W9-CNJ-962

CIGR Handbook of Agricultural Engineering Volume II

CIGR Handbook of Agricultural Engineering

Volume II
Animal Production & Aquacultural Engineering

Edited by CIGR–The International Commission of Agricultural Engineering

Part I Livestock Housing and Environment

Volume Editor:

El Houssine Bartali

Hassan II IAVM, Morocco

Co-Editors:

Aad Jongebreur

IMAG-DLO, Netherlands

David Moffitt

United States Department of Agriculture–NRCS, USA

Part II Aquaculture Engineering

Volume Editor:

Frederick Wheaton

University of Maryland, USA

Published by the American Society of Agricultural Engineers

LCCN 98-93767 ISBN 0-929355-98-9

For Information, contact:

Manufactured in the United States of America

Editors and Authors

Part I Livestock Housing and Environment

Volume Editor
El Houssine Bartali
Department of Agricultural Engineering, Hassan II Institute of Agronomy and Veterinary Medicine, B.P. 6202 Rabat Instituts, Rabat, Morocco

Co-Editors
Aad Jongebreur
Agricultural Research Department, Institute of Agricultural and Environmental Engineering (IMAG-DLO), Ministry of Agriculture, Nature Management and Fisheries, Mansholtlan 10-12, P.O. Box 43, NL-6700 AA Wageningen, The Netherlands

David Moffitt
United States Department of Agriculture–Natural Resources Conservation Service, South Central Regional Office, P.O. Box 6459, Fort Worth, TX 76115, USA

Authors
El Houssine Bartali
Department of Agricultural Engineering, Hassan II Institute of Agronomy and Veterinary Medicine, B.P. 6202 Rabat Instituts Rabat, Morocco

James M. Bruce
Scottish Agricultural College, 581 King Street, Aberdeen AB9 1UD, Scotland, United Kingdom

Carl-Magnus Dolby
Continuing Education, Swedish University of Agricultural Sciences, Box 45, 230 53 Alnarp, Sweden

Vincenzo Menella
Faculty of Agriculture, Institute of Agricultural Mechanisation and Land, University of Perugia, Borgo XX Giugno, 74, 06121 Perugia, Italy

David H. O'Neill
Wrest Park, Silsoe Research Institute, Beds MK 45 4 HS, United Kingdom

Soeren Pedersen
Danish Institute of Agricultural Sciences, Research Center Bygholm., Box 536, 8700 Horsens, Denmark

Krister Sallvik
Department of Agricultural Biosystems and Technology, Swedish University of Agricultural Sciences, P.O. Box 59, 230 59 Alnarp, Sweden

Shahab S. Sokhansanj
Department of Agricultural and Bioresource Engineering, University of Saskatchwan, 1 Building 57 Campus Drive AO3 ENG, Saskatoon, SK S7N 5A3, Canada

Jean Claude Souty
Ministère de l'Agriculture et de la pêche, DEPSE 13, 19 Avenue du Maine, 75015 Paris Cedex 15, France

Michel Tillie
6, rue Pasteur, 62217 Beaurains, France

Part II Aquaculture Engineering

Volume Editor

Frederick Wheaton
Department of Biological Resources Engineering, University of Maryland, College Park, MD 20742, USA

Authors

David E. Brune
Agr & Bio Engineering Department, Clemson University, 116 McAdams Hall, Clemson, SC 29634-0357, USA

John Hochheimer
NCQA, 2000 L Street, N.W., Suite 500, Washington, D.C. 20036, USA

Odd-Ivar Lekang
Department of Agricultural Engineering, Agricultural University of Norway, AS, Norway

John G. Riley
Bioresource Engineering Department, University of Maine, 5710 Bioresource, Orono, ME 04469-5710, USA

Sahdev Singh
Department of Biological Resources Engineering, University of Maryland, College Park, MD 20742, USA

Steven T. Summerfelt
Fresh Water Institute, P.O. Box 1746, Shepherdstown, WV 25443, USA

Michael B. Timmons
Agr & Bio Engineering Department, Riley Robb Hall, Cornell University, Ithaca, NY 14853, USA

Frederick Wheaton
Department of Biological Resources Engineering, University of Maryland, College Park, MD 20742, USA

Editorial Board

Contents

Part I Livestock Housing and Environment

Foreword xv
Acknowledgments xvii
Preface xix

1 Characteristics and Performances of Construction Materials 3
1.1 Concrete and Steel 3
1.1.1 Concrete 3
1.1.2 Reinforced Concrete 6
1.1.3 Steel Construction 7
1.2 Masonry and Blocks 9
1.2.1 Concrete Blocks 9
1.2.2 Earth Concrete Blocks 9
1.2.3 Burnt-Clay Bricks 10
1.3 Wood as a Construction Material for Farm Buildings 10
1.3.1 Wood as a Material 10
1.3.2 Structures in Timber 13
1.3.3 Joints 18
1.3.4 Panel Products 20
1.3.5 Wall Coverings 23
1.3.6 Thermal Insulation 28

2 Environment for Animals 31
2.1 Animal Environment Requirements 31
2.1.1 Thermal 32
2.1.2 Thermoneutral Zone 32
2.1.3 Heat Balance at Animal Level 33
2.2 Animal Heat and Moisture Production 41
2.2.1 Equations for Total Heat Production, Φ_{tot} 41
2.2.2 Proportion Between Sensible and Latent Heat Dissipation 44
2.2.3 Conversion of Latent Heat to Moisture Dissipation 46
2.2.4 Heat and Moisture Production at House Level 46
2.2.5 Diurnal Variation in Heat and Moisture Production at House Level 50
2.3 Environmental Control of Livestock Housing 54
2.3.1 Natural Ventilation 54
2.3.2 Forced Ventilation 68

3 Livestock Housing 89
3.1 Sheep Housing 89
3.1.1 Types of Holdings 89
3.1.2 Reference Background 90
3.1.3 Process and Product Characteristics 90

3.1.4 Criteria for the Defining of Building Systems 91
3.1.5 Building Systems for Intensive Milk-Production Holdings 92
3.1.6 Planning Parameters 96
3.1.7 Building Systems for Meat-Production Holdings 97
3.1.8 Planning Parameters 100
3.1.9 Conclusion 101
3.2 Pig Housing 101
3.2.1 Reference Scenario 101
3.2.2 Types of Pig Holdings 102
3.2.3 Criteria for Defining the Building System 103
3.2.4 Systems Planning and Integration Parameters 105
3.2.5 Closed-Cycle Pig-Breeding Center 106

4 Equipment and Control 115
4.1 Feed and Supply Distribution 115
4.1.1 Intensive Conditions 115
4.1.2 Extensive Conditions 128
4.2 Feed Mixers and Intake Control 130
4.2.1 Cattle Breeding 130
4.2.2 Sheep Production 133
4.3 Watering Equipment 133
4.3.1 Cattle Breeding [8] 133
4.3.2 Goats and Sheep 135
4.3.3 Pig Breeding [9] 138
4.3.4 Poultry [6] 138
4.4 Milk Storage and Control [11] 142
4.4.1 General Points 142
4.4.2 Influence of Refrigeration on the Microbial Flora 142
4.4.3 Refrigeration Systems 143
4.4.4 Tank Equipment 144
4.4.5 Control of Coolers 144

5 Storing Forages and Forage Products 147
5.1 Introduction 147
5.2 Losses in Swath 148
5.3 Losses During Storage of Square-Baled Hay 148
5.4 Losses During Storage of Round-Baled Hay 150
5.5 Storage of Cubes 152
5.6 Storage of Alfalfa Pellets 154
5.7 Storage of Loose Hay 156
5.8 Cube Spoilage During Transport 158

6 Waste Management and Recycling of Organic Matter 163
6.1 Waste Management 163
6.1.1 Effects of Manure on the Water Resource 163

	6.1.2	Effects of Manure on the Air Resource	167
	6.1.3	Effects of Manure on the Animal Resource	170
	6.1.4	Manure Characteristics	170
	6.1.5	Manure-Management Systems	178
6.2		Recycling of Organic Matter	187
	6.2.1	Land Application	187
	6.2.2	Energy Production	193
7		Draught Animals	197
7.1		Utilization	197
	7.1.1	Principles of Animal Traction	197
	7.1.2	Tasks Performed	200
	7.1.3	Species	202
	7.1.4	Work Performance	202
7.2		Husbandry and Work Performance	204
	7.2.1	Body Condition	204
	7.2.2	Work Schedules	205
	7.2.3	Multipurpose Animals	206
7.3		Draught Animals in Farming Systems	206

Part II Aquacultural Engineering

8		Aquacultural Systems	211
8.1		Introduction	211
8.2		System Types	211
	8.2.1	Ponds	212
	8.2.2	Raceways	213
	8.2.3	Net Pens and Cages	214
	8.2.4	Tanks and Recirculating Aquacultural Systems	214
9		Environmental Requirements	219
9.1		Primary Constraints in Aquacultural Systems	219
	9.1.1	Properties of Water	219
	9.1.2	Oxygen as a Constraint	220
	9.1.3	Other	221
9.2		Environmental Needs of Aquatic Organisms	223
10		Materials for Aquacultural Facilities	231
10.1		Considerations in Material Selection Process	231
	10.1.1	Weight of Water	231
	10.1.2	Corrosion	231
	10.1.3	Biofouling	232
	10.1.4	Ozone as a Constraint in Material Selection	235
10.2		System Components and Material Selection	236
	10.2.1	Tanks	236
	10.2.2	Raceways	236

10.2.3 Waterproof Lining 236
10.2.4 Screen Mesh 237
10.2.5 Nets 237
10.2.6 Ozone Unit 237
10.3 Advantages and Disadvantages of Various Materials 237
10.3.1 Masonry 238
10.3.2 Metals 239
10.3.3 Plastics/Rubber Compounds 240
10.3.4 Wood 242
10.3.5 Others 243

11 Facilities Design 245
11.1 Introduction 245
11.2 Ponds 246
11.2.1 Pond Photosynthesis 247
11.2.2 Diurnal Limits 247
11.2.3 Productive Systems and Polyculture 248
11.2.4 pH Limits 249
11.2.5 Nitrogen Control 249
11.2.6 Pond Types 249
11.3 Raceways 252
11.3.1 Design Densities and Loadings (Raceways or Tanks) 252
11.3.2 Raceway Length 254
11.3.3 Fish Growth 255
11.3.4 Design Principles and Considerations 256
11.4 Net Pens 263
11.4.1 Site Selection 264
11.4.2 Net-Pen Design and Construction 265
11.4.3 Net-pen summary 270
11.5 Tanks 271
11.5.1 Biomass Loading 271
11.5.2 Labor Requirements 271
11.5.3 Tank Shapes and Sizes 272
11.5.4 Water Inlet 274
11.5.5 Water Outlet 275

12 Equipment and Controls 281
12.1 Feeding Equipment 281
12.1.1 Hand Feeding 283
12.1.2 Automatic Feeders 284
12.1.3 Demand Feeders 287
12.2 Pumps 288
12.2.1 Types of Pumps 289
12.2.2 Power Source 293

12.3 Harvest Equipment 295
12.3.1 Types of Harvesting Equipment 296
12.4 Monitoring Equipment 300
12.4.1 Sensors 301
12.4.2 Monitoring and Control Systems 304

13 Waste-Handling Systems 309
13.1 Introduction 309
13.1.1 Effluent regulations 310
13.2 Materials to Remove 311
13.2.1 Nitrogenous Compounds 312
13.2.2 Solids 313
13.2.3 Dissolved Matter 314
13.2.4 Carbon Dioxide 314
13.2.5 Pathogens and Chemicals Used in Aquaculture 315
13.2.6 Nutrients 315
13.3 Methods to Remove Ammonia 316
13.3.1 Microbial Action 316
13.3.2 Media 320
13.3.3 Submerged Filters 321
13.3.4 Trickling Filters 322
13.3.5 Rotating Biological Contactors 322
13.3.6 Pressurized-Bead Filters 323
13.3.7 Fluidized-Bed Biofilters 323
13.3.8 Biofilter Comparison 326
13.4 Methods to Remove Solids 327
13.4.1 Settling Basins 327
13.4.2 Microscreen Filters 330
13.4.3 Granular Media Filters 332
13.4.4 Dissolved Air Flotation and Foam Fractionation 332
13.4.5 Ozonation 333
13.4.6 Discussion of Solids-Removal Options 334
13.5 Methods to Dispose of Solids 334
13.6 Methods to Remove Dissolved and Colloidal Organic Matter 335
13.7 Methods to Remove Carbon Dioxide 336
13.7.1 Air Stripping 336
13.7.2 Chemical Addition 337

Index 351

Foreword

This handbook has been edited and published as a contribution to world agriculture at present as well as for the coming century. More than half of the world's population is engaged in agriculture to meet total world food demand. In developed countries, the economic weight of agriculture has been decreasing. However, a global view indicates that agriculture is still the largest industry and will remain so in the coming century.

Agriculture is one of the few industries that creates resources continuously from nature in a sustainable way because it creates organic matter and its derivatives by utilizing solar energy and other material cycles in nature. Continuity or sustainability is the very basis for securing global prosperity over many generations—the common objective of humankind.

Agricultural engineering has been applying scientific principles for the optimal conversion of natural resources into agricultural land, machinery, structure, processes, and systems for the benefit of man. Machinery, for example, multiplies the tiny power (about 0.07 kW) of a farmer into the 70 kW power of a tractor which makes possible the production of food several hundred times more than what a farmen can produce manually. Processing technology reduces food loss and adds much more nutritional values to agricultural products than they originally had.

The role of agricultural engineering is increasing with the dawning of a new century. Agriculture will have to supply not only food, but also other materials such as bio-fuels, organic feedstocks for secondary industries of destruction, and even medical ingredients. Furthermore, new agricultural technology is also expected to help *reduce* environmental destruction.

This handbook is designed to cover the major fields of agricultural engineering such as soil and water, machinery and its management, farm structures and processing agricultural, as well as other emerging fields. Information on technology for rural planning and farming systems, aquaculture, environmental technology for plant and animal production, energy and biomass engineering is also incorporated in this handbook. These emerging technologies will play more and more important roles in the future as both traditional and new technologies are used to supply food for an increasing world population and to manage decreasing fossil resources. Agricultural technologies are especially important in developing regions of the world where the demand for food and feedstocks will need boosting in parallel with the population growth and the rise of living standards.

It is not easy to cover all of the important topics in agricultural engineering in a limited number of pages. We regretfully had to drop some topics during the planning and editorial processes. There will be other requests from the readers in due course. We would like to make a continuous effort to improve the contents of the handbook and, in the near future, to issue the next edition.

This handbook will be useful to many agricultural engineers and students as well as to those who are working in relevant fields. It is my sincere desire that this handbook will be used worldwide to promote agricultural production and related industrial activities.

Osamu Kitani
Editor-in-Chief

Foreword

Acknowledgments

At the World Congress in Milan, the CIGR Handbook project was formally started under the initiative of Prof. Giuseppe Pellizzi, the President of CIGR at that time. Deep gratitude is expressed for his strong initiative to promote this project.

To the members of the Editorial Board, co-editors, and to all the authors of the handbook, my sincerest thanks for the great endeavors and contributions to this handbook.

To support the CIGR Handbook project, the following organizations have made generous donations. Without their support, this handbook would not have been edited and published.

Iseki & Co., Ltd.
Japan Tabacco Incorporation
The Kajima Foundation
Kubota Corporation
Nihon Kaken Co., Ltd.
Satake Mfg. Corporation
The Tokyo Electric Power Co., Inc.
Yanmar Agricultural Equipment Co., Ltd.

Last but not least, sincere gratitude is expressed to the publisher, ASAE; especially to Mrs. Donna M. Hull, Director of Publication, and Ms. Sandy Nalepa for their great effort in publishing and distributing this handbook.

Osamu Kitani
CIGR President of 1997–98

Preface

The first part of this volume deals with livestock housing and environment. It is intended to be a useful tool in the hands of engineers, architects, educators, material and equipment suppliers, students, and others. There has been very significant progress in many areas of livestock housing and environment in the past years. This indicated a need for a volume of the *CIGR Handbook on Agricultural Engineering* devoted to this subject.

Information incorporated in this volume is of a nature that could be valuable in making decisions in this field: characteristics of construction materials, environment control, livestock housing design, silage storage, equipment and waste management, and draught animals husbandry. Frequent reference is made to other sources where additional detailed information can be obtained because of space limitations, a careful selection of topics has been made.

The chapters have been written by experts from research institutions, management departments, and universities recognized as outstanding authorities in their respective fields. The authors' wide experience has resulted in concise chapters geared toward practical application in planning, design, and management of livestock housing.

The editors are especially grateful to the contributors, not only because they appreciate the great value of their contributions but also because they are keenly aware of their considerable efforts in taking time to prepare their manuscripts.

The editor of Part II expresses his sincere appreciation to Dr. Sahdev Singh for his valuable assistance in editing Part II. Without Dr. Singh's time, expertise, and attention to detail, Part II would not have been completed.

El Houssine Bartali and Frederick Wheaton
Editors of the Volume II

PART I Livestock Housing and Environment

1 Characteristics and Performances of Construction Materials

1.1 Concrete and Steel

El Houssine Bartali

1.1.1 Concrete

Concrete as Material

Plain concrete is obtained by adequately mixing in specific proportions aggregates (gravel and sand), Portland cement, and water. Plain or reinforced concrete is used in livestock housing for structures, foundations, floors, and walls. It is a durable material that can resist attack by water, animal manure, chemicals, and fire. High-quality concrete is recommended for milk-, silage-, or manure-containing structures.

Properties of Concrete

Two main properties of concrete are strength and workability.

The strength of concrete depends on various factors, mainly the proportions and quality of the ingredients and the temperature and moisture under which it is placed and cured. The methods for proportioning and placing concrete to achieve a preset required strength can be found in the literature. Concrete can develop a very high compressive strength equivalent to two to five times that of wood [1]. Compressive strength of concrete increases with its age. This is measured by crushing cubes or cylinders of standard sizes. Concrete design is based on the characteristic strength values at 28 days of age. Its tensile strength remains weak, however, about one tenth of its compressive. For this reason, steel rods (rebars) are combined with concrete. In reinforced concrete, the area and positioning of steel bars determined according to applicable standard codes.

Workability of concrete relates to its ability to be poured in forms and to properly flow around steel bars. This measured by a slump test. For slump values less than 2 cm, concrete needs strong vibration in order to be properly put in place. For values of slump between 10 and 14 cm, concrete is very soft and may need slight stitching [2]. Concrete is placed in forms. This operation is undertaken either on the construction site or in a prefabrication plant.

Composition of Concrete

Cement

Portland cement is obtained from blending a mixture of calcareous (75%) and clayey (25%) materials, which are ground and burnt (at about 1450°C) to produce clinker. The latter is cooled and reground with the addition of gypsum into a very fine powder. The role of gypsum is to control the rate of set of concrete [3].

Two main categories of cement can be distinguished: artificial Portland cements, containing at least 97% clinker and blended Portland cements, which include pozzolona and contain not less that 65% clinker. This second category is obtained by adding pozzolanic materials to clinker. These materials include natural pozzolana, volcanic ash, fly ash, and blast-furnace slag.

Cement is sold in 40- or 50-kg paper bags or in bulk. Cement should be protected and kept in dry places away from ground moisture or damp air. If not lumps, may develop and reduce its strength. Main types of Portland cement include ordinary Portland cement; rapid-hardening Portland cement, which is very finely ground, develops strength more rapidly, and is suitable for early stripping of form work and early loading of buildings; and low heat Portland cement which avoids excess heat generated by chemical reactions and cracking in large structures. Ordinary Portland cement is suitable for most farm and all normal purposes.

Five classes of cement are distinguished and based on minimum compressive strengths at 28 days of age. These values vary from 32.5 MPa to 52.5 MPa.

When water is added to cement, the hydration process of cement starts. The strength of concrete is heavily dependent on cement–water ratio. Excess water makes concrete weak because it leaves voids after it evaporates.

Aggregates

Aggregates include gravel and sand. These are inert materials. They may be classified into the following size categories in millimeters: 0.08, 4, 6.3, 10, 20, 31.5, 40, 63, and 80 [4].

The size of a mesh screen that distinguishes between sand or small aggregate and gravel or coarse aggregate is about 6 mm. Larger particles should allow concrete to pass between reinforcement bars and must not exceed one quarter of the minimum thickness of the concrete member. Maximum particle size is usually 20 to 25 mm.

Nature, size distribution, and shape of aggregates affect strength, workability, and cost of concrete. Concrete strength is improved if sharp, flat, rough aggregates are used; however, this requires more cement paste. An adequate size distribution helps save cement. The proportion of cement needed varies with both total surface area of aggregates and volume of voids. A fairly even distribution of sizes with a well-graded aggregate leaves a minimum volume of voids to be filled with cement.

Aggregates are glued together by cured cement paste. It is important that aggregates be hard, strong, and clean, free of organic material and silt.

The presence of excessive quantities of silt and organic matter in the material will prevent cement from properly binding the aggregates. Silt tests and organic-matter tests are used to assess suitability of aggregates for concrete. These tests make it possible to know if material can be used as it is, or if it is necessary to wash material before using it or to reject it [5].

Naturally moist sand can contain 2.5% to 5.5% moisture content. This much water is taken into account when determining the concrete mixture.

In livestock housing one can use usual aggregates: natural deposit materials such as sand, gravel, hard limestone, silicolimestone, or crushed stone. Density of sand and stone varies from 2600 to 2700 kg/m^3. One can also use lightweight aggregates: expanded clay or expanded shale.

Water

The strength of concrete is very dependent on the amount of water (water–cement ratio) and on its quality. Enough water is needed to allow curing of cement based on chemical reactions. It is recommended that the water–cement ratio should not be lower than 0.4:1. On the other hand, any excess water is bound to evaporate and to induce cracking in concrete.

Clean water is needed. Attention should be given to its content of suspended material, organic matter, and salt. Suspended materials should not exceed a few grams per liter, and the amount of soluble salts should not exceed 30 g/L for plain or slightly reinforced concrete. Water fit for drinking is best. Sea water may be used but not for reinforced concrete. Sulfate-resistant cement may be needed for use with sulfate-containing water [3].

Admixtures

Admixtures are added in small quantities to concrete immediately after or before its mixing in order to improve some properties of the material. The list of such products includes accelerating products, retarding products, water-reducing products, air-entraining products to improve resistance to freezing and thawing, superplasticizers, and pozzolans.

Batching and Mixing

Proportions of concrete ingredients may vary according to the use, workability, and level of strength desired for the concrete. For ordinary concrete, the following quantities of ingredients may be used: per cubic meter, 800 L of gravel, 400 L of sand, 350 L of cement, and 150 L of water [6].

Other recommended trial mixes .of concrete are available in the literature [3]. Nominal mixes, which are represented by proportions of cement, sand, and stone may be used to designate a given grade of concrete. Specific grades used range from a grade as low as C7, presenting a characteristic crushing strength of 7 MPa, having a nominal mix 1:3:8, and suitable for strip footings and trench fill, to C60 used for prestressed concrete [4].

Mixing can take place on the construction site or in a factory, after which concrete must be placed within 30 min in forms either poured in place or used for precast units. Mixing can be made by power mixer or by hand depending on batch size. Enough mixing is needed in order to make a homogeneous distribution of ingredients. Excess mixing may induce a decrease in concrete strength or a loss in slump. A concrete mix with low workability will require more compaction. For most farm livestock constructions, manual compaction is used. Workability can be improved by using rounded aggregates with a suitable selection of sand and stone proportions.

Curing

In order to allow chemical reaction between cement and water to be completed, the concrete surface must be watered or protected by grass, sacks, sand, or polythene layer. These measures, which should take place within 10 h of casting, aim at preventing water from evaporating and may last for at least a week. They are particularly important in hot climates.

1.1.2 Reinforced Concrete

Reinforcing steel bars or mesh are systematically necessary to support traction in the tensile zone. Bars may also be needed in some cases to help concrete support excessive compression loads. Design of reinforced concrete structures is carried out in compliance with design codes specified in each country. Reinforced concrete presents the following advantages compared with steel: better rigidity and fire resistance. However, reinforced concrete structures are heavier than steel structures. Density of concrete is usually taken as about 24 kN/m^3 and that of reinforced concrete as 25 kN/m^3 (kilo-newtons per m^3).

Reinforcing Bars

Steel rods are available as plain bars or deformed bars; the latter have a better bonding with concrete. Welded mesh also is used particularly as reinforcement of flat slabs and is available in rolls of 30×1.5 m. Steel bars are designated by their diameter in millimeters or their number. They are available with an indication of their characteristic strength such as 400 MPa in FeE-400 or their minimum yield strength such as 41 MPa in type 40 steel.

In order to ensure resistance of structures, steel rods should be rust-free and dirt-free and properly surrounded by concrete. Cover thickness should be around 30 mm to 40 mm in order to avoid rusting due to liquids and air. Any cracks in concrete may allow corrosion and therefore expansion and weakening of bars. When assessing quality of reinforcement disposition, one has to check that bars are properly positioned in the tensile zone, adequately hooked, well overlapped over a joining distance of at least 40 times the bar diameter for adequate load transfer, well supported and tied together, and properly imbedded in concrete.

Concrete Floors

Foundation

A building should be located in a well-drained site with no risk of sliding of bearing soil layers. Foundations are the elements of construction in contact with bearing soil, to which they convey superstructure loads. They should be made of materials that resist pressure and humidity. Foundations under walls are about 0.30 m thick and 0.2 to 0.3 m high. In order to protect them from frost, they are placed at a depth of 0.5 m to 1 m.

Buildings should always be founded on good soil in order to avoid structural disorders that can be induced by differential settlement. Physical and mechanical characteristics of soil should be determined by adequate soil and site investigation. Careful attention should be given to the presence of the water table because it may generate loss in soil resistance, wash away fine particles of soil, and attack concrete. Design of foundation takes into account the combination of dead and live loads applied to the construction and the strength of bearing soil. Loads applied by livestock buildings to soils are usually not excessive. Such buildings usually rest on superficial concrete foundations under walls or columns.

Paving

Paving is reinforced concrete slabs (about 0.12 m thick) resting on a subgrade made of a layer of cement gravel, stones, or stabilized earth (0.2 to 0.40 m), meant to distribute applied concentrated loads or to act as a draining layer. A water-proof material can be incorporated beneath the slab to avoid soil water moving upward. This reduces the chance of cracking induced by temperature changes. Joints are added to keep the area of slab sections under 70 m^2 in order to reduce cracking due to concrete shrinkage [5].

Floor slab can be reinforced with welded mesh or chicken wire. Its design takes into account applied live concentrated or distributed loads and bearing capacity of soil. The floor surface is usually not flat. It is provided with slopes, gutters, and hallways. Floors should be waterproof, resistant enough to animal waste and easy to clean and to disinfect. The floor surface should not slippery.

1.1.3 Steel Construction

Barns may be designed as steel warehouses. The latter do not usually have a symmetric shape unless they are not used to store straw and hay. The cost per square meter of a warehouse increases generally with the building span. For a span less than 15 m, steel warehouses may be more cost-competitive than timber warehouses.

Steel is used in livestock housing as hot-rolled or cold-formed shapes. It is used in the form of iron sheet for walls, partitions, or roofing. It is also found as the structural element for beams, columns, or trusses in various forms such as I-beams, angles, or pipes. Because of their thinner elements, cold-formed shapes are more susceptible to local buckling than hot-rolled sections.

In agricultural buildings in general and in livestock housing in particular, light-gauge steel FeE-24 and A36 are the most commonly used, with a limit stress of 24 MPa.

Steel-bearing elements are generally made of hot-rolled steel. The most common shapes are IPN, IPE (standard I shape sections) corner, and tubular (circular, square, rectangular). Other shapes may be obtained by welding and formage of flat products. All these elements are assembled through bolts or welding.

Steel for Roofs and Walls

Iron sheets, corrugated or nervurated, used for roofs or walls are commonly 0.75 to 0.80 mm thick for roofing and 0.65 to 0.75 mm thick for walling. These iron sheets need to be protected against corrosion using galvanization, paint, or lacquer. The main types found are steel galvanized by hot immersion, galvanized steel prelacquered in oven, painted galvanized steel, and painted steel [5].

Selection of shapes and their types of protection is made in terms of type of roofing used and weather encountered. Nervuration provides iron sheets with resistance to bending, which is variable with shape, height, and spacing. Selection of nervuration takes into account distance between rafters and climatic loads of the region under consideration.

In low-height and close warehouses, the use of corrugated galvanized iron sheet should be avoided due to risks of condensation.

It may be more advantageous to use aluminum, which presents several advantageous qualities: light weight, lack of alterability, high sunlight-reflecting potential, and low cost. Disadvantages of aluminum include that it is noisy under hail or heavy rain or temperature changes, which frightens animals.

Steel Protection Against Corrosion

Steel surfaces are sensitive to outside conditions. Weather aggressively induces oxidation of the surface of steel. This progressive alteration reduces strength, which may cause failure of steel elements if no measure of protection is taken. Aggressiveness of the environment is variable with hygrometry, temperature, and presence of dirt and some gases in ambient air.

Thus, in order to ensure adequate behavior of steel, it is necessary to adopt a protection system against corrosion adapted to the nature and intensity of the alteration to be avoided. The lower edges of steel columns and frames are inserted in concrete pipes in order to protect them against corrosion from manure. The space between steel columns and concrete pipes are filled with cement. Steel rusts when exposed to atmospheres above a critical relative humidity of about 70%. Serious corrosion occurs at normal temperatures only in the presence of both oxygen and water, both of which must be replenished continually. To select a paint system for corrosion protection, therefore, it is necessary to begin with the function of the structure, its environment, maintenance practices, and appearance requirements [7].

The manager of livestock housing needs to proceed about every 2 years to a thorough visual control of all steel elements in order to detect any corrosion. Then it is necessary to brush all attacked zones and to apply a new protection to them immediately.

Various painting protections are available on the market. To allow a reliable and durable protection such paints should include a primary layer, an intermediate layer, and one or more finishing layers. Before any application of paint, the steel should be brushed by hand or power to remove loose mill scale, loose rust, weld slug, flux deposit, dirt, and foreign matter. Oil and grease should be removed by solvent. Paint application should follow immediately because unprotected steel is very sensitive to oxidation.

Protection by zinc often is used for agricultural buildings. It may be applied through galvanization by immersion of steel in hot zinc (450°C), metalization by projection of finely pulverized zinc onto the surface to be protected, or paints rich in zinc, which represent a good base for fixing finishing paints.

Prelacquered Shapes

Prelacquered shapes provide important supplementary protection against corrosion. It is important to make an adequate choice of lacquer to ensure sufficient protection against aggressive atmosphere. Availability of different colors of lacquers makes it possible to make a building better fit in a given landscape [5]. Prelacquering is in general done in a continuous manner. It consists of applying on a surface initially galvanized a lining made of polymerized plastic binders that have been oven-cooked. The main types of lacquers are acrylics, siliconed acrylics, and syliconed polyesters.

1.2 Masonry and Blocks

El Houssine Bartali

1.2.1 Concrete Blocks

Concrete blocks are economical, adaptable, and readily available. However, they are difficult to insulate. Blocks may be used in buildings and manure storage tanks, among other places. They can be hollow or solid and load-bearing or non–load bearing and are available in various shapes and sizes. Lightweight aggregates such as pumice, volcanic cinders, and scoria or ordinary heavy weight aggregate can be used for their production [1]. Compressive strength levels at 28 days of age for heavyweight blocks can vary from 4 MPa for hollow blocks to 16 MPa for solid blocks. These levels range respectively from 2.5 MPa to 7 Mpa for lightweight concrete blocks.

Mortar is needed to bond blocks together and provide strength and waterproofing to a wall. Good water retention is an important characteristic for a mortar. It is influenced by the quality of sand, binder, and dosage. Recommended mortar mixes are available in the literature [5].

Mortars are masonry cement, made of a mixture of Portland cement, lime or hydrated lime, and sand. Hydrated lime improves water-retention capacity, workability, and adherence of mortar. It is recommended to avoid excess cement in mortar. The strength of mortar should be in harmony with the strength of the block. To insure desired compressive and tensile strengths of a masonry wall, one should specify a full bed of mortar, with each course well hammered down, and all joints completely filled with mortar.

Such blocks are made of the following ingredients, for which common proportions used are indicated: cement (50 L); gravel (120 L) and sand (90 L). Sizes in centimeters usually encountered include $5 \times 20 \times 40$, $10 \times 20 \times 40$, $15 \times 20 \times 40$, $20 \times 20 \times 40$, and $25 \times 20 \times 20$. In livestock housing, blocks may be used over the full area or a section of walls. Concrete blocks are used instead of iron sheets wherever walls have to withstand impact loads or pressure from animals or where walls are in contact with stacked manure. Walls of barns may be composed of concrete blocks about 1 m high at their bottom edge, on top of which iron sheets are used.

In order to allow slight movement of walls and avoid random cracking, vertical control joints are used. Such joints are efficient in relieving stress caused by expansion, contraction, or differential settlement. These joints are placed around doors and windows and at the intersection of bearing walls.

Reinforcing is used to increase the strength of concrete-block walls. Horizontal reinforcing can be achieved with horizontal bars placed in mortar joints or with bond beams. Vertically reinforcing steel is incorporated in blocks. Pilasters are needed to support hollow-block walls.

1.2.2 Earth Concrete Blocks

Such blocks present the advantages of low cost and better heat insulation properties. The fabrication process includes mixing, which can be done by hand or mechanically and should create a homogeneous mix; and moulding and compacting, which can be undertaken by compaction material. Rate of daily block production varies with equipment used. It can be in the range of 600 blocks per day for 15 × 23 × 29 cm blocks. Earth concrete blocks need be dried in a covered and well-ventilated area and over a suficient period of time of up to 4 weeks.

As an indication composition of earth concrete blocks may be as follows: laterite earth, 0.5 m^3; clay, 0.5 m^3; sand, 0.2 m^3; and cement, 50 kg. Percentage of cement in the mix may vary from 3% to 12% depending on the percentage of clay used. One cubic meter of mix can yield 83 blocks of 15 × 20 × 40 or 125 blocks of 10 × 20 × 40. Joints are made of cement mortar. Because of the vulnerability of earth concrete blocks to rain water, walls should be protected on their external surface with a cement mortar, and foundations and lower sections of walls should be made of concrete or stones.

1.2.3 Burnt-Clay Bricks

Such bricks are made of clay earth free of any plant material. Ingredients used include clay, silt, and fine sand. They can be locally made or produced in a factory. The latter case ensures better quality bricks. Clay paste is burnt at a temperature ranging from 550°C to 1200°C. Sizes of bricks in centimeters commonly found include 22 × 11 × 5.5, 30 × 15 × 8, and 30 × 11 × 11.

Advantages offered by burnt-clay bricks include affordable cost, good thermal insulation, and resistance to moisture, erosion, and insects. Their compressive strength is medium to high. Livestock housing built with such bricks has lasted for several years in various locations.

References

1. Midwest Plan Service. 1983. *Structures and Environment Handbook*, 11 ed. Ames, IA: Iowa State University.
2. Renaud, H., and Letertre, F. 1992. *Ouvrages Béton Armé Ed.* Paris: Foucher.
3. Lindley, J. A., and Whitaker, J. H. 1996. *Agricultural Buildings and Structures*, rev. ed. St. Joseph, MI: ASAE.
4. FAO. 1986. *Farm Structures in Tropical Climates*. Rome: Author.
5. CATED. 1982. *Les Batiments Agricoles*. Paris: Author.
6. Institut de l'Elevage and ITAVI. 1996. *Bâtiments d'Elevage Bovin, Porcin et Avicole. Réglementation et préconisations relatives à l'environnement*. Paris: Author.
7. Merritt, F. S. 1976. *Standard Handbook for Civil Engineers*. New York: McGraw-Hill.

1.3 Wood as a Construction Material for Farm Buildings

Carl-Magnus Dolby

1.3.1 Wood as a Material

Wood Structure

Wood is the only major building material that is a renewable resource. It is constantly renewed by natural methods every 60 to 120 years. Wood is a cellular material principally

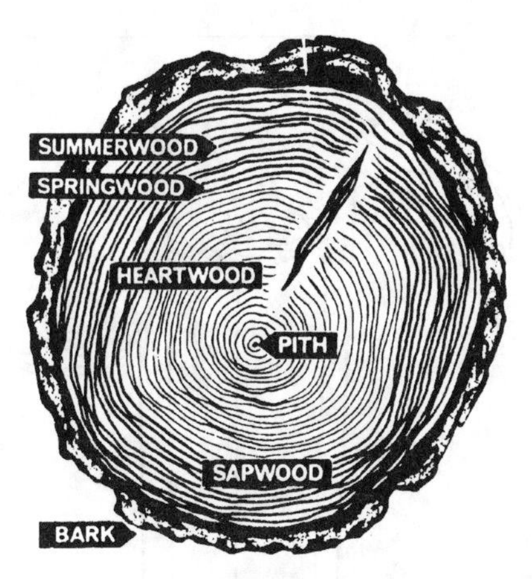

Figure 1.1. Cross-section of softwood log showing bark, wood, and pith.

comprising long, tube-like cells or fibers. These cells are mainly cellulose and bonded together with lignin. Most cells are oriented vertically in the tree. The new layers of cells produced at the outer active region of the tree make up the sapwood; the inner region of dying cells make up the heartwood (Fig. 1.1). Throughout its life in the tree, wood remains moist or "green," with the amount of moisture depending on the species, the part of the tree and whether it is sapwood or heartwood.

Species of trees are divided into two broad classes, coniferous trees (softwoods), which have needle-like leaves, and broad-leaved trees (hardwoods). Coniferous trees (pine and spruce) are the principal timbers for construction.

Advantages of Timber

Timber is the most easily worked of all structural materials. There is a reason why timber has remained a primary construction material for thousands of years. The reason is simply that no comparative material has all the advantages of timber.

Timber is light. The density of coniferous trees is approximately 500 kg/m^3, which, compared with the density of steel and concrete, is just about 1/16 and 1/5 respectively. The lightness of timber means that timber buildings do not require such solid foundations as buildings constructed of heavier materials.

Timber is strong. For its weight timber is stronger than any other building material. For example, stress-graded timber is available with a greater strength/weight ratio than mild steel.

Timber withstands impact. It is excellent at absorbing impact and usually only suffers local indentation. Timber is therefore suitable for uses such as external and internal boarding, flooring, and partition walls.

Timber is not a fire hazard. Contrary to popular belief timber is an excellent structural material when exposed to fire. Large timber members burn slowly and form char on the surface. Their strength is reduced gradually during a fire, and collapse does not take place until a very advanced stage of the fire.

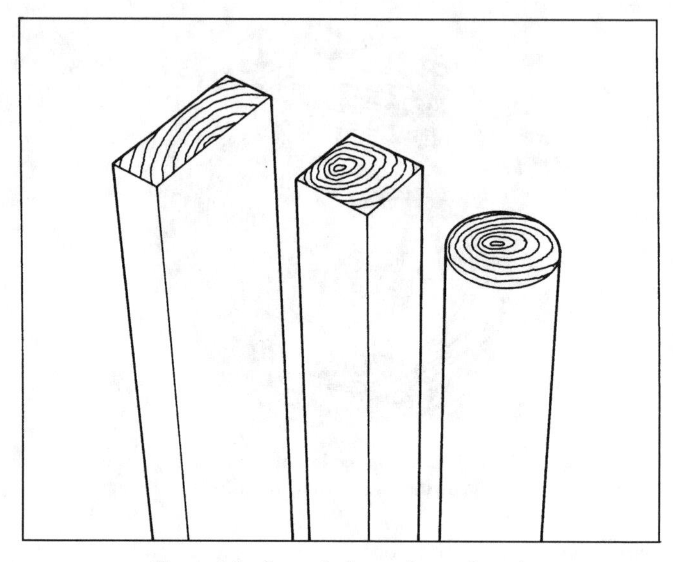

Figure 1.2. Sawn timber and roundwood.

Timber is easily worked. There are many simple ways of assembling timber parts and of joining timber to other materials. The material suits do-it-yourself builders very well. Alterations and additions are simple.

Timber is durable. A correctly designed and detailed timber structure is extremely durable. There are timber buildings in existence today that are over 1000 years old. In a well-designed timber structure there is little risk of excessive moisture movements or decay. Resistance to chemicals makes timber a valuable structural material in severely exposed environments.

Timber is attractive. It has a natural association with life and warmth and has appeal. The texture and characteristics are highly expressive. The attraction is enhanced with age.

Sawn Timber

Timber is sawn wood, a material obtained from trees. Timber has been used by builders and craftsmen for centuries and has been an important building material throughout history across the whole world. Timber can provide a structural frame as strong and durable as steel and concrete.

For a complete specification of sawn timber (Fig. 1.2) the following should always be stated:

- Purpose for which the timber is used
- Species
- Condition of seasoning
- Sizes
- Grade
- Surfacing
- Preservative treatment

- Special requirements
- Packing

Roundwood

Round timber poles are strong, versatile, and economic elements of building construction and have been one of the most valuable building materials throughout history. Round timber is stronger than sawn timber of equal cross-sectional area because fibers flow smothly around natural defects and are not terminated as sloping grain at cut surfaces. A round pole possesses a very high proportion of the basic strength of its species, because knots have less effect on the strength of naturally round timbers compared with sawn sections. The cost and wastage of sawing are eliminated.

Roundwood is strong parallel to its grain, relatively light, and economic and uses very little energy in its processing. For instance, the cost of sawn timber is approximately two to three times the cost of unsawn round timber of equivalent strength because a larger tree must be selected for a sawn and planed rectangular section. Its performance over time depends on its natural durability or suitability for preservation and on its inherent strength. These properties vary from species and in case of strength within species. Round timber poles are available in a wide range of sizes. The diameter is usually called "small-end diameter," with common sizes from 100 to 350 mm.

Glue-Laminated Timber

Glue-laminated timber (glulam) is frequently used in place of sawn timber for beams, columns, frames, and arches. In relation to weight and price it is practically the strongest material used. Glue-laminated timbers are manufactured of three or more layers of wood glued together with the grain of all layers or laminations approximately parallel. The laminations may vary depending on species, number, size, shape, and thickness. Glue-laminated timbers made of laminations of a certain (Fig. 1.3) grade have generally higher allowable unit stresses than solid members of the same grade. These higher stresses result especially from the dispersion of defects in a laminated member and the advantage of higher strength of dried wood in certain types of service. Proper fabrication of glulam members requires special equipment and manufacturing facilities, skilled workmanship, and a high standard of quality control.

Glulam can have a straight or curved shape or can be made with a variable section as in tapered beams or portal frames. For exposed conditions glulam can be treated with preservative in a manner similar to solid sections. Glulam is resistant to chemical attack and is often used in structures involved with corrosive substances such as fertilizers.

Depending on specific loading conditions a steel beam may be 20% heavier and a reinforced concrete beam 600% heavier than an equivalent glulam beam to carry the same load. The resulting lighter structure can lead to significant economy in foundation construction.

1.3.2 Structures in Timber

Structural Systems

Timber has been used for rural structures for centuries. In prehistoric time wood was the only material capable of spanning large distances. Development of the construction

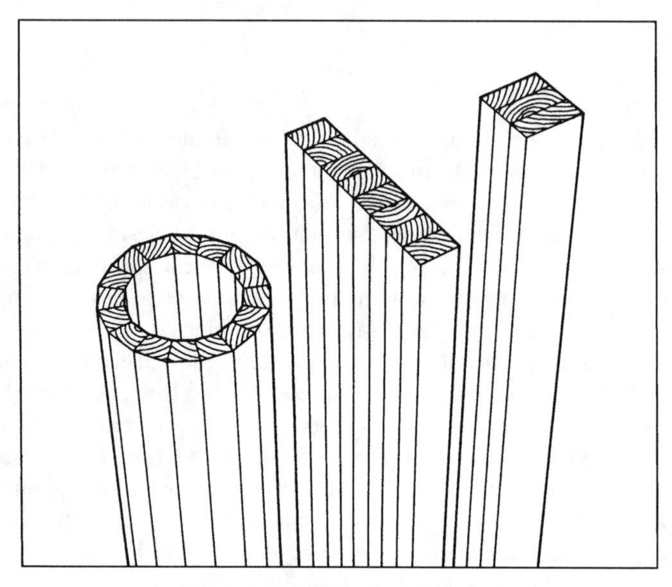

Figure 1.3. Glue-laminated timber.

techniques in wood was done by generations of carpenters who continually discovered efficient systems of support. Nowadays steel and concrete compete hard with timber as structural materials and current conditions for timber are quite different from conditions in the past. However, timber has maintained its position very well in agriculture and fits quite satisfactorily in the rural environment. See reference 1.

In general, three categories of load-carrying structural systems can be distinguished, namely free-span trusses supported by wall framing, post and beam constructions, and portals and arches. Each structural system can be composed of different structural components, which is evident from Fig. 1.4. In all timber structures the structural form and use of material depends on the vision of the designer, the technical constraints, the motivation to overcome imposed conditions, and ultimately the cost.

Factors Influencing Choice of Structural System

The choice of a particular structural system will generally be determined by the cost and fitness for purpose. In agriculture the design of a new production building and hence the selection of a structural system is governed by following factors:

- The farmer's intentions
  - Type of production
  - Size of production building
  - Technical installations
  - The site of building
  - Financial solutions
- Codes, regulations, guidelines
  - Loads

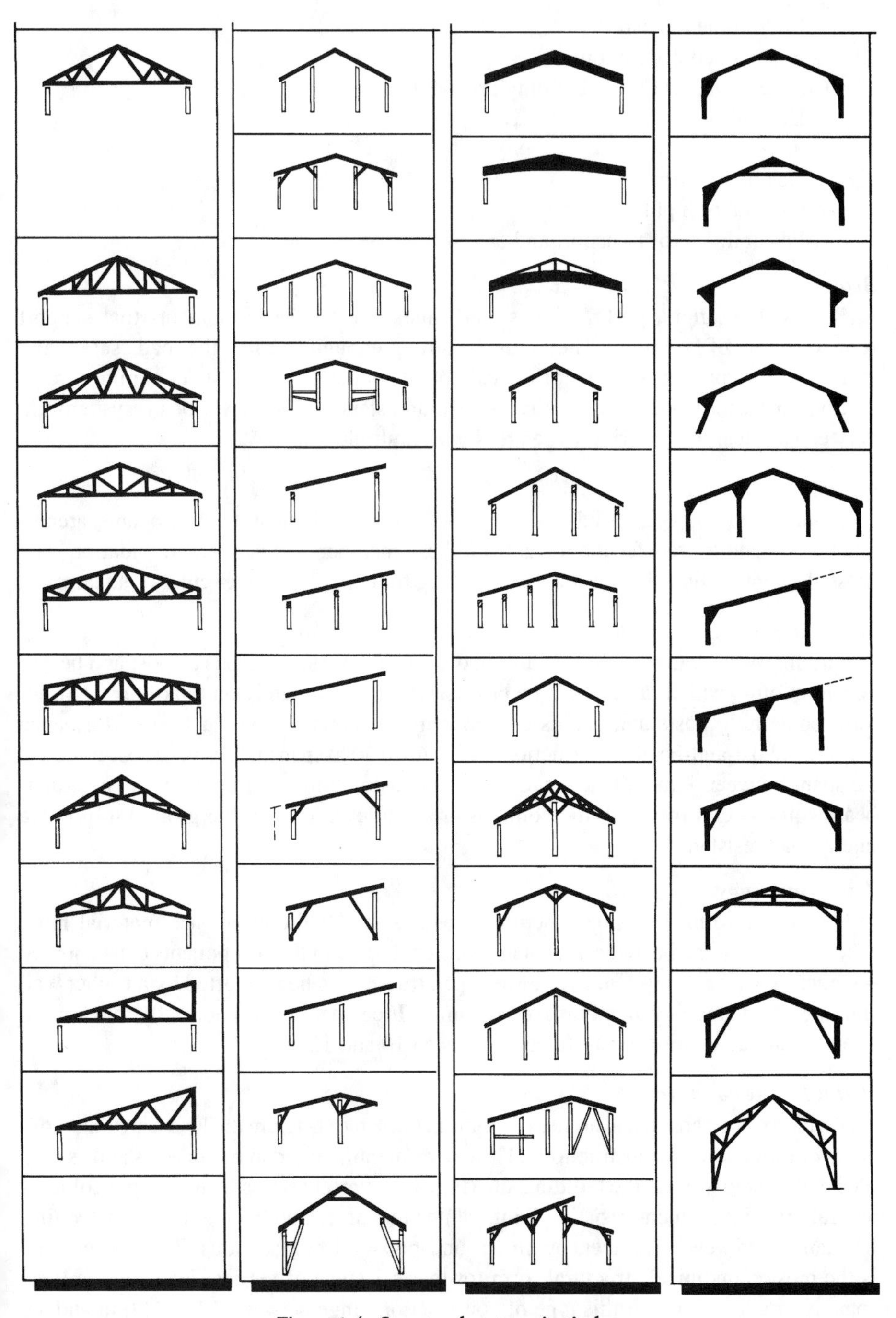

Figure 1.4. Structural systems in timber.

Environmental aspects
Fire-risk considerations
- Choice of materials and building procedure
  Durability
  Price
  Use of own timber
  Own labor input
  Possibilities for transport and erection

Beams

Beams like rafters, purlins, joists, and lintels are the basic elements that support roofs or floors of buildings. They must be strong enough to carry the loads safely and stiff enough to prevent sagging or undesirable movement. The most economical cross-section for a timber beam is one that is deep and slender. Cantilever beam systems are very economical for covering large areas with multiple spans.

Columns

Columns (or posts) are used to support roof and floor elements such as beams, arches, and trusses and to transfer the loads from these members down to the foundation. The most efficient section shape for a free-standing timber column is circular or square.

Post and Beam Construction

The simplest arrangement of beams and columns to resist loads is a "post and beam" construction in which rafters bear on beams, which rest squarely on top of posts. Structures containing posts and beams comprise many levels of construction performance (Fig. 1.5). Post-and-beam constructions range up to 30 m span and are very cost-effective for spans between 7 and 20 meters. Cantilever construction is a simple form of post-and-beam construction in which the columns are cantilevered from the ground to provide lateral load resistance.

Truss Buildings

Trusses perform the same function as beams in a building but use material more efficiently. They can be made almost any size or shape and the components can be joined together in various ways (Fig. 1.6). Free-span trusses can be supported by a timber wall framing consisting of hinged or fixed columns. Free-span trusses normally range up to about 20 m but are most cost-effective between 10 and 15 m.

Portal Frames and Arches

Portals (rigid frames) and arches combine in the same structure the load-carrying properties of roof and wall constructions. The can be manufactured in many ways and assembled at the factory or on the building site (Fig. 1.7). Portal frames from straight columns and rafters with moment-resisting joints can be manufactured from glue-laminated timber, laminated veneer lumber, or timber and plywood box sections. The curved arch is the most economical structural form for the resistance of vertical loads over a large span. Normally, spans for this type of construction range between 12 and 30 m and are cost-effective for over 20 meters.

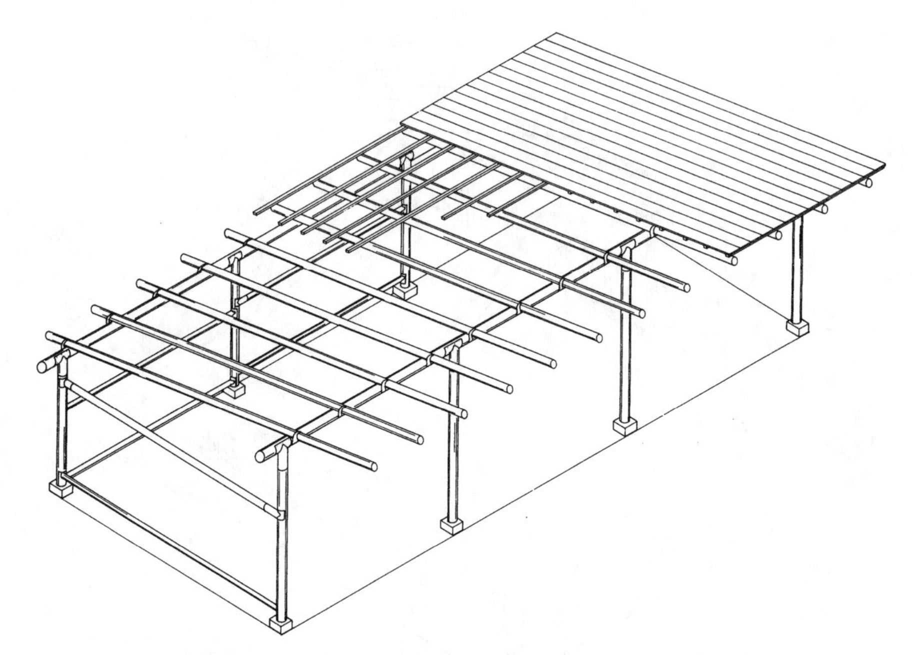

Figure 1.5. Example of beam and post construction in roundwood.

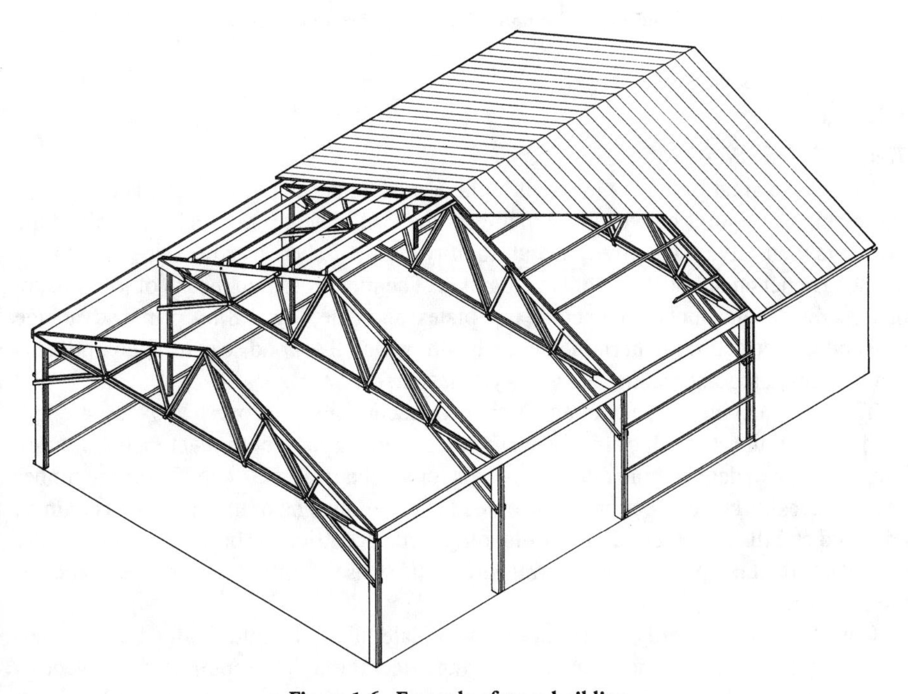

Figure 1.6. Example of truss building.

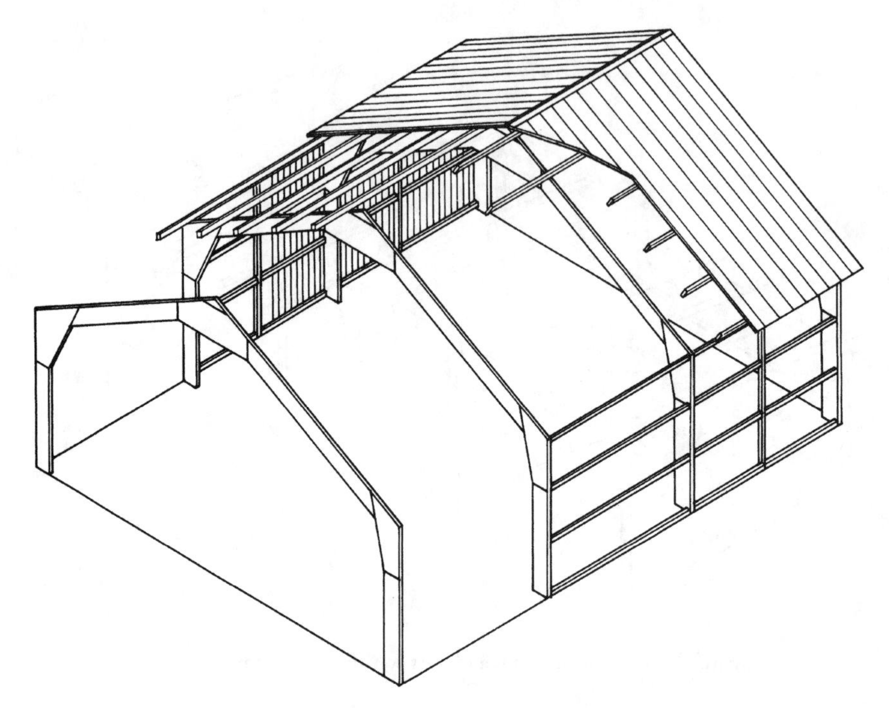

Figure 1.7. Example of structure of rigid frames.

1.3.3 Joints

Timber Connections

One of the great problems inherent in the use of timber in construction is the design and making of a joint between two members that will develop as much strength as the members themselves. The strength and stability of any timber structure depends heavily on the fasteners that hold its members together. The most common means of jointing are nails, screws, bolts, metal connectors, nail plates, and glue (Fig. 1.8). A prime advantage of wood as a structural material is the ease with which the wood structural components can be jointed together using this variety of fasteners.

The load-carrying capacity of any timber connection may be governed by the strength of the timber, the strength of the fastener, or a combination of both. Factors that require consideration in determining allowable loads for mechanically fastened joints are timber species, density, critical section, angle of load to grain, spacing of mechanical fastenings, edge and end distances, conditions of loading, and eccentricity. The species and density of timber affect the permissible bearing stress and consequently the lateral resistance of the fastener.

Connections are usually loaded in two ways, laterally or longitudinally. Lateral loading, which is more common, causes a shearing effect through the connector. Connectors must be designed to handle the shear loads at joints. The other type of loading on a connector is withdrawal loading.

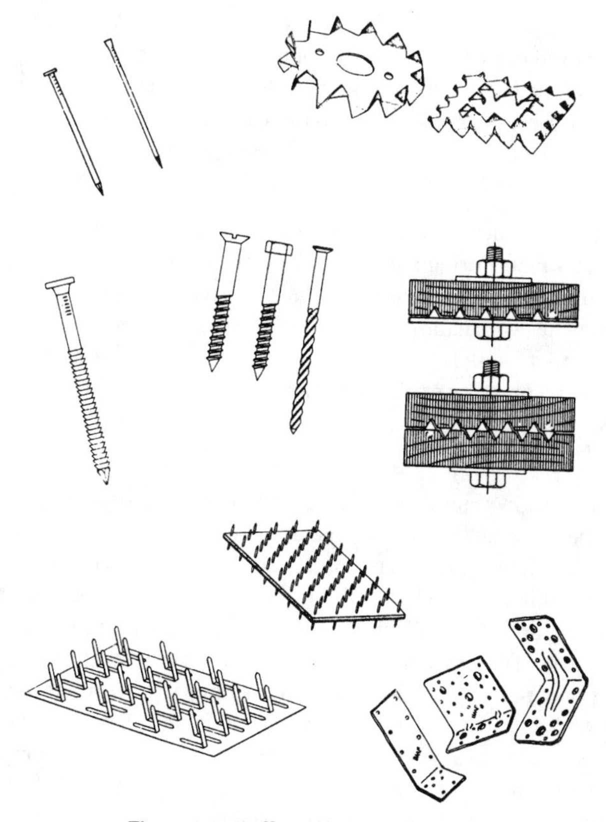

Figure 1.8. Different kinds of fasteners.

Selection of Fasteners

There are many interrelated factors that affect the selection of fasteners for a particular construction. The main considerations are the load-carrying capacity required, the thickness of timber, the method of manufacture, the type of loading, and assembly or erection procedures. The use of nails is very well suited to do-it-yourself constructions, but for heavy framed assemblies the choice of bolts and metal connectors is preferable. The size of members in a connection is often determined not by stress considerations but by the spacing and edge-distance requirements of the fixings. The use of glue in particular requires high-quality workmanship and strict quality-control procedures.

Nails

Nails are the most common mechanical fasteners in rural constructions. They are used to join solid timber members together or wood-based sheathing or steel plates to solid wood. There are many types, sizes, and shapes of nails, which can be used in a variety of applications. In farm buildings, nails with corrosion-resistant coatings are recommended because of moisture conditions. Gun nails or air-driven nails permit an economic fixing of a large number of nails in shearwalls or moment-resisting joints. The ultimate strength of a nailed joint depends on the nail penetration, coating, and diameter.

Bolts

Bolts are often used in roof trusses or in connections with beams of large dimensions. They can be used alone but are commonly used together with metal connectors in order to increase the load-carrying capacity and to reduce deformations. The metal connectors are installed only in the outer fibers and do not bend or deform.

Nail Plates

Connectors of sheet metal with or without punched nails are mostly fixed to the outside of wood members during assembly. They are used in the manufacture of trussed rafters or on building sites. Punched metal plates consist of 1.0- to 1.5-mm thick hot dip galvanized metal plates with nails punched out from the plate by a stamping process. The nails are short, slender, and closely spaced. The teeth of the nail plates are forced into the wood members by pressure equipment. Roundwood members can be joined together by using molded pre holed hand nail plates.

Glued Joints

The fabrication of glued joints should be done under factory-controlled conditions. This is because there are several factors that affect the strength of the glued joint. Wood members to be glued should have clean, machined, and dry surfaces that fit well together. The moisture content in the wood should not exceed 15%. The glued pieces are pressed together by clamping or nailing and kept in a room with suitable temperature and humidity for the curing period. The advantages of glued joints are that, if properly made, they are quite strong and rigid. A well-manufactured glue joint exhibits fatigue behavior similar to the wood that it is joining. When choosing adhesive for rural constructions some essential factors such as weather resistance, resistance to microorganisms, effect of preservative treatments, effect of heat (fire), durability, and cost factors should be taken into account. Two types of synthetic resin adhesives can be expected to satisfy the mentioned requirements, namely resorcinol formaldehyde and phenol formaldehyde.

1.3.4 Panel Products

Types of Panel Product

Wood-based sheet materials can satisfactorily be used in new farm buildings as well as in rebuilding, maintaining, or repairing old ones. The many uses include cladding, flooring, sheet bracing in walls and ceilings, and components such as web in I-beams and box beams. Panel products consist of processed wood material (veneers, sawdust, chips, strips, shavings, flakes, and fibers) of various sizes, geometries, and species bound together to form sheets. The categories usually recognized within this group of panel products are plywood, particleboards, and fiberboards.

Plywood

Plywood is an assembled product comprising thin layers of wood (veneers or plies) bonded together with the grain usually at right angles (Fig. 1.9). Selected logs are rotary-peeled in a lathe to form a ribbon of veneer, which is dried, clipped to sheet widths, graded, and glued together. At production two types of adhesives are used,

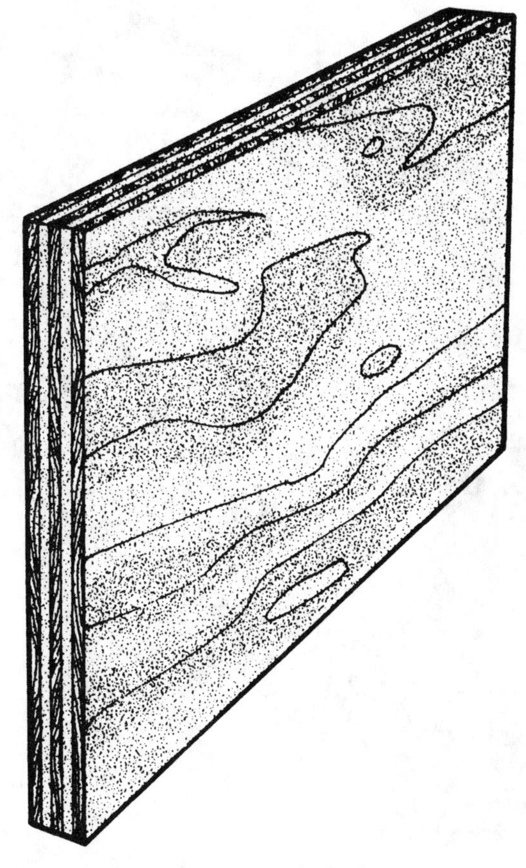

Figure 1.9. Plywood.

urea-formaldehyde resin glue and phenol-formaldehyde resin glue. The latter type of adhesive predominates in rural constructions. Grading rules for plywood are dependant on the country of manufacture and species. Grade names in general are based on the quality of the veneers used for the face and back of the panel.

Particleboards and Fiberboards

Particleboards, waferboards, strandboards, and fiberboards are composed of particles of various sizes or fibers obtained from refinement of wood chips (Fig. 1.10). Their proporties and performance are closely related to the type of particles used. There are basically two ways of making these materials. Particleboards are made with external bonding agents (adhesives) in a "dry" process, whereas fiberboards use water as a processing medium to form hydrogen bonding (softboard) or lignin bonding (hardboard). Adhesives are sometimes used to enhance the quality of wet-process boards.

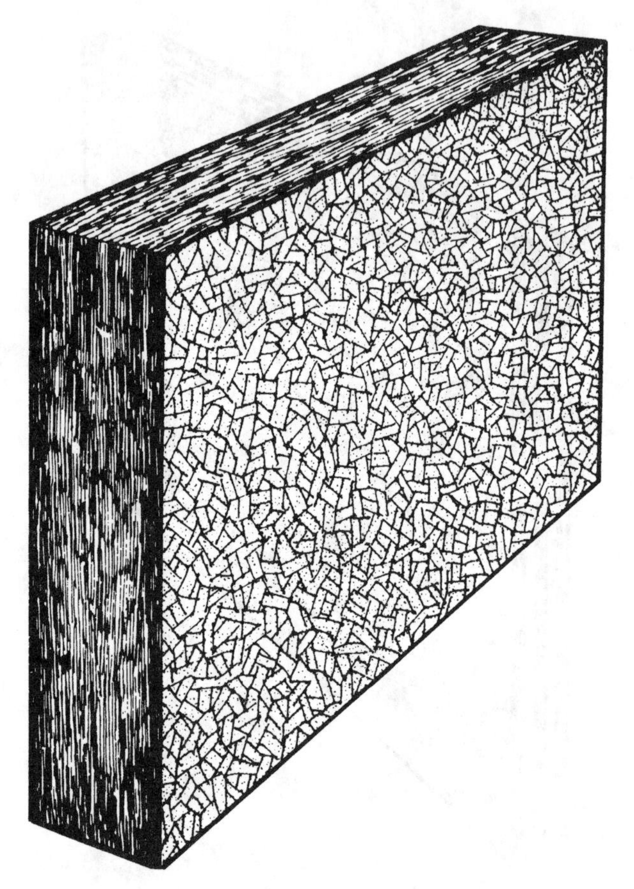

Figure 1.10. Particleboard.

Properties

Wood-based sheet materials are classified according to climate, strength, and surface fineness. Plywood has in relation to its weight high strength and stiffness. Its ability to withstand concentrated and impact loads is better than other wood material. The board properties are influenced by particle size and orientation, distribution of particles in the board, binder type, additives such as wax or emulsifiers, final density, and a range of pressing variables. Panel products are in some respects less sensitive to moisture than wood, but dimensional changes such as swelling and shrinking should be noticed.

Applications

Wood-based sheet materials can be used for a wide variety of applications in agriculture. They are most often used as flooring or sheathing to carry face loads that are applied perpendicular to the surface. The selection of material depends on a lot of factors including demands, material properties, and costs. A complete specification for purchasing wood-based sheet materials should include the following items: purpose, construction, thickness, size, quality, type of bond, and method of finishing. Fiberboards are as a rule the cheapest material, followed by particleboards. Plywood is expensive but is often the most cost-effective panel product due to its strength in load-carrying constructions, durability, and maintenance.

1.3.5 Wall Coverings

Aspects of Design

Today a lot of different cladding materials with a wide spectrum of colors, profiles, and textures are available on the market. It is necessary to choose cladding materials with skill and care so that they are in harmony with the surrounding landscape and existing buildings. A badly chosen exterior cladding material can devastate the appearance of the entire farmstead. It is important to remember that some materials weather well and their appearances improve with age. Others become faded and blotchy. Nonwood materials such as metal sheets, fibercement sheets and similar are available and commonly used.

Exterior wood cladding is light and durable. A correctly applied wooden facade has a long lifespan even with poor maintenance. The lifetime of a wooden facade depends on construction methods, surface treatment, timber quality, and climate. It is important that the facade is protected from permanent moisture and earth contact is well ventilated. The timber cladding also is easily repaired and replaced.

Weather Boarding

There are different ways of covering the wall with timber dependent on tradition and design, for example vertical and horizontal boarding (Fig. 1.11), diagonal sidings, and application of plywood or particleboard. Interesting effects can be achieved with the use of a combination of profiled boards. Plywood sheetings are easy to use and give interesting surface lines. A timber cladding combined with other materials such as concrete or bricks is common in farm building.

Vertical boarding is traditional in many countries and vertical spaced boarding especially is used in buildings in which good ventilation is needed. The spaced board is then often combined with a concrete or masonry wall. Whether horizontal or vertical boarding is technically preferable may be less important than correct detailing and nailing. The nails should be corrosion-resistant, that is, galvanized, to insure spot-free siding under adverse conditions.

Finishing

Wood-based sheet materials and weather boarding are best protected through a combination of constructive protection and painting. All surfaces to be finished should be clean and dry. Painting in intense sunshine should be avoided. Surfaces of wood claddings are given a prime coat as early as possible after setting up. The primer should be spread abundantly on the surface. Alkyd paints can be used as a priming paint. Two coats of good-quality latex, alkyd, or oil-based paint should be used over the nonporous primer. This is particularly important for surfaces that are fully exposed to weather or wet conditions. In order to avoid future separation between coats of paint or intercoat peeling the first topcoat should be applied within 2 weeks of the application of the primer and the second coat within 2 weeks of the application of the first coat.

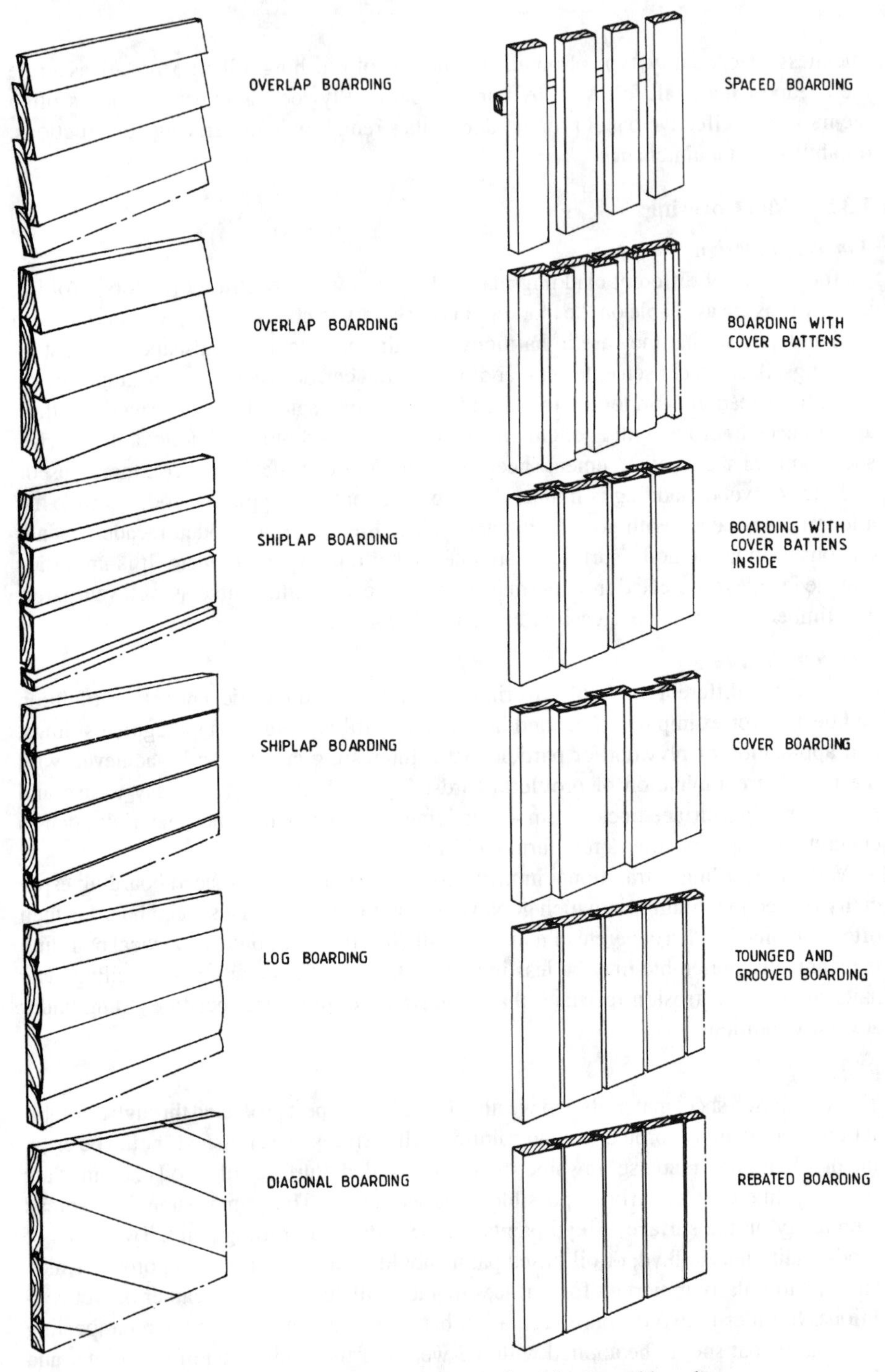

Figure 1.11. Different types of vertical and horizontal boarding.

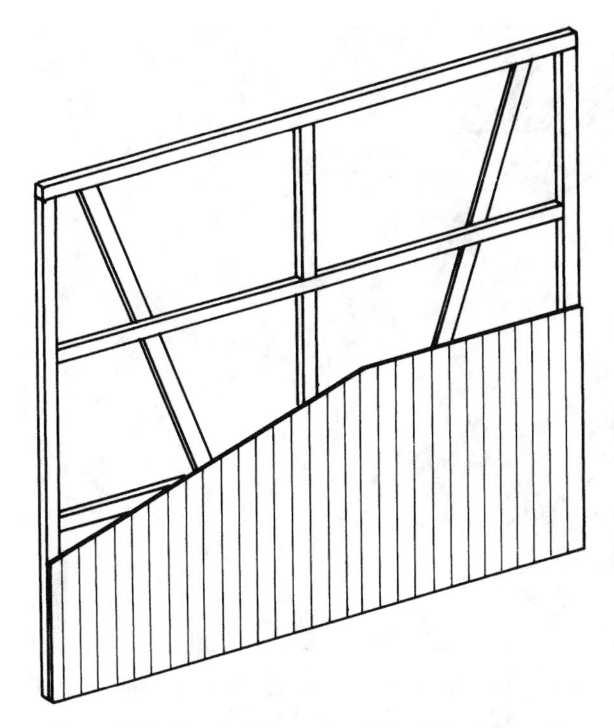

Figure 1.12. Doors and gates for rural applications can be manufactured in timber.

Doors and Gates

Doors and gates for rural application can be manufactured in timber (Fig. 1.12). Several types of doors are used such as hinged, sliding, folding, and overhead doors. They can be insulated or not and easily given a different cladding and thus can provide a variation in the facade. Doors with diagonal sidings fit in well with facades of horizontal cladding. As doors are frequently exposed to damage from vehicles the siding should be impact-resistant and easy to repair. Wooden panels are durable and easy to replace but heavy in weight. Plywood is strong and has an excellent bracing effect. Corrugated steel and aluminium sheets have low weight but are easily damaged and may be difficult to repair.

Roof Coverings

Roof coverings should provide a long-lived water-proof finish that will protect the building and its contents from rain, snow, and wind. Materials used for pitched roofs (Fig. 1.13) include metal sheets, fibercement sheets, wood sheathing, and tiles. Perhaps the most-used roof covering in agriculture is metal sheathing with materials such as aluminium and galvanized iron. The choice of roofing materials is usually influenced first by cost but also by local code requirements, building design, and preferences based on past experience.

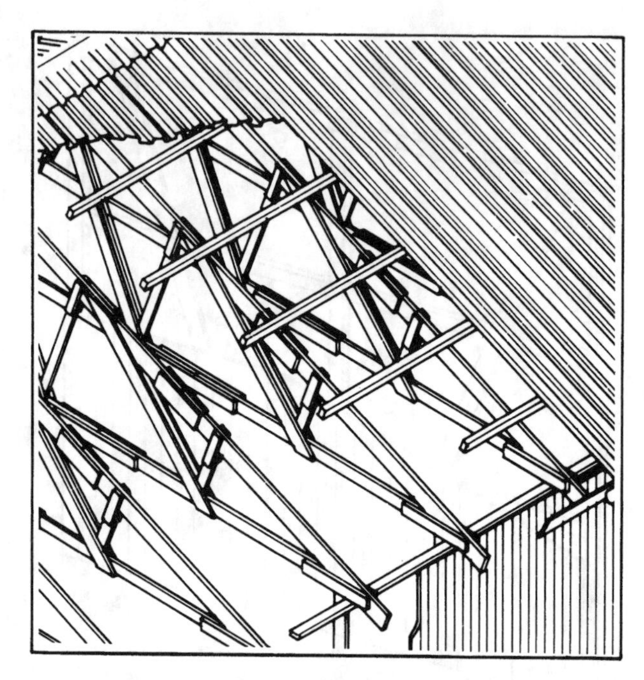

Figure 1.13. Uninsulated roof with purlins and cladding fixed to trusses.

The detailing is important and a large roof overhang not only gives the building a certain expression but also protects the timber facade against wind, rain, and sun.

Wood Protection

Wood is an organic material that is affected by its surroundings. The purpose of wood protection is to protect the timber from degradation. The main causes of degradation are fungal decay, insect attack, and degradation due to weathering. Wood protection can be achieved in two ways, by constructive design and by applying chemical preservatives (Fig. 1.14). The recommendation is to design against degradation before using wood treated with chemicals.

Wood Protection by Design

Timber building design, both in its concept and detail, should aim to

- Protect untreated timber from direct sun and rain
- Avoid details that trap moisture
- Avoid condensation points for moisture
- Insulate or isolate timber from sources of moisture
- Provide mechanical barriers to termites (if this hazard exists).

In realizing these aims, detailed design considerations for untreated timber should involve the following. Building sites should always be graded to provide positive drainage away from foundation walls. All exposed wood surfaces should be pitched to assure

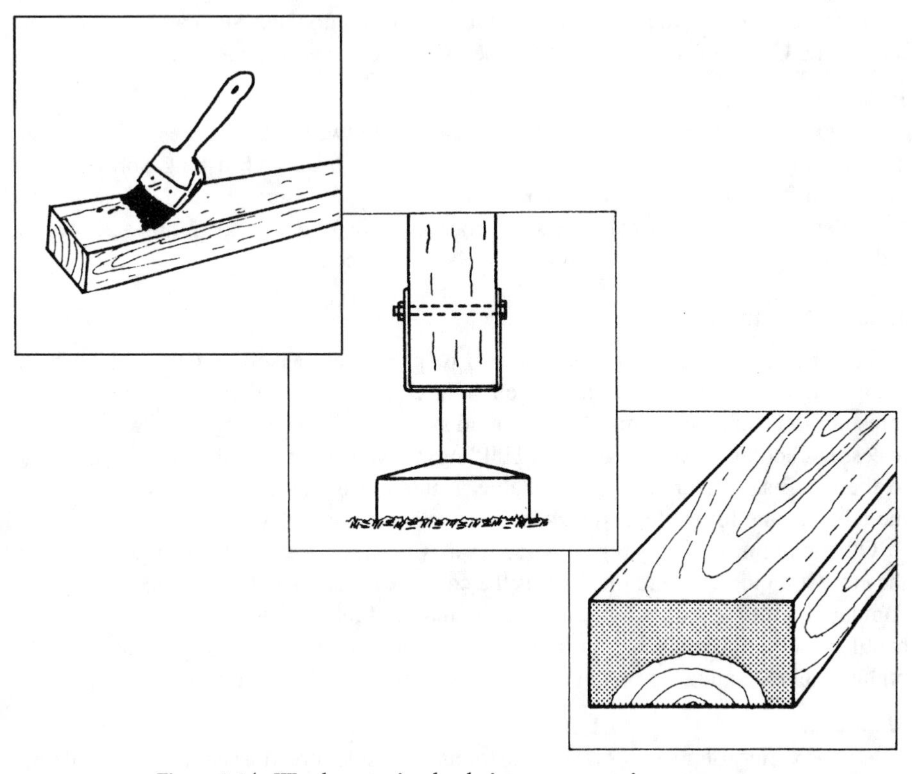

Figure 1.14. Wood protection by design or preservative treatments.

rapid runoff of water. Construction details that tend to trap moisture in end grain must be avoided. A fairly wide roof overhang with well designed gutters and downpipes is desirable. Wood in contact with concrete near the ground should be protected by a moisture-proof membrane such as heavy asphalt paper. In most cases preservative treatment of wood in actual contact with the concrete is advisable. Vapor barriers, if installed, should be near the warm face of insulated walls and ceilings. Roofs must be weathertight and attic cross-ventilation is desirable.

Chemical Wood Preservation

Wood can be protected from attack by fungi and insects by means of treatment with preservatives. The level of protection can be varied to match particular hazards by altering the type of preservative, its concentration in the wood, and its penetration into the wood. There are various methods of applying wood preservatives, some providing better penetration than others. The penetration also depends on the wood species, and in most cases heartwood is more resistant to treatment than sapwood.

Types of Preservatives

Wood preservatives are generally divided into three main groups, water-borne preservatives, organic-solvent preservatives, and creosotes or coal-tar types. Examples of types of water-borne preservatives are boric (mixtures of compounds such as boric acid and borax) and CCA (salts or oxides of Copper, Chronium, and Arsenic).

Preservation Processes

Wood-preservation treatments can be divided into two types, nonpressure processes and pressure processes. The nonpressure methods, for instance brushing, spraying, and dipping, differ widely in the penetration and retention of preservatives and consequently in the degree of protection. The pressure process obtain a deeper penetration and a higher retention, and accordingly the wood gets a better protection.

1.3.6 Thermal Insulation

In many farm buildings it is desirable to keep the heat transmission through the roof, walls, and floor to a minimum. Insulation of the constructions will help to reduce the heat transmission from the warm side to the cold side. It will keep the building warmer or, if necessary, cooler in hot climates. For buildings in which an internal temperature similar to the external temperature is acceptable, insulation is not necessary. Such buildings include housing for cattle, calves, and sheep and storage of hay and straw.

Other buildings housing pigs and poultry may need supplementary heating to maintain the correct temperature. In these cases the cost of insulation and its installation must be compared with the expenditure on energy, that is, fuel costs and the efficiency of the heating and control system. As long as the value of savings in energy expenditure is higher than the annual cost of insulation it is profitable to insulate.

Materials

A wide variety of materials is suitable for use as insulation. Most common is insulation made from low-density materials such as glass or mineral wool, expanded polystyrene, and polyurethane foam. The insulation material can be rigid, flexible, loose-fill, or reflective. Rigid insulation is usually made of polystyrene or polyurathane and available in board form. Flexible insulation consists of either blankets or batts.The materials used are mineral wools, which are fibrous materials made from glass or rock. Because flexible insulation has no rigidity it must be applied in spaces between other structural members. Loose-fill insulation is made from materials similar to those for rigid and flexible insulations. Loose-fill material requires a complete enclosure such as ceilings or wall cavities to hold it in place (Fig. 1.15). This insulation is available in bags. Relective insulation consists of metal foil or other reflective surfaces.

Selection of Materials

For most applications the insulation will only be effective if the insulation material is

- Unaffected by high levels of humidity
- Fire resistant
- Easy to clean
- Easy to fix over large areas
- Strong enough to withstand rough handling during and after installation.

These requirements should be considered in addition to the insulation value and cost.

Figure 1.15. Thermal insulation of glass or mineral wool in stud frame wall.

Reference

1. Dolby, C.-M., A. Hammer, and K.-H. Jeppsson. 1988. *Rural Constructions in Timber*. Alnarp, Sweden: Swedish University of Agricultural Sciences.

2 Environment for Animals

Some terms used in the livestock environment field are more conceptual than precise. Some authors have used a very broad sense of the term *environment* to include all non-genetic factors influencing the animal. In this case we have used a restricted definition to include only thermal factors, aerial factors, floor, and fittings influencing the animals. Housing of livestock is originally motivated to reduce climatic impacts on production and to facilitate management including control of diseases and improvement of health. Since the 1960s the general concern about the environment and pollution has made desirable control of the runoff or emission originating from animal wastes. Under real circumstances there are two main bases for practices: strict economic and animal welfare. Functional relationships between animal performance and weather parameters are therefore necessary to establish.

2.1 Animal Environment Requirements

Krister Sallvik

In many countries there has been a lot of work done to state "animal rights" or "welfare codes of practice," [1, 2]. These rights are both ethical and scientific. Within our definition of environment some key issues are

- Adequate fresh air
- Food, sufficient and of type to keep the animal in full health and vigor
- Freedom of movement (including the ability freely turn around and exercise limbs)
- Comfort of immediate environment (e.g., freedom from draughts, a bedded area)
- Freedom to follow innate behavior patterns except if there denial cannot reasonable be avoided.

Most of the codes of practice issued can be divided into four basic sections:

- Housing, which includes control of ventilation, air speed, temperature, humidity, gases, dust, noise, and lighting
- Space allowance and other critical measures for animals depending on species, age, sex, and system of husbandry
- The provision of food, water, and bedding in both intensive and extensive systems
- Management and planning, which include isolation facilities for culling, medical treatment, isolation, and outside shelters in extensive systems

2.1.1 Thermal

All farm animals are homeothermic, exhibiting a relatively constant long-term body temperature under the most varying conditions with large variations of internal heat production due to metabolism (for maintenance and production) and variation in the surrounding thermal environment (temperature, air velocity, radiation, bedding, etc). The surrounding environment can also add heat to an animal if the ambient air temperature exceeds its body temperature or by direct or indirect solar radiation. Consequently it is essential for agricultural engineers to understand animal heat balance from the animal and physiological view. It is recognized [3], however, that short-term fluctuations in body temperature occur because of metabolic processes, physiological changes, and environmental stressors. There is a diurnal rhythm (circadian) that largely reflects activity and feeding during the day and rest at night. There is even a rhythm of body temperature for ad-lib fed animals that is monophasic, with the maximum near midnight and the minimum in late morning to early afternoon. During hot conditions the range can be around 2°C. The brain substance cannot be exposed to temperature higher than 45°C without severe damages.

2.1.2 Thermoneutral Zone

The thermoneutral zone is a very important concept. The most-used graphical model presented by Mount 1973 is idealized and presented in Fig. 2.1. The definition of the

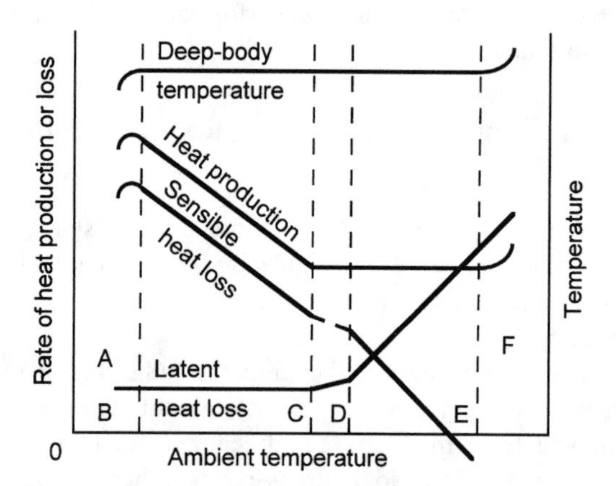

Figure 2.1. Diagrammatic representation of relationships between heat production, evaporative and nonevaporative heat loss, and deep-body temperature in a homeothermic animal. A: zone of hypothermia; B: temperature of summit heat metabolism and incipient hypothermia; C: critical temperature; D: temperature of marked increase in latent heat loss; E: temperature of incipient hyperthermal rise; F: zone of hyperthermia; CD: zone of least thermoregulatory effort; CE: zone of minimal metabolism; BE: thermoregulatory range. (*Source:* [4])

thermoneutral zone is the range of environmental temperatures within which the metabolic rate is minimum and independent of temperature. The temperatures that bound this zone are known as *upper* and *lower critical temperatures*. The lower is well defined (and also marked in Fig. 2.3). For the upper critical temperature there does not exist any agreement as to an absolute definition [5].

In a thermoneutral environment, animal total heat production can be calculated as the difference between the input of metabolizabled energy and energy retained in products and growth of body tissue [6]:

$$\phi_{tot} = [E_m + (1 - K_y) \cdot E_y + (1 - K_p) \cdot E_p + (1 - K_g) \cdot E_g] \cdot 11.57 \qquad (2.1)$$

where

ϕ_{tot} = Thermoneutral total heat production of the animal (W)
E_m = Metabolizable energy required for maintenance (MJ/d)
K_y = Efficiency of utilization for milk production
E_y = Metabolizable energy required for milk production (MJ/d)
K_p = Efficiency of utilization for pregnancy
E_p = Metabolizable energy required for pregnancy (MJ/d)
K_g = Efficiency of utilization for growth
E_g = Metabolizable energy required for growth (MJ/d).

The coefficients of utilization depend on the metabolizability of the feed. To calculate the total heat production equations from "the nutrient requirements of ruminant livestock" (ARC, 1980 or NRC, 1981) can be used.

There are two different situations for the animal thermoregulatory system, short and medium/long-term. Intensity and duration of thermal stressors as well as cycling or constant thermal environment must be considered. A dramatic change in thermal conditions must be taken care of by an alarm reaction known as stress [9]. For some species behavioral patterns are very important. The long-term responses to thermal environment include adaptation of feed and water intake, change of metabolism, and fur or coat insulation.

The concept of a variable setpoint temperature is found useful by physiologists to explain regulation of body temperature. The setpoint temperature apparently is modified by such factors as microbial infections and the thermal environment.

Until now the calculations of the thermoneutral zone by heat balance equations is static and assume steady-state conditions. The thermoregulatory system are, in the animals, however, much more sophisticated than that. These systems have been developed to guarantee survival of animals under very sever weather conditions, which means that the control system both integrates thermal signals and must prepare for even more severe conditions. This is shown in both beef and dairy cattle, in which increased feed intake occurs at higher temperatures than indicated by calculating lower critical temperature.

2.1.3 Heat Balance at Animal Level

The metabolic heat production of a homeothermic animal must be balanced by heat losses of the same magnitude. Thus, when examining an animal by heat-balance

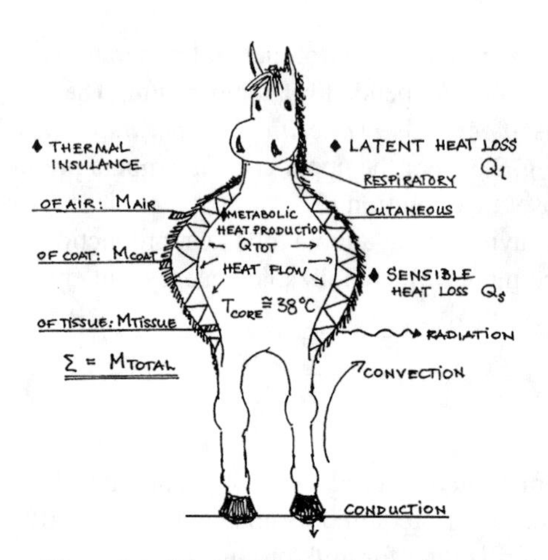

Figure 2.2. The principle of heat production and heat loss from an animal (horse). The three layers of thermal insulation are also shown. (***Source:*** **[10]**)

calculation and the consequences one must start from the inside of the animal (Fig. 2.2). This is very fundamental.

Thermoregulatory reactions are governed by the relation between the physiological requirement for heat dissipation and the physical potential for heat loss. The requirement for heat dissipation depends on the rate of internal heat production and exterior heat load. The potential for heat dissipation is determined by the thermal properties of the animal and its behavioral repertoire in relation to the thermal environment (including wet surfaces, bedding, and other animals). When characterizing the thermal environment, all physical factors affecting heat transfer and heat production in any form must be considered.

From the principle of energy conservation follows that metabolic heat production inside the animal must be balanced by an equal rate of sensible and latent heat dissipation to the environment together with the change in stored heat:

$$\phi_{\text{tot}} = \phi_s + \phi_l + \phi_{\text{sto}} \tag{2.2}$$

where

ϕ_{tot} = Thermoneutral total heat production of the animal (W)
ϕ_s = Sensible heat loss (W)
ϕ_l = Latent heat loss (W)
ϕ_{sto} = Rate of heat storage (W).

Because the thermoregulatory system of a homeotherm keeps the body temperature almost constant over a wide range of thermal environments the influence of heat storage on the heat balance is small. As a consequence of steady-state conditions the effect of heat storage can be ignored.

Sensible Heat Loss

The rate of sensible heat loss for a standing animal depends on the difference between the body temperature and the ambient temperature together with the thermal resistance of the body:

$$\phi_s = \frac{t_{\text{body}} - t_{\text{amb}}}{R_{\text{body}}} \tag{2.3}$$

where

ϕ_s = Sensible heat loss (W)
t_{body} = Deep body temperature (°C)
t_{amb} = Ambient temperature (°C)
R_{body} = Total thermal resistance of the body (K/W).

The thermal resistance of the body depends on the surface area of the body together with the thermal insulances of the tissue, coat, and boundary layers:

$$R_{\text{body}} = \frac{m_{\text{tiss}} + m_{\text{coat}} + m_a}{A_{\text{body}}} \tag{2.4}$$

where

R_{body} = Total thermal resistance of the body (K/W)
m_{tiss} = Thermal insulance of the body-tissue layer (m^2K/W) (Table 2.1)
m_{coat} = Thermal insulance of hair–coat layer (m^2K/W) (Table 2.2)
m_a = Surface thermal insulance of the surface–air boundary layer (m^2K/W) (see Eq. 2.4)
A_{body} = Surface area of the body (m^2) (Table 2.3).

The surface thermal insulance of the surface-air boundary layer is depending on radiant and convective conditions. Physicial properties in surroundings and for the animal influencing this is air velocity, surface temperatures and emission factors and the trunk diameter of the animal. The latter a function of body weight. Gustafsson [22] analysed

Table 2.1. Thermal insulance of body-tissue layer

	m_{tiss} (m^2K/W)		
Animal	Maximum	Minimum	Source
Cattle	$0.03 \cdot m^{0.33}$		[11]
Cattle	$0.017 \cdot m^{0.41}$	0.02	[12]
Bull	0.14	0.039	[13]
Calf	0.094	0.039	[13]
Horse	0.100		[10]
Pig	$0.02 \cdot m^{0.35}$	$0.01 \cdot m^{0.35}$	[14]
Piglet	0.065	0.029	[15]
Weaner	0.10	0.057	[15]
Pig adult	0.30		[15]
Sheep	0.2	0.078	[16]

Note. m = body weight (kg).

Table 2.2. Different classes of coats and their thermal insulance

Type of Coat	Depth of Coat (mm)	Thermal Insulance (m^2K/W)
Cattle, kept outdoors		
Very heavy winter coat	30–40	0.25
Heavy winter coat	20–30	0.20
Winter coat	15–20	0.15
Cattle kept indoors		
Very heavy coat	15–20	0.15
Heavy dense coat	10–15	0.12
Medium to dense coat	10–15	0.10
Light coat	5–10	0.06
Very light, close-cut coat	0.03	0.03
Cattle, general		$\frac{d \cdot 10^{-3}}{0.15}$
Horse		0.12
Sheep (Merino)	10	0.22
	20	0.31
	30	0.40
Sheep (Cheviot)	10	0.12
Chicken	5	0.10
	10	0.20
	20	0.40

Note. Data for windspeed < 0.2 m/s.
Sources: [10, 12, 13, 16].

Table 2.3. Expressions for body surface area of A_b, different animals

Animal	Surface area, A_b (m^2)	Source
General	$0.06 \cdot m^{0.67}$	[7]
Cow	$0.12 \cdot m^{0.60}$	[9]
Beef cow	$0.13 \cdot m^{0.56}$	[17]
Pig	$0.09 \cdot m^{0.66}$	[18]
Pig	$0.097 \cdot m^{0.633}$	[17]
Growing pig	$0.09 \cdot m^{0.633}$	[19]
Horse	$1.09 + 0.008m$	[20]
Chicken	$0.10 \cdot m^{0.67}$	[21]

Note. m = body weight (kg).

the radiative conditions and proposed the heat transfer coefficient for radiation to be 5.4 W/m^2K under normal housing conditions. Bruce and Clark [19] proposed following expression for the heat-transfer coefficient for convection:

$$\alpha_{\text{conv}} = \frac{15.7v^{0.6}}{m^{0.13}} \ (\text{W/m}^2\text{K}) \tag{2.5}$$

where v = air velocity (m/s).

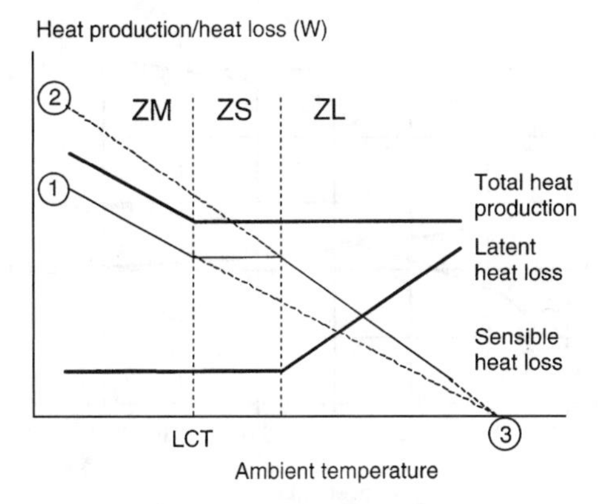

Figure 2.3. Basic thermoregulatory responses in relation to ambient temperature. ZM, ZS, and ZL are temperature zones with different types of regulation (metabolic, sensible, and latent regulation, respectively). LCT: lower critical temperature. (*Source:* [23])

The total thermal insulance of the surface–air boundary layer is then

$$m_a = \frac{1}{5.4 + 15.7\left(\frac{v^{0.6}}{m^{0.13}}\right)} \ (\mathrm{m^2\ K/W}). \tag{2.6}$$

Figure 2.3 shows how the thermoregulatory reactions of an animal are restricted by the limits imposed by the laws of physics. The line between points 1 and 3 corresponds to the lowest possible rate of sensible heat loss that corresponds to maximum thermal resistance of the body. Similarly, the line between 2 and 3 shows the highest possible rate of sensible heat loss with minimum thermal resistance. No sensible loss is possible if ambient (air and radiant) temperature is equal to body temperature.

Convective heat transfer depends on external windspeed (forced convection) or thermal buoyancy (free convection). Radiative heat losses are influenced by surface temperatures, radiating areas, and emission factors. There are many references with experiments and calculation of the heat transfer by convection and radiation, [12, 13, 19, 22, 24, 25].

Figure 2.4 shows how convective heat loss increases with increasing air speed. In the calculation both free and forced convection are taken into account. The convective heat dissipation lowers the coat surface temperature and the net result is that the increasing convective loss is partly compensated for by radiant heat dissipation. This compensating mechanism also moderates the effect of latent heat loss from the skin.

In Tables 2.1, 2.2, 2.3 some of these basic parameters are given for different types of animals.

For a lying or recumbent animal one must also consider conduction heat losses to the floor and the thermal properties of the floor including the bedding. Animal behavioral patterns and the number of animals in a group, resulting in reduced exposed area, must also be considered.

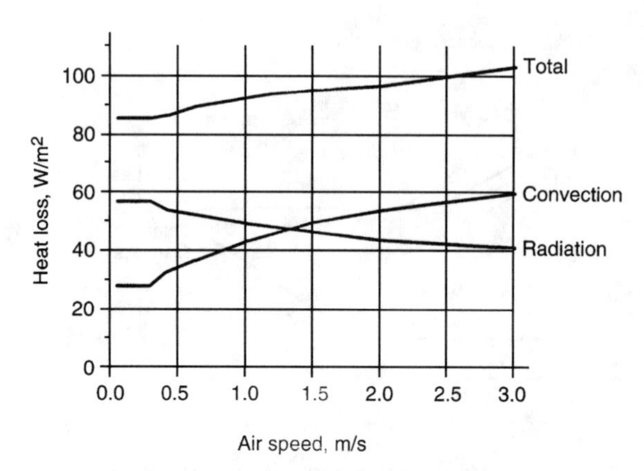

Figure 2.4. Convective, radiant, and total sensible heat dissipation from the surface of an animal. Convective heat loss considers both natural and forced convection. (*Source:* [12])

If ambient temperature rises above the limit at which the thermal resistance of the body is at its minimum (upper limit of ZS in Fig. 2.3), sensible heat loss decreases linearly because the governing temperature difference between the environment and the body decreases and the requirement of heat balance can only be met by increasing latent heat dissipation.

In a cooler environment the thermal resistance of the body increases (by increasing the thermal insulance or decreasing the exposed surface area) until it reaches its maximum level. The ambient temperature at which the thermal resistance is maximal corresponds to the lower limit of the thermoneutral zone. In Fig. 2.3 the range of temperatures within which the thermal resistance of the body gradually changes from its maximum to its minimum level is called the zone with sensible thermoregulatory control (ZS).

Even in a cool environment, some heat is lost through evaporation to the air from the respiratory tract and water-vapor diffusion through the skin. In the following subsections the latent loss at lower critical temperature is called *minimum latent heat loss*.

Latent Heat Loss

The theory of thermodynamics shows that water-vapor pressure differences rather than temperature differences govern latent heat transfer. The water-vapor pressure at the skin surface depends on the rate of diffusion through the skin and the rate of secretion from the sweat glands. Water also evaporates from the respiratory tract. Increasing the air exchange rate ("thermal panting") can increase this evaporation rate. The thermoregulatory system has a wide range of control over both sweat-gland activity and respiratory rate. Thus, from a physiological viewpoint the rate of latent heat loss is governed by the heat-dissipation requirement rather than the water-vapor pressure of the ambient air.

Experimental data (e.g., [26, 27]) indicate that the thermoregulatory system uses regulation of sensible heat loss instead of latent heat loss whenever possible. Thus, in the lower range of the thermoneutral zone heat balance is attained mainly through sensible regulation, while latent heat-loss regulation dominates the upper range. Mount [4] refers to the lower range as the "zone of least thermoregulatory effort."

The thermal resistance of the body depends only on its size and the thermal properties of the insulating layers and is not related to the requirement of cooling caused by the total heat production. Thus, in a specific thermal environment (in the upper range of the thermoneutral zone) animals with the same rate of sensible heat loss, the rate of latent heat loss will depend on their total heat production. This has been demonstrated experimentally by Blaxter et al. [26], who made measurements on sheep at different feeding levels and over a wide range of temperatures.

Minimal Latent Heat Loss

This parameter is needed to calculate the lower critical temperature for the animal, and it determines the level of evaporative heat loss in the zone with metabolic thermoregulatory control (ZM) and (ZS). Based on a literature review [12, 28], the minimal latent heat loss from cattle can be calculated as

$$\phi_{\mathrm{l\,min}} = 10 \cdot A_b + 0.07 \cdot \phi_{\mathrm{tot}} \tag{2.7}$$

where

$\phi_{\mathrm{l\,min}}$ = Latent heat loss (W)
A_b = Surface area of the body (m^2)
ϕ_{tot} = Thermoneutral total heat production of the animal (W).

This equation was chosen because it models both moisture diffusion through the skin (which is related to the body surface area) and respiratory heat loss. The respiratory heat loss depends on the rate of air exchange, which is related to the metabolic rate. The equation gives a minimal latent heat loss of 10% to 20% of the total heat production depending on the relation between body surface area and total heat production.

For pigs Bruce and Clark [19] give the following expression for minimum latent heat loss:

$$\phi_{\mathrm{l\,min}} = 0.09(8.0 + 0.07m)m^{0.67}\ \mathrm{(W)}. \tag{2.8}$$

Cold Environment

When ambient temperature decreases below the thermoneutral zone (below lower critical temperature), sensible loss increases due to the increasing temperature difference between the body and the environment. This induces an increase in metabolic heat production. Obviously, this increase in metabolic rate requires an adequate supply of metabolisable energy. The temperature range in which body temperature is controlled by regulating the metabolic rate is called the zone with metabolic thermoregulatory regulation (ZM) (see Fig. 2.2).

Hot Environment

Acute heat stress often increases the total heat production due to the muscle activity associated with panting [29]. After some time, the heat-stressed animal adapts by decreasing its metabolism and consequently its total heat production. At the present, it does not seem possible to calculate the upper limit of the thermoneutral zone using only the basic thermal properties of the body and the metabolic rate.

Models

Bruce and Clark [19] have presented a fundamental and widely used models for heat production and critical temperature for growing pigs. Sterrenburg and Ouwerkerk [18] have presented a computerized model known as "BEZOVA" for pigs. For cattle Ehrlemark and Sallvik [23] have presented "ANIBAL." Morgan [10] deals with horses; Bruce [11] with beef cattle; and Bruce, Broadbent, and Topps [30] with lactating and pregnant sows. Information on parameters, assumptions, and computer programs for calculating animal heat balances is many times available directly from the authors. However, there are very few new basic studies on physical and physiological parameters.

The models can be used to illustrate lower critical temperature under differences in conditions such as production or feed intake level, air velocity, coat condition, floor and bedding, or number of animals in a group.

Tables 2.4 and 2.5. Illustrate for cattle and pigs respectively very big differences in lower critical temperature.

Table 2.4. Lower critical temperature (LCT) for cattle

Animal	Body Weight (kg)	Production (kg/d)	LCT (°C)
Milking cow 240-d pregnant	500	0	−18
Milking cow	500	15	−30
Milking cow	500	35	−50
Beef cattle	75	0.7	10
Beef cattle	150	0.9	−5
Beef cattle	500	0.9	−20
Replac heifer	250	0.55	0
Replac heifer 240-d pregnant	500	0.5	−20
Calf	50	0.5	10

Note. Assumptions: air speed 0.1 m/s coat depth for cattle 10 mm.
Source: [23]

Table 2.5. Lower critical temperature (LCT) for piglets

Pig Weight (kg)	LCT (°C), Concrete Floor		LCT (°C), Expanded Metal Floor		LCT (°C), Straw Bedding	
	1 Pig/Pen	10 Pigs/Pen	1 Pig/Pen	10 Pigs/Pen	1 Pig/Pen	10 Pigs/Pen
2*maintenance						
1	31	29	28	26	27	23
5	28	23	26	24	24	20
10	27	24	25	22	23	18
4*maintenance						
1	27	24	23	20	22	17
5	23	20	20	17	18	12
10	21	17	18	15	16	9

Source: [31].

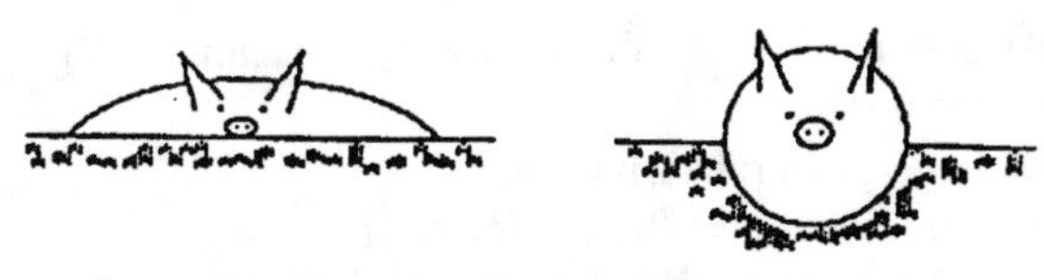

Figure 2.5. A pig with 50% of its body surface in contact with a solid floor compared with a pig on deep straw bedding according to the concept by Bruce [33]. The figure shows the unrealistic physiological consequences. (*Source:* [32])

Within the same kind of animal (e.g., milking cows) there is a very big influence of the lactation stage. Lower critical temperature for a dry but highly pregnant cow is −18°C and for a high yielding cow −50°C. For pigs there is shown the importance of natural behavior with access to bedding, littermates as well as of feed intake. A single pig on concrete has a lower critical temperature of 31°C compared to a litter of 10 pigs eating four times maintenance kept on straw having a lower critical temperature of 9°C. For horses Morgan [10] estimated the lower critical temperature as 18°C if they were acclimatized to indoor conditions and fed for maintenance and light work.

When using some of these models in new types of husbandry systems such as deep-straw bedding one must be aware of great errors in the results [32]. This is illustrated in Fig. 2.5.

2.2 Animal Heat and Moisture Production

Krister Sallvik and Soren Pedersen

All farm animals are homeothermic and must keep their body temperatures constant. The animals dissipate heat partly to maintain essential functions (Φ_m maintenance) and partly due to production. The total heat dissipation from an animal, Φ_{tot}, is mainly dependent on:

- Body mass
- Production (milk, meat, egg, foetus)
- The proportion between lean and fat tissue gain
- Energy concentration in the feed

2.2.1 Equations for Total Heat Production, Φ_{tot}

Equations for total heat production, Φ_{tot}, presented below follow the general Eq. (2.1) and are based on references [6–8, 34–37]. The first part of the equations gives the heat dissipation due to maintenance, Φ_m, and is a function of the metabolic weight; for example, for a cow Φ_m is $5.6 \cdot m^{0.75}$.

Notations

Φ_{tot} = animal total heat dissipation in the barn (W)
Φ_d = daily feed energy intake (W)
Φ_m = heat dissipation due to maintenance (W)

Φ_{prod} = heat dissipation due to the production by the animal (W)
m = animal weight (kg)
n = daily feed energy in relation to Φ_m
K = factor for additional heat dissipation from horses
K_t = coefficient for proportioning between sensible and latent heat dissipation
K_y = coefficient of efficiency at weight gain
Y_1 = milk production (kg/d)
Y_2 = meat and egg production (kg/d)
M = concentration of energy in feed, MJ/$kg_{dry\ matter}$
p = number of days pregnant

Cattle

Calf:

$$\Phi_{tot} = 6.44m^{0.70} + \frac{13.3Y_2(6.28 + 0.0188m)}{1 - 0.3Y_2}$$

Y_2 is normally 0.5 kg/d.

Veal calf, beef cattle:

$$\Phi_{tot} = 7.64m^{0.69} + Y_2\left[\frac{23}{M} - 1\right]\left[\frac{57.27 + 0.302m}{1 - 0.171Y_2}\right]$$

$Y_2 = 0.7 - 1.1$ kg/day
M for roughage $= 10$ MJ/$kg_{dry\ matter}$
M for concentrate $= 11 - 12$ MJ/$kg_{dry\ matter}$

Cow replacement:

$$\Phi_{tot} = 7.64m^{0.69} + Y_2\left[\frac{23}{M} - 1\right]\left[\frac{57.27 + 0.302m}{1 - 0.171Y_2}\right] + 1.6 \cdot 10^{-6}p^3$$

$Y_2 = 0.6$ kg/d. M is according to veal calf and beef cattle

Cow:

$$\Phi_{tot} = 6.6m^{0.75} + 22Y_1 + 1.6 \cdot 10^{-6}p^3$$

Pigs

Piglet:

$$\Phi_{tot} = 7.4m^{0.66} + (1 - K_y)(\Phi_d - \Phi_m)$$

$K_y = 0.47 + 0.003m$
$\Phi_d = n \cdot \Phi_m$

Fattening pig:

$$\Phi_{tot} = 5.09m^{0.75} + (1 - K_Y)(\Phi_d - \Phi_m)$$

K_y and Φ_d are according to piglet.

Sow, boar, gilt:

$$\Phi_{tot} = 5.5m^{0.76} + 28Y_1 + 2 \cdot 10^{-6}p^3 + 76Y_2$$

Y_2, pregnant sow = 0.18 kg/d
Y_2, pregnant gilt = 0.62 kg/d

Nursing sow:

$$\Phi_{tot} = 5.5m^{0.76} + 28Y_1$$

$Y_1 = 6$ kg/d

Other Kinds of Animals

Horse:

$$\Phi_{tot} = 6.1m^{0.76} + K \cdot \Phi_m$$

$K = 0$ for horses in little work/training
$K = 0.25$ for horses in moderate work/training
$K = 0.50$ for horses in hard work/training

Laying hen, in cage:

$$\Phi_{tot} = 6.28m^{0.76} + 25Y_2$$

Laying hen, on floor:

$$\Phi_{tot} = 6.8m^{0.76} + 25Y_2$$

Broiler:

$$\Phi_{tot} = 10m^{0.75}$$

Sheep:

$$\Phi_{tot} = 6.4m^{0.76} + 33Y_1 2.4 \cdot 10^{-5} \cdot p^3$$

Y_1 nursing ewe = 1 – 1.5 kg/d

Lamb:

$$\Phi_{tot} = 6.4m^{0.76} + 145Y_2$$

$Y_2 = 0.25$ kg/day

Heat Producing Unit

In the previous text, the animal total heat production refers to a single animal under thermoneutral conditions. Sometimes, it is more convenient to refer to a specific *heat production unit*, e.g. as used below, defined as a quantity of animals producing 1000 W in total heat at 20°C [6]. To illustrate an animal's adjustment of feed intake due to low and high temperatures respectively, and its effect on the total heat production, Eq. (2.9) according to [6] can be used. However it should be understood that Eq. (2.9) does not take into account neither size of animal nor production level i.e. the actual upper and lower critical temperature.

$$\Phi_{tot} = 1000[1 + 4 \cdot 10^{-5}(t + 10)^4] \quad (2.9)$$

In this way Eq. (2.9) expresses the animal total heat production at different indoor temperatures.

2.2.2 Proportion Between Sensible and Latent Heat Dissipation

It is essential to separate sensible and latent heat dissipation when calculating ventilation demand and judging animal comfort. Experimental results on sensible and latent heat production are rare because normally experiments are focused on total heat. In Fig. 2.3 it is shown that

$$\Phi_{tot} = \Phi_s + \Phi_l. \quad (2.10)$$

Φ_s is dissipated proportional to the temperature difference between the animal surface and the ambient air. Consequently Φ_s is zero if the surrounding air temperature is equal to animal surface temperature, that is, around 40°C.

Φ_l is dissipated in the form of moisture. To maintain animal heat balance and body temperature, Φ_l increases with increasing temperature. The proportion between Φ_s and Φ_l is affected also by kind of animal, production, surface area, fur, and dryness of the skin.

To calculate heat and moisture dissipation in the barn, Φ_{tot} must be distributed between Φ_s and Φ_l. The proportional sensible heat is given by

$$\Phi_s = \Phi_{tot}[0{,}8 - \alpha(t_x - K_t)^4] \quad (2.11)$$

where

K_t = coefficient when proportioning between sensible and latent heat dissipation could be the same numerical value as lower critical temperature
t_x = measured or assumed ambient temperature
α = nondimensional factor
$\alpha = \frac{0.8}{(40 - K_t)^4}$

If K_t is unknown, one should choose

$K_t = -10$
$\alpha = 1.28 \cdot 10^{-7}$

One can get a model of how an animal dissipates sensible heat by using normal heat-transfer equations as shown in Fig. 2.3. The following example shows this.

Example

Assumptions for one fattening pig are as follows:
Body mass = 45 kg
Daily feed intake = 3.5 times maintenance

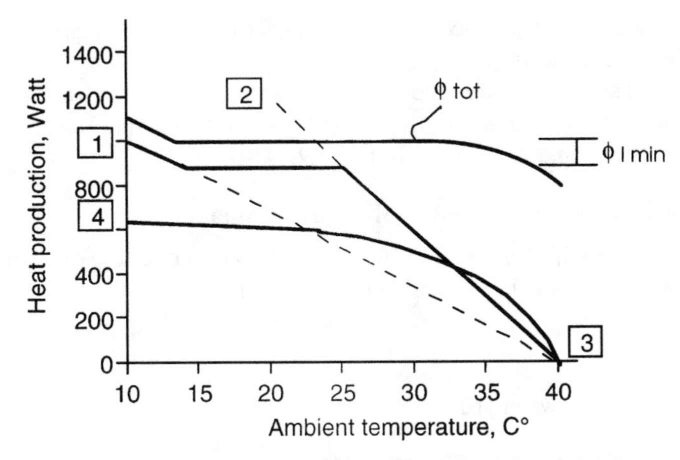

Figure 2.6. Distribution of sensible and latent heat dissipation from a single 45-kg pig (Φ_{tot} converted to 1000 W at 20°C) at house level (line 3-4) and at animal level (line 3-1).

$\Phi_{tot} = 174$ W (according to equation for Φ_{tot} for fattening pigs presented previously)

$A_b = 1.11\ m^2$ (according Table 2.3)

Pig standing up, no conduction heat losses Total A_b takes part in the heat and moisture dissipation

Radiant temperature = ambient air temperature

m_{tiss} maximal = 0.077 (m^2K/W) (according Table 2.1)

m_{tiss} minimal = 0.038 (m^2K/W) (according Table 2.1)

$m_a = 0.12$ (m^2K/W) (air velocity 0.15 m/s and according to Eq. [2.6])

$R_{max} = 0.177$ (m^2K/W) (according to Eq. [2.4])

$R_{min} = 0.144$ (m^2K/W) (according to Eq. [2.4])

$\Phi_{l\,min} = 23{,}28$ (W) (according to Eq. [2.7])

Converting these results for any animal producing. Φ_{tot} is 1000 W at 20°C, these assumptions are illustrated practically in Fig. 2.6, explaining how animals physiologically regulate their body temperatures within the laws of thermal physics, as generally shown in Fig. 2.3 previously.

From the total heat production, Φ_{tot}, one must first deduct the minimum latent heat dissipation $\Phi_{l\,min}$, and thus the remaining part of the Φ_{tot} is available for sensible heat dissipation. The line between points 1 and 3 represents the sensible heat dissipation at R_{max}. The point (temperature) at which the heat dissipation according to R_{max} equals ($\Phi_{tot} - \Phi_{l\,min}$) is the lower critical temperature. At temperatures below the lower critical temperature Φ_{tot} must increase so that the pig can maintain body temperature.

At the point (temperature) at which the sensible heat loss at R_{min} represented by the line between 2 and 3 is not enough to balance the heat production, the latent heat must increase. This temperature however, is not equivalent with the upper critical temperature.

In reality animals perform a much smoother transfer between these principles for the thermoregulation of their body temperature.

At house level some of the sensible heat is used to evaporate water from wet surfaces, feed, and manure. This changes the proportion between Φ_s and Φ_t. This is exemplified by the line between 3 and 4. See also Section 2.2.4.

2.2.3 Conversion of Latent Heat to Moisture Dissipation

The latent heat dissipation, Φ_l, in watts is transferred to moisture dissipation, F, in grams per hour and is calculated according to following:

$$F = \frac{\Phi_l}{r} \tag{2.12}$$

where $r = 0.680$ Wh/g water at 20°C

2.2.4 Heat and Moisture Production at House Level

The heat and moisture production at house level is based on animal level of heat and moisture production and then the influences of various factors. These factors mean that more moisture evaporates with heat from the animal's sensible heat. Factors include flooring system, stocking, density, watering, moisture content of the feed and feeding system, animal activity, and relative humidity.

In Fig. 2.6 the line between 1 and 3 represents Φ_s at animal level at vasoconstriction; the line between 2 and 3 represents Φ_s at vasodilation. The upper curve is Φ_{tot}.

The difference between the lines for Φ_{tot} and $\Phi_{l\,min}$ is Φ_l, which with increasing temperature increases according to the line between 2 and 3. At house level Φ_s is dissipated according the line between 3 and 4.

Equation 2.13 gives a general expression of the sensible heat dissipation at house level within the ambient temperature range from 10°C to 40°C under housing conditions similar to what are standard in Northern Europe

$$\Phi_s = a + bt_{amb} + ct_{amb}^2 + dt_{amb}^3 \tag{2.13}$$

where a, b, c, and d are coefficients (see Table 2.6) and t_{amb} is ambient temperature in centigrades or °C. The result is shown in Figs. 2.7 through 2.12.

The figures are mainly based on results with indoor temperatures in the normal range of 15 of 30°C.

Table 2.6. Coefficients for Eq. (2.13) on sensible heat

				Layers		
Coefficients	Basic	Cattle	Pigs	Cages	Floor	Broiler
a	884	752	876	982	681	805
b	−7,404	−6,294	−42,597	−46,984	−1,190	−39,216
c	−0,0381	−0,0324	2,4200	2,6788	−0,3111	2,2288
d	−0,0082	−0,0070	−0,0475	−0,0529	−0,0021	−0,0438

Note. Coefficients defined for ambient temperatures in the range of 10° to 40°C

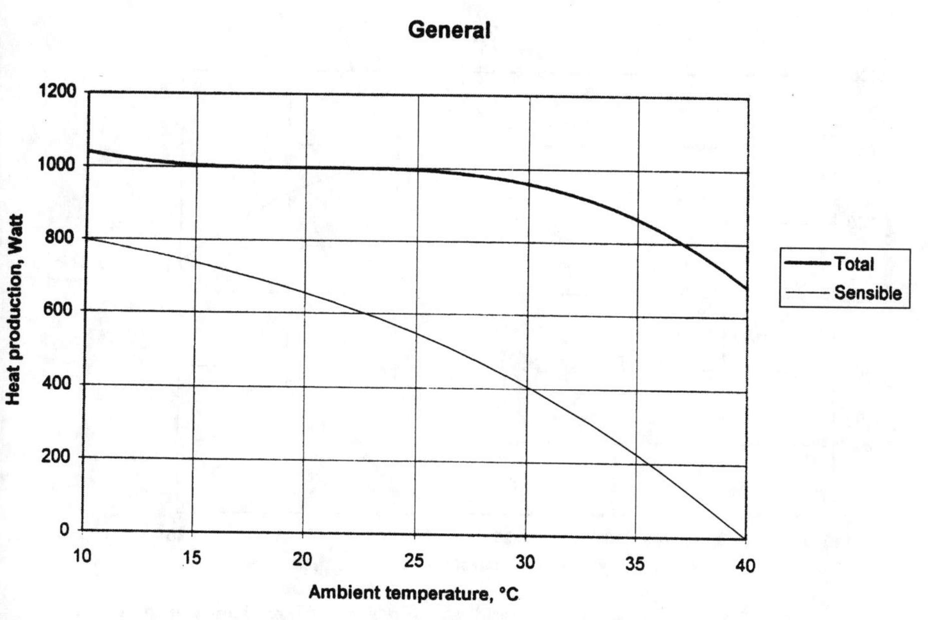

Figure 2.7. Sensible and latent heat dissipation at house level. General animal house.

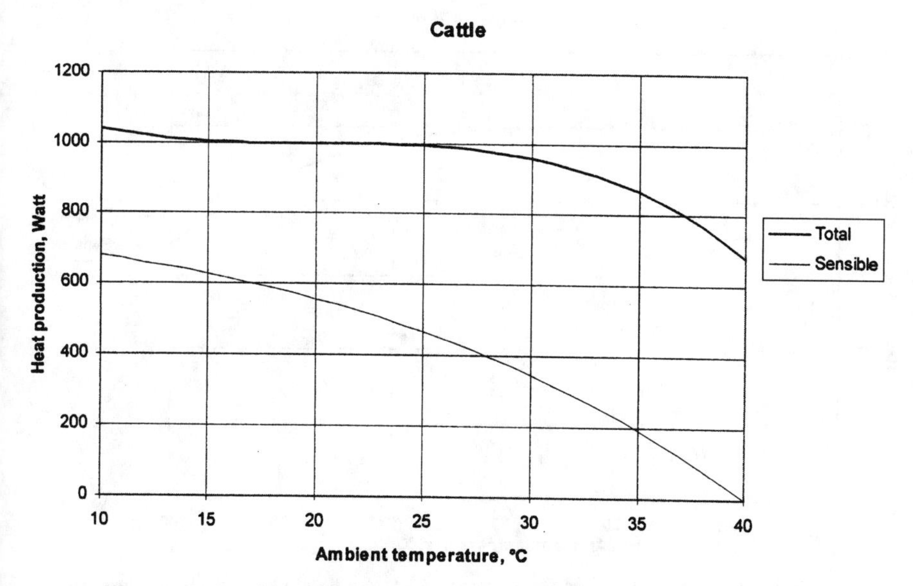

Figure 2.8. Sensible and latent heat dissipation at house level. House for cattle.

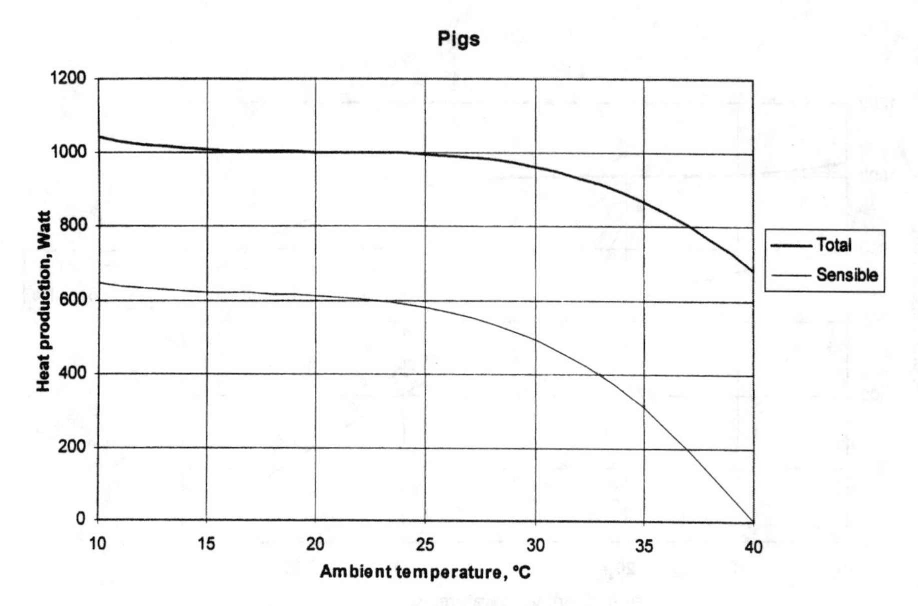

Figure 2.9. Sensible and latent heat dissipation at house level. House for pigs.

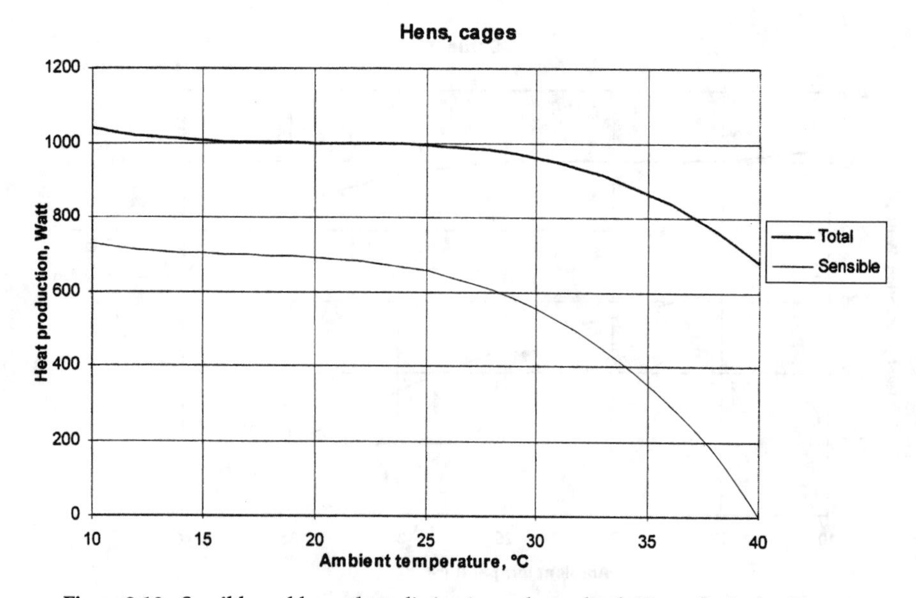

Figure 2.10. Sensible and latent heat dissipation at house level. House for laying in cages.

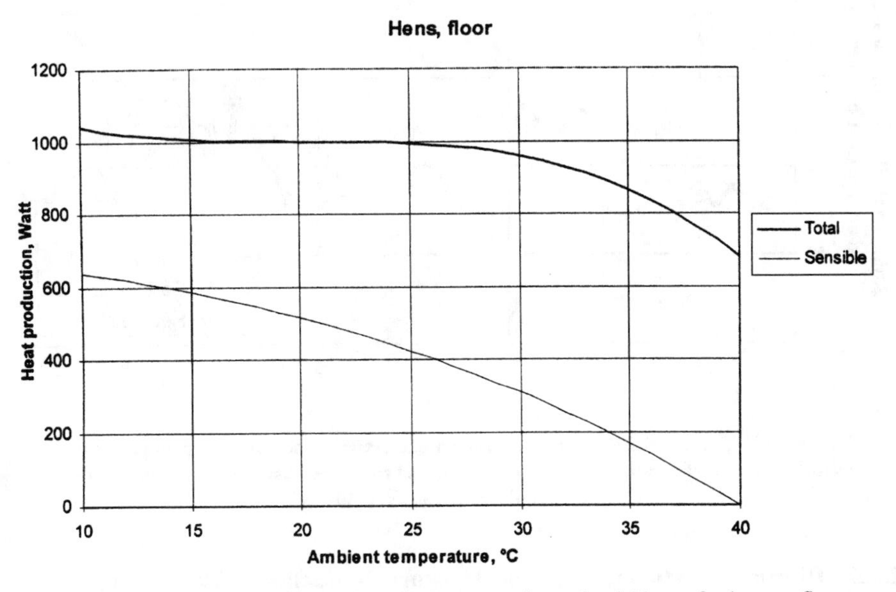

Figure 2.11. Sensible and latent heat dissipation at house level. House for hens or floor.

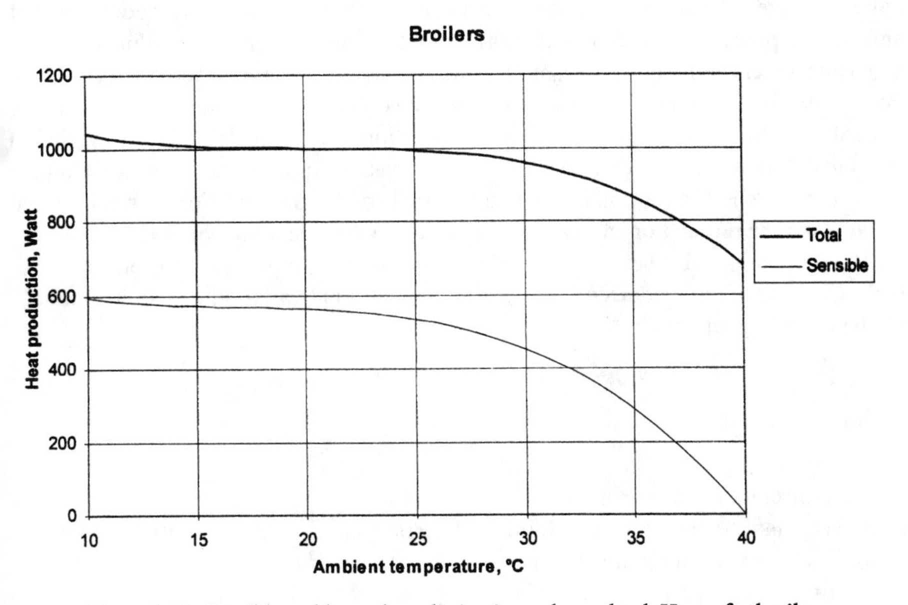

Figure 2.12. Sensible and latent heat dissipation at house level. House for broilers.

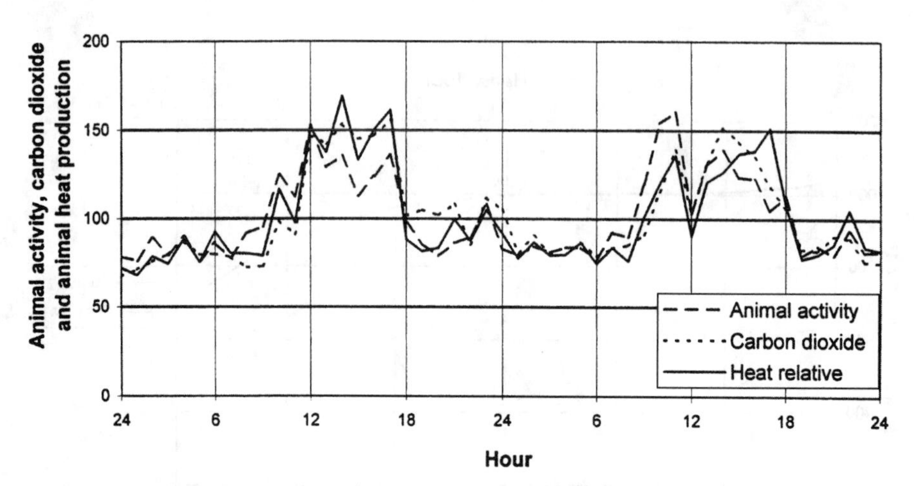

Figure 2.13. The diurnal variation between animal activity (measured by passive infrared detector) in a house for fattening pigs and the animal total heat dissipation and production of carbon dioxide. (*Source:* [39])

2.2.5 Diurnal Variation in Heat and Moisture Production at House Level

Normally, animal heat production is expressed by steady state equations, as shown previously. In commercial animal houses, there are various reasons why heat production varies during the 24-h cycle, for example because of feeding intervals, working routines, or light regimes. It is not possible at present to give worldwide equations for adjustment of animal heat production according to diurnal rhythm, but for some well-defined housing and management situations information is available based on animal activity recordings. For other specific production systems it is expected to be possible in the future to make diurnal corrections. The variation in heat dissipation is very closed related to variation in animal activity, which can be registered by passive infrared detector, a technique presented by Pedersen [38]. In Fig. 2.13 an example of the close relation between animal activity and the production of heat, Φ_{tot}, and carbon dioxide is shown [39].

The correlation, R^2, between animal activity and heat production for pigs is about 60%. Ouwerkerk and Pedersen [40] expressed the diurnal variation in activity and Φ_{tot} by the following equation:

$$A = a + b \cdot \sin[(2 \cdot \pi/24) \cdot (t + 6 - h_{min})] \tag{2.14}$$

where

A = animal activity
a = constant (average over 24 h)
b = constant (maximum and minimum deviation from average activity)
h_{min} = time with minimum activity
t = time
time displacement = 6

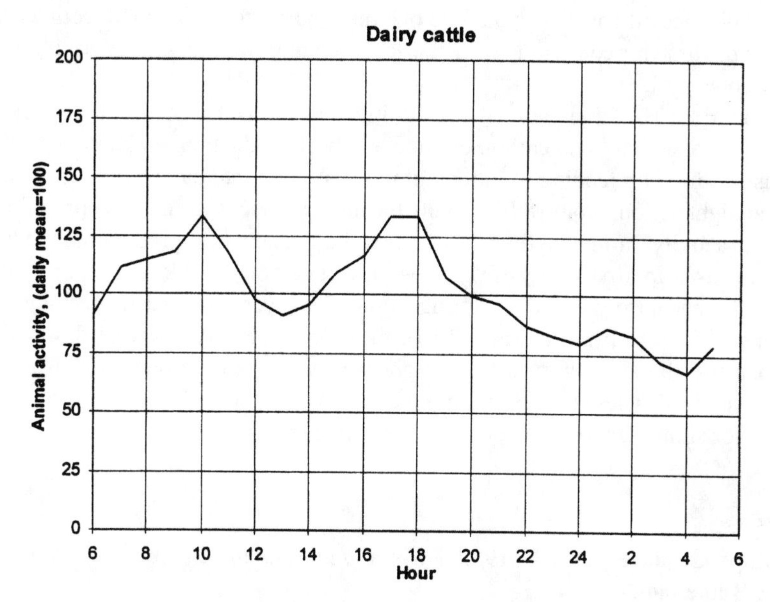

Figure 2.14. The diurnal variation of animal activity (measured by passive infrared detector technique) in houses for dairy cows. (*Source:* [41])

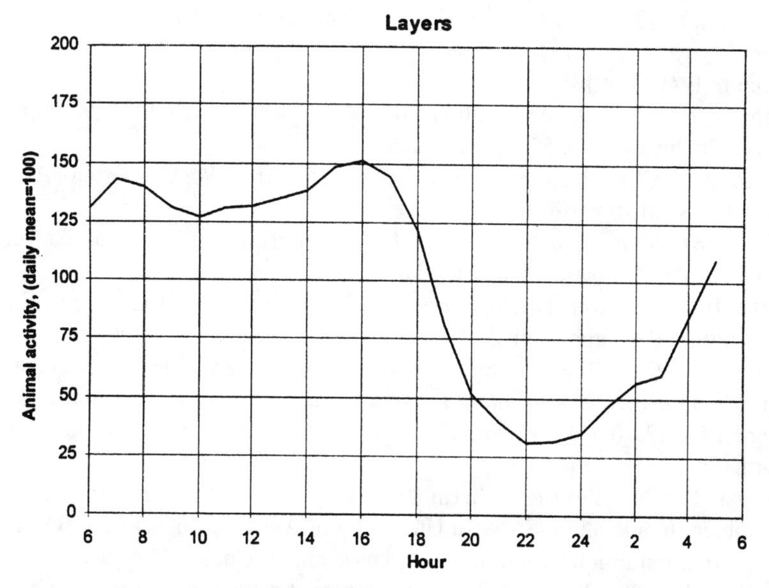

Figure 2.15. The diurnal variation of animal activity (measured by passive infrared detector technique) in houses laying hens. (*Source:* [41])

A more sophisticated equation would be one that takes into account two activity peaks during the daytime for conditions with restricted feeding two times a day, as in Fig. 2.14 for dairy cows [36].

The importance of taking animal activity into account is that there is a good relation between animal activity and carbon dioxide and heat production as shown in Fig. 2.13 in a house with ad-lib feeding of fattening pigs. Dividing into day periods from 6:00 to 18:00 and night periods from 18:00 to 6:00, the night activity accounts for approximately 70% of the activity during daytime. For carbon dioxide and heat production, the figures are in both cases approximately 75%, showing that carbon dioxide and heat production are much higher during day than during night. For layers with ad-lib feeding, based on groups of identical houses (Fig. 2.15), the animal activity over the day period is more constant; the activity during the night sometimes is only one third of the 24-h mean activity [36]. These two examples clearly show that steady-state equations are not sufficient for simulation of climatic conditions in animal houses.

References

1. Phillips, C., and Piggins, D. 1992. *Farm Animals and the Environment*. Wallingford C.A.B International cop.
2. Moss, R. (ed.). 1992. *Livestock, health and welfare*. Harlow Longman.
3. Hahn, G. L., Nienaber, J. A., Klemcke, H. G., and Gose, G. L. 1986. Body temperature fluctuations in meat animals. ASAE Paper No. 86-4009. St. Joseph, MI: ASAE.
4. Mount, L. E. 1973. The concept of thermal neutrality. In *Heat Loss of Animals and Man—Assessment and Control* (ed. J. L. Monteith and L. E. Mount) pp. 424–435. London: Butterworths.
5. Webster, A. J. F. 1991. Metabolic responses of farm animals to high temperature. EAAP Publication No. 55, pp. 15–22.
6. CIGR. 1984. Climatization of animal houses. Report of Working Group. Aberdeen, Scotland: Scottish Farm Buildings Investigation Unit.
7. *NRC. 1981. Nutrient requirements of beef cattle*. National Research Council. National Academic Press, Washington DC.
8. ARC. 1980. *The Nutrient Requirement of Ruminant livestock*. Slough England: Agricultural Research Council, Commonwealth Agricultural Bureaux.
9. Webster, A. J. F. 1976. The influence of the climatic environment on metabolism in cattle. In *Principles of Cattle Production. Easter School in Agricultural Science, Proceedings 23* (ed. H. Swan and W. H. Broster), pp. 103–120. London: Butterworths.
10. Morgan, K. 1996. Short-term thermoregulatory responses of horses to brief changes in ambient temperature. Swedish University of Agricultural Sciences, Department of Agricultural Engineering, Report 4, Dissertation, Uppsala, Sweden.
11. Bruce, J. M. 1986. Lower critical temperatures for housed beef cattle. *Farm Building Progress* April:23–28. Scottish Farm Building Investigation Unit. Aberdeen, Scotland.

12. Ehrlemark, A. 1988. Calculation of sensible heat and moisture loss from housed cattle using a heat balance model. Swedish University of Agricultural Sciences, Department of Farm Buildings, Report 60, Uppsala, Sweden.
13. Blaxter, K. L. 1969. *The Energy, Metabolism of Ruminants*, 2 ed. London: Hutchinson.
14. Bruce, J. M. 1979. *Heat Loss from Animals to Floors*. Farm Building Progress No. 55. Scottish Farm Building Investigation Unit. Aberdeen, Scotland.
15. Stombaugh, D. P. 1976. Thermoregulation in swine. ASAE Paper No. 76-5009. St. Joseph, MI: ASAE.
16. McArthur, A. J. 1981. Thermal insulation and heat loss from animals. In *Environmental Aspects of Housing for Animal Production*, pp. 37–60. London: Butterworths.
17. Brody, S. 1945. *Bioenergetics and Growth*. New York: Reinhold Publishing Corporation.
18. Sterrenburg, P., and Ouwerkerk, E. N. J. 1986. Rekenmodel voor de bepaling van de thermische behaagliijkeidszone van verkens (BEZOVA). Instituut voor mechanisatatie, arbeiden gebouwen, IMAG, Rapport 78, Wageningen, The Netherlands.
19. Bruce, J. M., and Clark, J. J. 1979. Models of heat production and critical temperature for growing pigs. *Animal Production* 28:353–369.
20. Hugdson, D. R., McCutcheon, J. L., Byrd, S. K., Brown, W. S., Bayly, W. M., Brengelmann, G. L., and Gollnick, P. D. 1993. Dissipation of metabolic heat in the horse during exercise. *Journal of Applied Physiology*. 74:1161–1170.
21. Leighton and Siegel, 1966.
22. Gustafsson, G. 1988. *Air and heat balances in animal houses*, Swedish University of Agricultural Sciences, Department of Farm Buildings, Report 59. Dissertation. Lund, Sweden.
23. Ehrlemark, A. G., and Sällvik, K. G. 1996. A model of heat and moisture dissipation from cattle based on thermal properties. *Transactions of the ASAE* 39(1):187–194.
24. Mount, L. E. 1977. The use of heat transfer coefficients in estimating sensible heat loss from the pig. *Animal Production* 25:271–279.
25. Wiersma, F., and Nelson, G. L. 1967. Nonevaporative convective heat transfer from the surface of a bovine. *Transaction of the ASAE* 10:733–737.
26. Blaxter, K. L., McGraham, N. C., Wainman, F. W., and Armstrong, D. G. 1959. Environmental temperature, energy metabolism and heat regulation in sheep: II. The partition of heat losses in closely clipped sheep. *Journal of Agricultural Science*. 52:25–49.
27. Gebremedhin, K. G., Cramer, C. O., and Porter, W. P. 1981. Predictions and measurements of heat production and food and water requirements of Holstein calves in different environments. *Transactions of the ASAE* 24(3):715–720.
28. Ehrlemark, A. 1991. Heat and moisture dissipation from cattle. Measurements and simulation model. Swedish University of Agricultural Sciences, Department of Farm Building, Dissertation, Report 77, Uppsala, Sweden.
29. Yousef, M. K. 1985. Heat production: Mechanisms and regulation. In *Stress Physiology of Livestock, vol. 1: Basic Principles* (ed. M. K. Yousef). Boca Raton, FL: CRC Press.

30. Bruce, J. M., Broadbent, P. J., and Topps, J. H. 1984. A model of the energy system of lactating and pregnant cows. *Animal Production* 38:351–362.
31. Bruce, J., and Blaxter, 1979.
32. Sallvik, K., and Wejfeldt, B. 1993. Lower critical temperature for fattening pigs on deep straw bedding. ASAE IV Livestock Environment Symposium, University of Warwick, Coventry, U.K.
33. Bruce, J. M. 1977. *Conductive heat loss from the recumbent animal.* Farm Building R & D studies, vol. 8 February, pp. 9–15.
34. Swedish Standard. 1992. *Farm Buildings - Ventilation, heating and climatic analysis in insulated animal houses - Calculations.* SS 95 10 50. Stockholm, Sweden.
35. CIGR. 1992. Climatization of animal houses. 2nd Report of Working Group. Faculty of Agricultural Sciences, State University of Ghent, Belgium.
36. Pedersen, S., Takai, H., Johnsen, J. O., Metz, J. H. M., Groot Koerkamp, P. W. G., Uenk, G. H., Phillips, V. R., Holden, M. R., Sneath, R. W., Short, J. L., White, R. P., Hartung, J., Seedorf, J., Schroeder, M., Linkert, K. H., and Wathes, C. M. 1998. *A Comparison of Three Balance Methods for Calculating Ventilation Flow Rates in Livestock Buildings.* Journal of Agricultural Engineering Research. Volume 70, Number 1, Special Issue, pp. 25–37, May 1998.
37. Pedersen, S., and Thomsen, M. G. 1998. *Heat and Moisture Production for Broilers on Straw Bedding.* In progress.
38. Pedersen, S. 1993. Time based variation in airborne dust in respect to animal activity. Fourth International Livestock Environmental Symposium, Coventry, U.K.
39. Pedersen, S., and Rom, H. B. 1998. *Diurnal Variation in Heat Production from Pigs in Relation to Animal Activity.* AgEng International Conference, Oslo 98. Paper 98-B-025.
40. van Ouwerkerk, E. N. J., and Pedersen, S. 1984. *Application of the carbon dioxide mass balance method to evaluate ventilation rates in livestock buildings.* XII CIGR World Congress Agricultural Engineering, Milan. Proceedings, vol. 1, 1994:516–529.
41. Pedersen, S. 1996. Døgnvariationer i dyrenes aktivitet i kvæg-, svine- og fjerkræstalde. Delresultater fra EU-projekt PL 900703. *Diurnal variation in animal activity in livestock buildings for cattle, pigs and poultry. Subresults from EU-project PL 900703.* (In Danish). National Institute of Animal Science, Denmark, Intern rapport Nr. 66, 33 pp., 1996.

2.3 Environmental Control of Livestock Housing

2.3.1 Natural Ventilation

James M. Bruce

All ventilation systems once were natural but, with the availability of electricity on farms and a blossoming control technology, mechanical or fan-powered ventilation took over. This shift from natural to mechanical ventilation took place mainly because building designers felt that the performance of mechanical systems was predictable, controllable, dependable, and effective and that natural ventilation was none of these. In recent decades, studies of the theory and practice of natural ventilation have shown that natural ventilation can be all of these things—and very much cheaper. We have progressed to the point, in many applications, at which natural ventilation has been accepted as preferable to mechanical ventilation. For example, most new pig buildings in the United Kingdom have automatically controlled natural ventilation (ACNV). In the past, the reputation of natural ventilation suffered due to ignorance of theory and consequent inconsiderate design. Far too often, natural ventilation has been blamed for the failures of designers.

Natural ventilation of buildings is generated by two distinct sources. Buoyancy or gravity effects are present, and in large part these are due to temperature differences between the outside and the inside air. The second source is wind blowing over a building, generating pressures and suctions at different points, which can force air in and out of the building.

Theoretical developments have been instrumental in the understanding of the phenomena and the development of practical applications [1, 2], and decisive in the interpretation of experimental data [3].

Natural Ventilation due to Thermal Buoyancy

Consider a set of n openings, distributed vertically, in a building in which the air density is lower than outside. A neutral plane is defined at height z such that no pressure difference exists due to a gravity or buoyancy effect.

$$p_o(z) = p_i(z) \tag{2.15}$$

The pressure difference at height h is then given by

$$\Delta p(h) = g(\rho_o - \rho_i)(z - h) \tag{2.16}$$

This pressure difference is available to overcome resistance to air flow and to provide kinetic pressure.

$$g(\rho_o - \rho_i)(z - h) = \left(f\frac{L}{D} + \Sigma\zeta\right)\rho v^2/2 \tag{2.17}$$

This gives

$$v = \frac{|z-h|}{z-h}\left[2g\frac{\Delta\rho}{\rho}|z-h| \Big/ \left(f\frac{L}{D} + \Sigma\zeta\right)\right]^{\frac{1}{2}}. \tag{2.18}$$

For the continuity of mass flow

$$\sum_{j=1}^{n}\int_{A_j}\rho v\,dA = 0. \tag{2.19}$$

Assuming ρ is approximately constant,

$$\sum_{j=1}^{n}\left(f\frac{L}{D} + \Sigma\zeta\right)_j^{-\frac{1}{2}}\int_{A_j}\frac{|z-h|^{\frac{3}{2}}}{z-h}\,dA = 0. \tag{2.20}$$

It is necessary to solve Eq. (2.20) to find z. The ventilation rate for the building is then found from

$$V = \left(2g\frac{\Delta T}{T}\right)^{\frac{1}{2}} \sum_{j=1}^{n} \left(f\frac{L}{D} + \Sigma\zeta\right)_j^{-\frac{1}{2}} \int_{A_j(z)} \frac{|z-h|^{\frac{3}{2}}}{z-h}\, dA \tag{2.21}$$

where $\Delta\rho/\rho$ has been replaced by $\Delta T/T$. By treating air as a perfect gas it is readily shown for approximately constant pressure and small ΔT that $\Delta\rho/\rho = \Delta T/T$.

The following relationship can be used to evaluate the integral in Eq. (2.21):

$$\int |a-y|^p (a-y)^q\, dy = -\frac{1}{p+q+1}|a-y|^p (a-y)^{q+1} \tag{2.22}$$

for $p + q \neq -1$

For a given set of openings

$$V = K\Delta T^{1/2}, \tag{2.23}$$

K being found from Eq. (2.21).

For a building the heat balance can be written

$$\Phi = (\rho c V + \Sigma k A)\Delta T \tag{2.24}$$

Substituting Eq. (2.23) into Eq. (2.24) and rearranging gives

$$V^3 + \frac{\Sigma k A}{\rho c} V^2 - \frac{K^2 \Phi}{\rho c} = 0 \tag{2.25}$$

which must be solved to give V, the ventilation rate.

Example 1

Consider a single vertical rectangular opening of height H and area A. From Eq. (2.20) or from symmetry $z = H/2$. Evaluating Eq. (2.21) gives

$$V = \zeta^{-\frac{1}{2}} \frac{A}{3}\left(g\frac{\Delta T}{T} H\right)^{\frac{1}{2}}. \tag{2.26}$$

Typically $\zeta^{-1/2} \approx 0.6$ for sharp-edged openings.

Therefore

$$V = 0.2A\left(g\frac{\Delta T}{T} H\right)^{\frac{1}{2}} \approx 0.04A(\Delta T H)^{\frac{1}{2}}. \tag{2.27}$$

For a doorway 0.9 m wide and 2.1 m high with a temperature difference of 10 K the ventilation rate would be

$$V = 0.04 \times 0.9 \times 2.1(10 \times 21)^{1/2} = 0.35\,\mathrm{m^3 s^{-1}}. \tag{2.28}$$

From Eq. (2.18) the velocity profile can be derived to be

$$v = 0.50\frac{\left|\frac{H}{2} - h\right|^{\frac{3}{2}}}{\frac{H}{2} - h} = 0.50\frac{|1.05 - h|^{\frac{3}{2}}}{1.05 - h}. \tag{2.29}$$

Equation (2.29) is shown in Fig. 2.16.

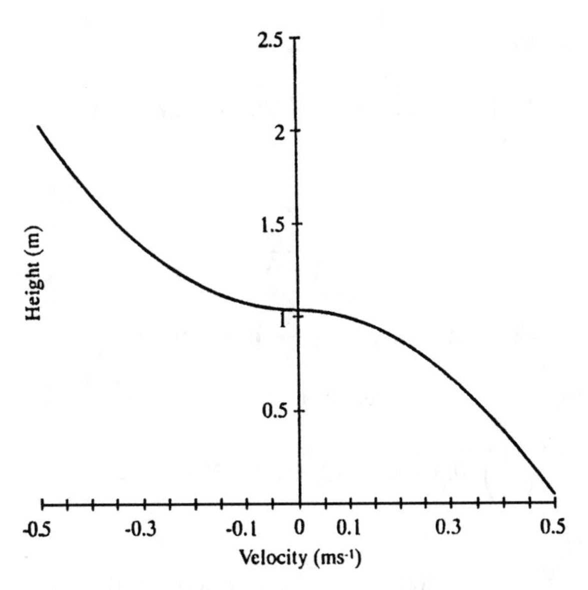

Figure 2.16. Velocity profile at the vertical opening.

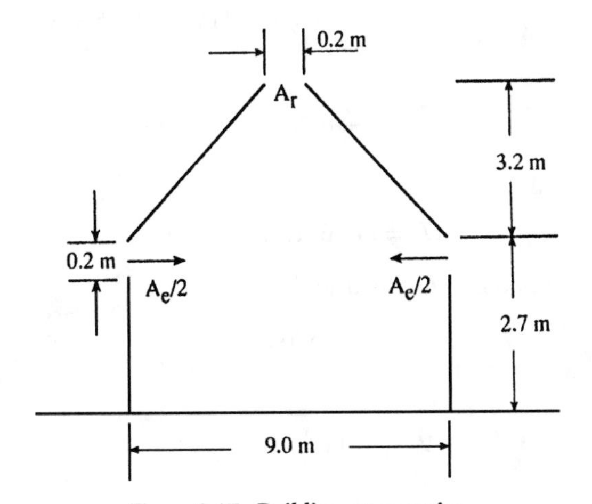

Figure 2.17. Building cross-section.

Example 2

Consider a building 30 m long with a cross-section as shown in Fig. 2.17. The building contains 400 pigs producing 100 W of sensible heat each. The heat-loss rate from the building is given by

$$\Sigma kA = 500\,\text{WK}^{-1} \tag{2.30}$$

Assuming that the height between the openings is 3.3 m and that the height of the eaves openings is negligible, the height of the neutral plane above the eaves openings is found using Eq. (2.20):

$$\int_{A_e} \frac{|z-h|^{\frac{3}{2}}}{z-h} dA + \int_{A_r} \frac{|z-h|^{\frac{3}{2}}}{z-h} dA = 0. \tag{2.31}$$

Integrating gives

$$\frac{|z|^{\frac{3}{2}}}{z} A_e + \frac{|z-3.3|^{\frac{3}{2}}}{z-3.3} A_r = 0 \tag{2.32}$$

$$z = 3.3/1 + \left(\frac{A_e}{A_r}\right)^2 = 0.66\,\text{m} \tag{2.33}$$

The constant in Eq. (2.21) is evaluated thus:

$$V = \left(\frac{2 \times 9.81}{273}\right)^{\frac{1}{2}} 0.6 \times 0.66^{\frac{1}{2}} (2 \times .02 \times 30) \Delta T^{\frac{1}{2}} = 1.6 \Delta T^{1/2}. \tag{2.34}$$

Substituting into Eq. (2.25) gives

$$V^3 + \frac{500}{1.2 \times 1010} V^2 - \frac{1.6^2 \times 400 \times 100}{1.2 \times 1010} = 0$$

$$V^3 + 0.41 V^2 - 84.5 = 0$$

from which

$$V = 4.26\,\text{m}^3\text{s}^{-1} \tag{2.35}$$

and

$$\Delta T = (4.26/1.6)^2 = 7.1\,\text{K}. \tag{2.36}$$

If the building were poorly insulated so that

$$\Sigma kA = 5000\,\text{WK}^{-1} \tag{2.37}$$

then

$$V^3 + 4.1 V^2 - 84.5 = 0$$

from which

$$V = 3.36\,\text{m}^3\text{s}^{-1} \tag{2.38}$$

and

$$\Delta T = 4.4\,\text{K} \tag{2.39}$$

If all the openings were reduced by half then the ventilation rates and temperature differences for the insulated and poorly insulated cases would be

$$V = 2.63\,\text{m}^3\text{s}^{-1} \quad \text{with} \quad \Delta T = 10.8\,\text{K} \tag{2.40}$$

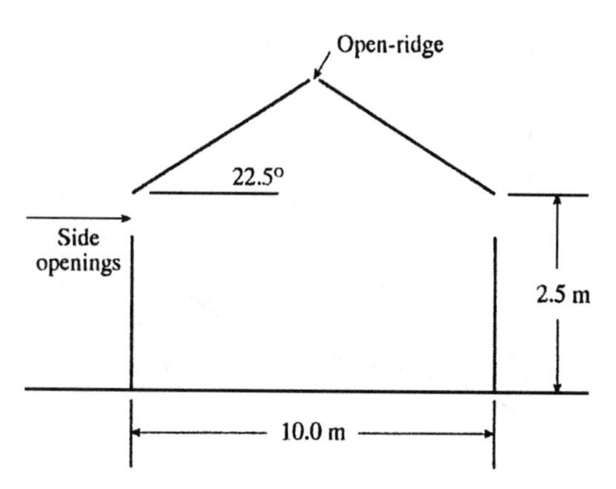

Figure 2.18. Building cross-section.

and

$$V = 1.88\,\mathrm{m^3 s^{-1}} \quad \text{with} \quad \Delta T = 5.5\,\mathrm{K} \tag{2.41}$$

We see from this that the ventilation rate is not halved in proportion to the openings and the temperature lift is not doubled. The effort required to carry out the calculations for a range of opening sizes and insulation levels is great. The use of a computer is highly desirable.

Example 3

Consider a building with the cross-section shown in Fig. 2.18.

The open-ridge and side openings can be of various sizes. The theory outlined allows us to examine the effect on the ventilation rate of different combinations of open-ridge and side openings. For this purpose we will consider a part of the building that is 1.0 m long. In this case the height of the side openings is not considered negligible.

Figure 2.19 shows the resultant effect of some combinations of openings. The important points to note are that with small side openings there is a very large effect on ventilation rate caused by using an open ridge, but increasing the open ridge does not have a proportional effect; and with large side openings there is a small effect caused by having an open ridge and increasing its size.

Natural Ventilation due to Wind

The pressure drop across an opening with airflow through it is given by

$$\Delta p = \left(f\frac{L}{D} + \Sigma\zeta\right)\frac{1}{2}\rho v^2. \tag{2.42}$$

The pressure difference due to wind across an opening in a building is given by

$$\Delta p = (C_{po} - C_{pi})\frac{1}{2}\rho v_w^2. \tag{2.43}$$

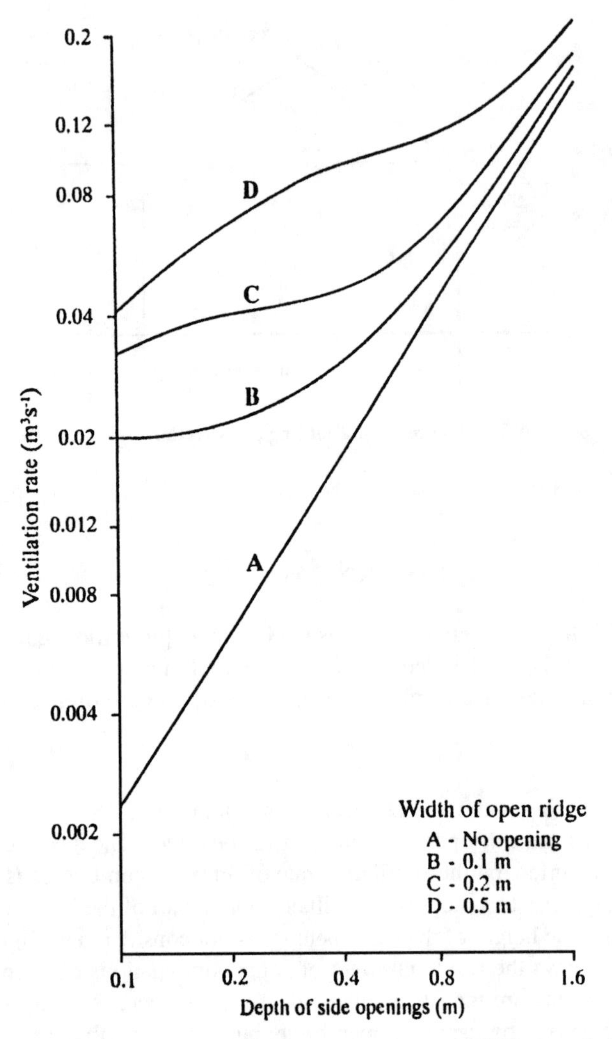

Figure 2.19. Ventilation rate due to a temperature difference of 1 K.

Combining Eqs. (2.42) and (2.43) gives

$$v^2 = v_w^2 \frac{C_{po} - C_{pi}}{f\frac{L}{D} + \Sigma\zeta}. \tag{2.44}$$

Taking account of the direction of flow through the opening, the solution is

$$v = v_w \frac{|C_{po} - C_{pi}|^{\frac{3}{2}}}{C_{po} - C_{pi}} \left(f\frac{L}{D} + \Sigma\zeta \right)^{-\frac{1}{2}}. \tag{2.45}$$

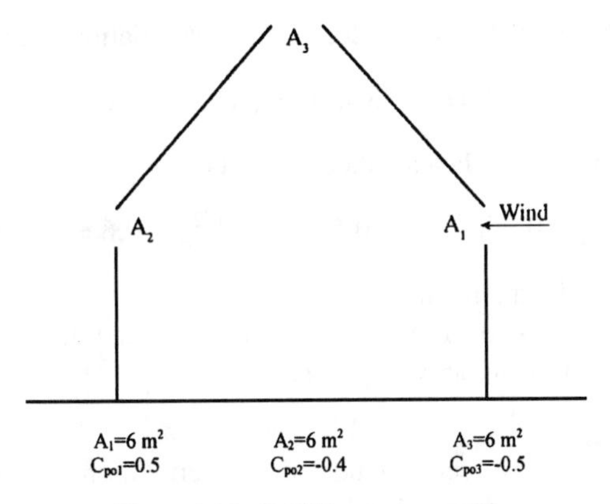

Figure 2.20. Building cross-section.

For continuity of mass flow in a building with n openings,

$$\sum_{j=1}^{n} A_j v_j = 0 \tag{2.46}$$

when $\rho \approx$ constant.

Given the external pressure coefficients, the areas, and resistance characteristics at each opening, the problem reduces to solving Eq. (2.46) to find C_{pi}. Once C_{pi} is found the ventilation rate can be calculated from

$$V = \sum_{j=1}^{n} A_j v_j \quad \text{for } v_j > 0 \text{ only.} \tag{2.47}$$

For more than a few openings the calculations are tedious and a computer is essential.

Example 1

Assume a wind of 3 ms^{-1} blowing across the same building shown in Fig. 2.17. Figure 2.20 illustrates the case.

The openings are sharp-edged, so $\zeta^{-1/2} = 0.6$. Equation (2.46) becomes

$$\frac{|0.5 - C_{pi}|^{\frac{3}{2}}}{0.5 - C_{pi}} + \frac{|-0.4 - C_{pi}|^{\frac{3}{2}}}{-0.4 - C_{pi}} + \frac{|-0.5 - C_{pi}|^{\frac{3}{2}}}{-0.5 - C_{pi}} = \Sigma = 0 \tag{2.48}$$

Successive guesses for C_{pi} give

C_{pi}	Σ
−0.1	−0.41
−0.2	−0.16
−0.3	+0.13
−0.26	+0.01.

This is nearly zero; therefore $C_{pi} = -0.26$. The inlet ventilation rate is therefore given by

$$A_1 v_w (C_{po1} - C_{pi})^{1/2} \zeta^{-1/2} = 6 \times 3 \times (0.5 + 0.26)^{1/2} \times 0.6 = 9.4\,\mathrm{m^3 s^{-1}}. \quad (2.49)$$

The outlet ventilation can also be calculated as a check:

$$6 \times 3 \times (|-0.4 + 0.26|^{1/2} + |-0.5 + 0.26|^{1/2}) \times 0.6 = 9.3\,\mathrm{m^3 s^{-1}}. \quad (2.50)$$

This is sufficiently good agreement.

This ventilation rate is two to three times that calculated due to thermal buoyancy. The wind can easily dominate buoyancy effects.

Example 2

Figure 2.21 illustrates the effect on wind-induced ventilation of different combinations of openings in the building described in Fig. 2.18.

Again, a unit length of 1.0 m is taken. The effect of the size of the open ridge is greater with small side openings than with large side openings. This was observed with buoyancy effects. Generally, however, the ventilation rate due to buoyancy shows a greater response to the size of the open ridge than does the wind-induced ventilation.

Combined Effects of Thermal Buoyancy and Wind [4]

For the sake of simplicity $(f\frac{L}{D} + \Sigma\zeta)$ is replaced by ζ alone.
Following Eqs. (2.17), (2.42) and (2.43) gives

$$\zeta \frac{1}{2}\rho v^2 = \frac{1}{2}\rho v_w^2 (C_{po} - C_{pi}) + g(\rho_o - \rho_i)(z - h). \quad (2.51)$$

This results in

$$v = \zeta^{-\frac{1}{2}} \left[\frac{\left|(C_{po} - C_{pi})v_w^2 + 2g\frac{\Delta T}{T}(z-h)\right|^{\frac{3}{2}}}{(C_{po} - C_{pi})v_w^2 + 2g\frac{\Delta T}{T}(z-h)} \right] \quad (2.52)$$

or

$$v = \zeta^{-\frac{1}{2}} \left[\frac{(C_{po} - C_{pi})v_w^2 + 2g\frac{\Delta T}{T}(z-h)}{\left|(C_{po} - C_{pi})V_w^2 + 2g\frac{\Delta T}{T}(z-h)\right|^{\frac{1}{2}}} \right]. \quad (2.53)$$

The same continuity constraint and heat balance apply as before but the solution is much more difficult and a computer is required even for a simple case.

Example 1

The same building shown in Figs. 2.17 and 2.20 will be used. First consider the effect of temperature lift with a constant wind of 1 $\mathrm{ms^{-1}}$. Figure 2.22 shows the relationship between ventilation rate and temperature difference. There appear to be two distinct domains. At low temperature differences only the windward opening, A_1, is an inlet, with A_2 and A_3 acting as outlets. At some temperature difference between 3 K and 4 K both A_1 and A_2 become inlets, so a new relationship develops. For the building concerned,

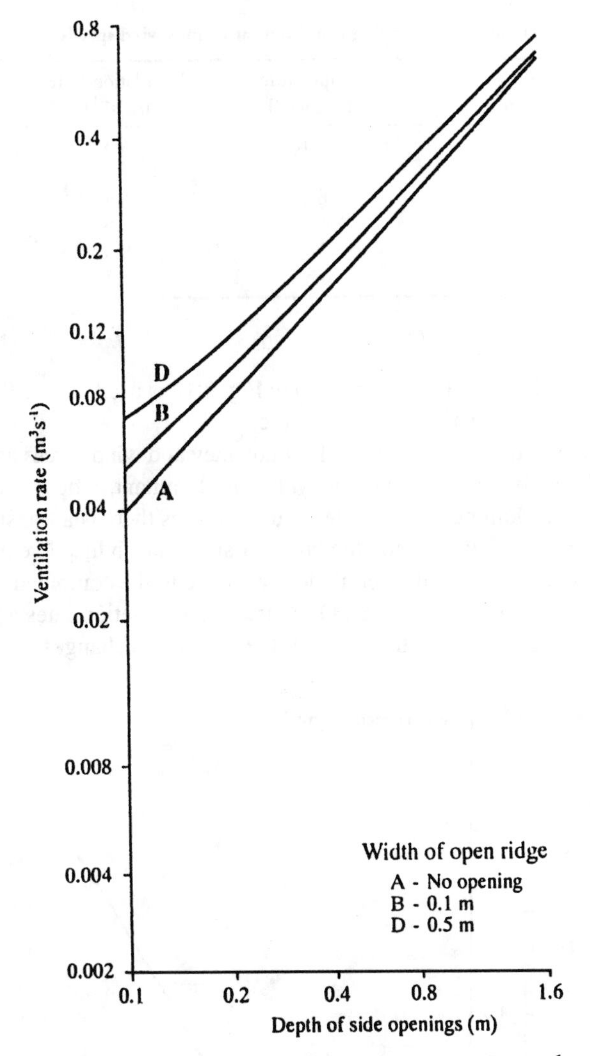

Figure 2.21. Ventilation rate due to a wind speed of 1 ms^{-1}.

with the assumed sensible heat production, the operating temperature difference is about 6.8 K with a ventilation rate of 4.4 m^3s^{-1}.

Table 2.7 shows the operating conditions at various wind speeds. So long as the wind speed is not greater than about 1 ms^{-1}, the temperature difference and ventilation rate do not change by much. The ventilation rate is, however, more than doubled by a wind speed of 3 ms^{-1} and is little affected by buoyancy. Over a small range of wind speeds the ventilation changes from being dominated by buoyancy effects to being dominated by wind.

Table 2.7. Operating conditions at various wind speeds

Wind Speed (ms^{-1})	Temperature Difference (K)	Ventilation Rate (m^3s^{-1})
0	7.1	4.3
0.5	7.1	4.3
1.0	6.8	4.4
2.0	4.7	6.6
3.0	3.3	9.6
3.0	No heat	9.4

Example 2

In this example, the same building shown in Fig. 2.18 is used, where the height of the side openings is not considered to be negligible.

Figure 2.23 shows the ventilation rate for buoyancy and wind forces acting alone and combined. With small side openings, buoyancy is dominant, whereas with large side openings the wind is dominant. With intermediate values there is a transition. It is clear that it would be wrong simply to add the buoyant and wind-induced ventilation rates.

Figures 2.24 and 2.25 show the ventilation rates due to the combined action of buoyancy and wind for small side openings. Of course the ventilation rates are much higher with the large side openings, but the shape of the graph also changes.

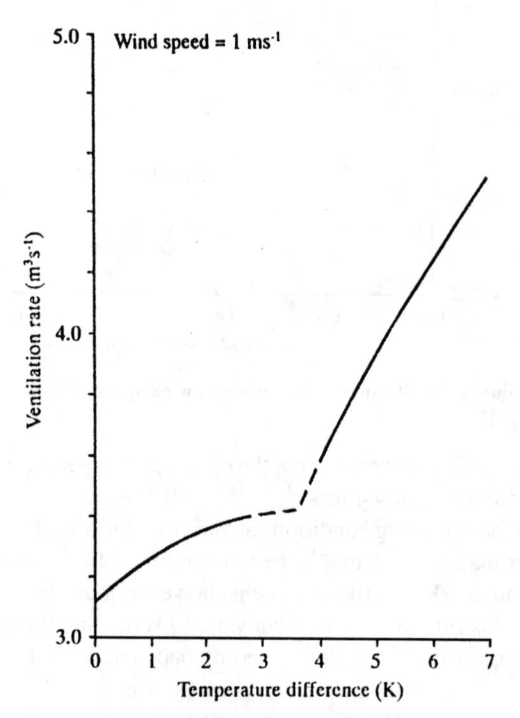

Figure 2.22. Ventilation rate for combined wind and thermal buoyancy effects.

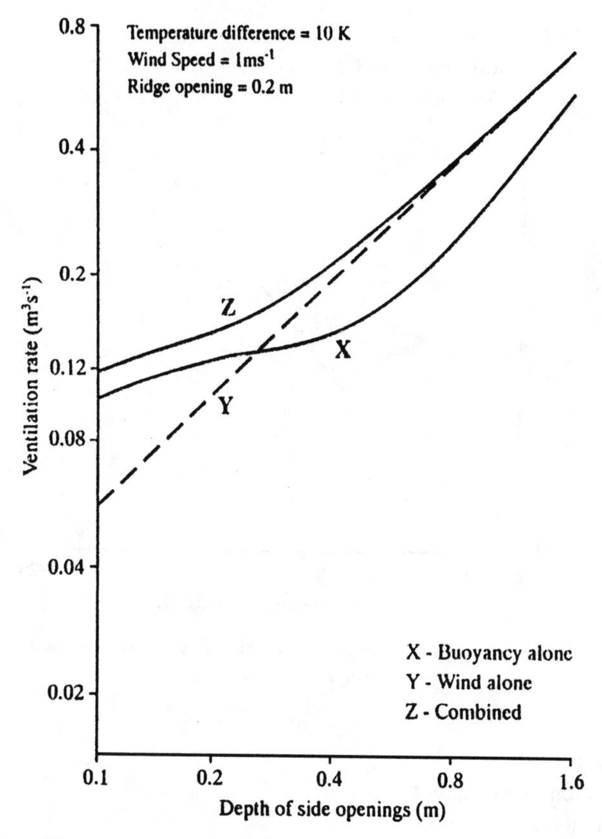

Figure 2.23. Ventilation rate due to buoyancy, wind, and combined effects.

Conclusions

Theories for the calculation of natural ventilation due to thermal buoyancy and wind have been presented and their use illustrated by examples. Neither theory is simple to apply. The combined effects of thermal buoyancy and wind can be calculated, but for even very simple cases a computer is essential. The combined effect of buoyancy and wind can be complex even for simple systems. From a practical point of view it would seem desirable to base design procedures either on thermal buoyancy or on wind acting separately. In this way manageable design methods are feasible.

Notation

A area
$A(z)$ area below the neutral plane
C_p pressure coefficient
c specific heat
D equivalent diameter

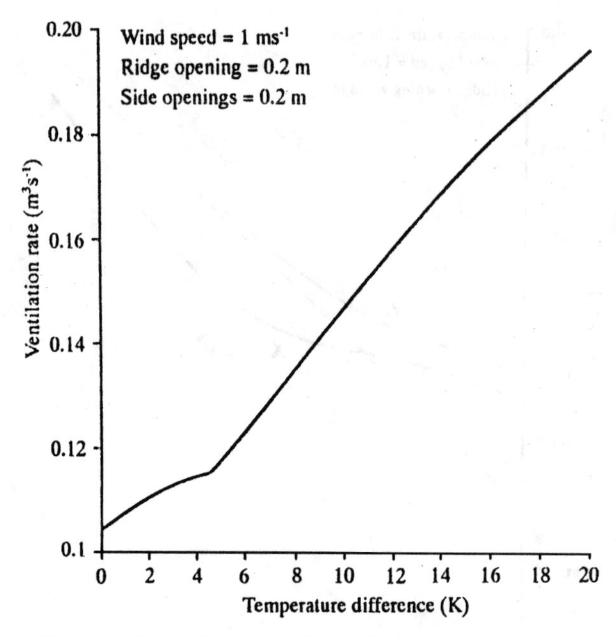

Figure 2.24. Ventilation rate for combined wind and thermal buoyancy effects.

f	friction factor
g	acceleration due to gravity
h	variable height
K	coefficient
k	heat transfer coefficient
L	length
n	number of openings
p	pressure
T	absolute temperature
V	ventilation rate
v	velocity
z	height of neutral plane

Subscripts:

e	eaves
i	inside
j	jth element
o	outside
r	ridge
w	wind
ρ	density

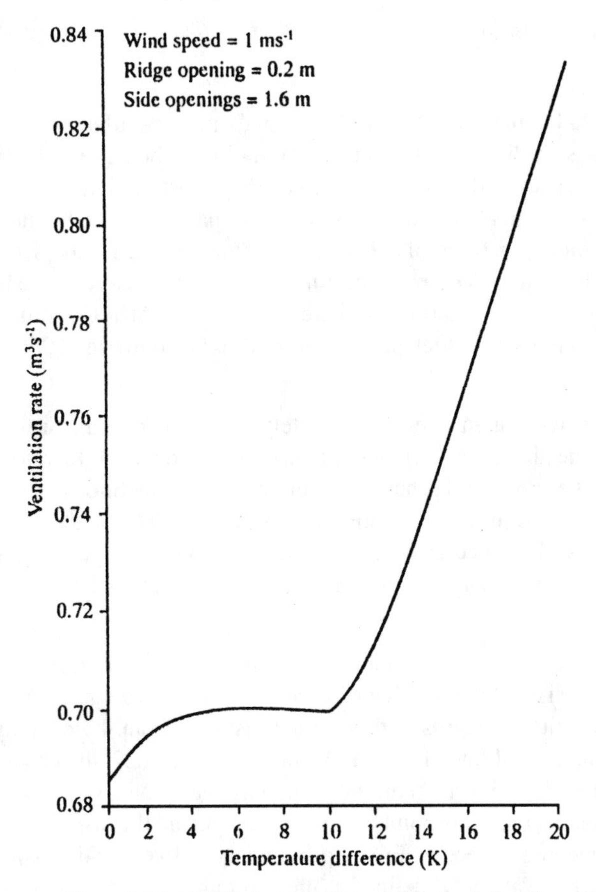

Figure 2.25. Ventilation rate for combined wind and thermal buoyancy effects.

ζ	resistance factor
1, 2, etc.	identifying number

All the dimensions of the variables are standard SI.

References

1. Bruce, J. M. 1975. A computer program for the calculation of natural ventilation due to wind. *Farm Building R&D Studies* November.
2. Bruce, J. M. 1978. Natural convection through openings and its application to cattle building ventilation. *J. Agric. Eng. Research* 23:151–167.
3. Bruce, J. M. 1982. Ventilation of a model livestock building by thermal buoyancy. *Trans. ASAE* 25:1724–1726.
4. Bruce, J. M. 1986. Theory of natural ventilation due to thermal buoyancy and wind: Pig, rabbit and small birds housing. CIGR Section II Seminar, Rennes, France.

2.3.2 Forced Ventilation

Soeren Pedersen

Forced ventilation means that the air-exchange rate in ventilated rooms is established by means of fans. It differs from natural ventilation in being less dependent on wind velocity and buoyancy and in the continued energy consumption.

A ventilation system always includes two main parts: inlet and outlet. In the case of forced ventilation at least one of them contains fans. If the fan is placed in the outlet system, it is called a *negative-pressure system*. If the fan is placed in the inlet, it is called a *positive-pressure system*, and if there are fans in both inlet and outlet it is called an *equal-pressure system*. The three principles are illustrated in Fig. 2.26.

Air Inlets

The airflow pattern at an inlet is completely different from the airflow pattern at an outlet, because the air stream from an inlet is unidirectional and therefore able to travel a long distance before it is discharged, contrary to the conditions at the outlet, where the air is sucked in from all directions. Normally, air inlets are shaped as inlet valves or nozzles with well-defined jets. They can also be shaped as porous surfaces, such as mineral wool ceilings, where the incoming air will have a very low velocity.

Free Jet

Figure 2.27 shows a free jet, which in principle means a jet thrown into an infinite room. The jet consists of a cone-shaped core with constant velocity and a mixing zone in which the velocity decreases both with the distance from the opening and with the distance from the central line of the jet. As shown in Fig. 2.27, the cone-shaped mixing zone has a summit somewhere behind the opening, depending on the shape of the inlet opening. For practical use this point can be neglected, and the distance can be measured just from the opening. There is no sharp boundary between the mixing zone and the room air, but normally the jet is defined as the part that is inside a mixing cone with a top angle of 24 degrees. At this distance the velocity is reduced to about 8% of the velocity at the central line [1].

For a jet from a circular opening the air velocity at different distances can be calculated by the following equation:

$$v_c = C\frac{v_o}{x}\sqrt{\frac{A}{\alpha\varepsilon}} \tag{2.54}$$

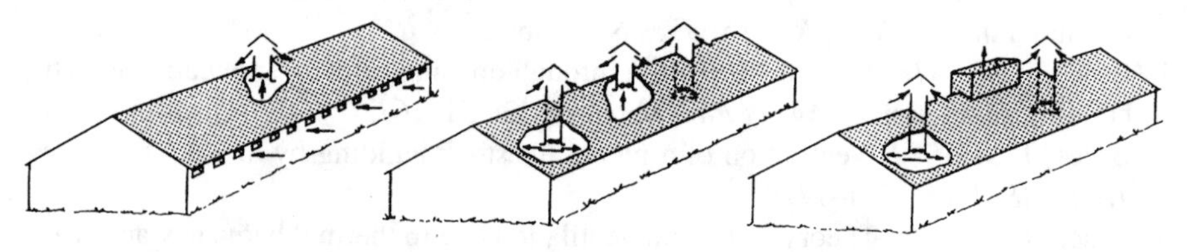

Figure 2.26. Ventilation principles.

Table 2.8. The constant C used in Eq. 1.

Shape of Inlet Opening		C-value
Circular	$2.5 < v_o < 5$	5.7
	$10 < v_o < 50$	7.0
Rectangular		
	width/height = 1	6.5
	width/height = 5	6.2
	width/height = 10	6.0
	width/height = 20	5.6

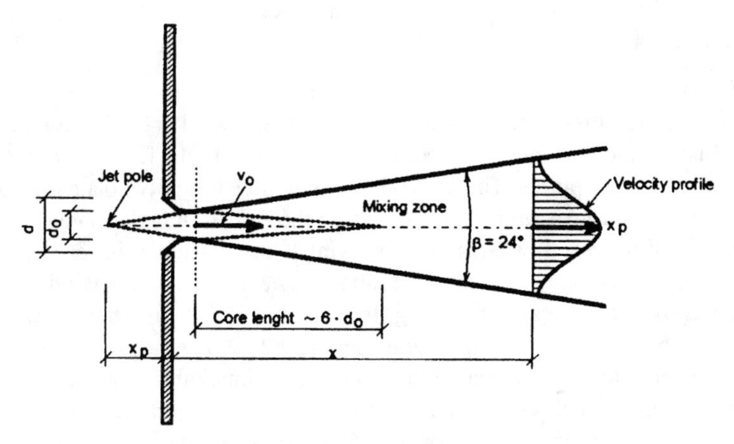

Figure 2.27. Free jet.

where

v_c = central velocity (m/s)
v_o = the average velocity in the inlet opening (m/s)
x = distance from the opening (m)
A = total opening area (m^2)
α = the contraction coefficient; for a sharp-edged opening α is 0.61, and for circular openings with well-rounded (bell-shaped) inlets it is close to 1.0
ε = the ratio between the free surface and the total surface of the opening; this means that $\varepsilon = 1$ if there is no grill in the opening
C = constant, which can be found in different tables (e.g., for circular openings at velocities from 2.5 to 5.0 m/s it is 5.7.)

For other inlet shapes, some C values are shown in Table 2.8.

The following is an example: for a sharp edged circular opening of 0.1 m and the incoming air with an average velocity of 5 m/s, the velocity at a distance of 4 m is:

$$v_c = 5.7\frac{5}{4}\sqrt{\frac{\pi 0.1^2/4}{0.61 \times 1}} = 0.81\ \text{m/s} \qquad (2.55)$$

To avoid draught in the living zone of the animals, the central velocity has to be discharged to below 0.2 m/s. In the example given, the velocity was reduced to 0.2 m/s at a distance of about 16 m.

Wall Jet

For a wall jet, which means a jet blown along and close to a plane, the velocity of the jet is $\sqrt{2}$ higher than that of a free jet, as shown in the following equation:

$$v_c = \sqrt{2}C\frac{v_o}{x}\sqrt{\frac{A}{\alpha\varepsilon}} \tag{2.56}$$

For the wall jet, the angle of the mixing zone has been increased to 33 degrees. For other types of jets, see reference [1].

Diffuse Inlets

Diffuse inlet systems use porous materials like mineral wool or textiles for the air-inlet system. This method was initially used in Norway in the 1970s [2], and later on it was adopted in many other places. The advantage is that there is nearly no air velocity when the air leaves the porous material, and therefore the risk of draught is negligible. This is for instance of interest in weaner and farrowing houses, where it is very important to avoid draught. On the other hand, it is not appropriate under housing conditions where it is advantageous to have air-jet cooling, for example in fattening houses with indoor temperatures above that of the comfort zone. Figure 2.28 shows the pressure drop needed across the material to obtain different flow rates. It is remarkable that the pressure drop for porous materials such as mineral wool increases in proportion to the increase in volume flow. For the aluminum sheet with 4.5 mm slots [3], the pressure drop increases

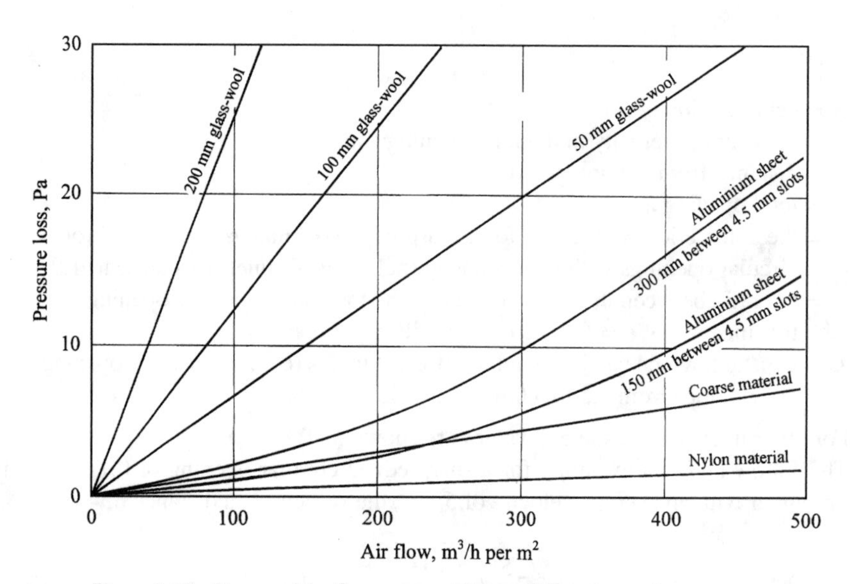

Figure 2.28. Air capacities for porous ceiling at different negative pressures.

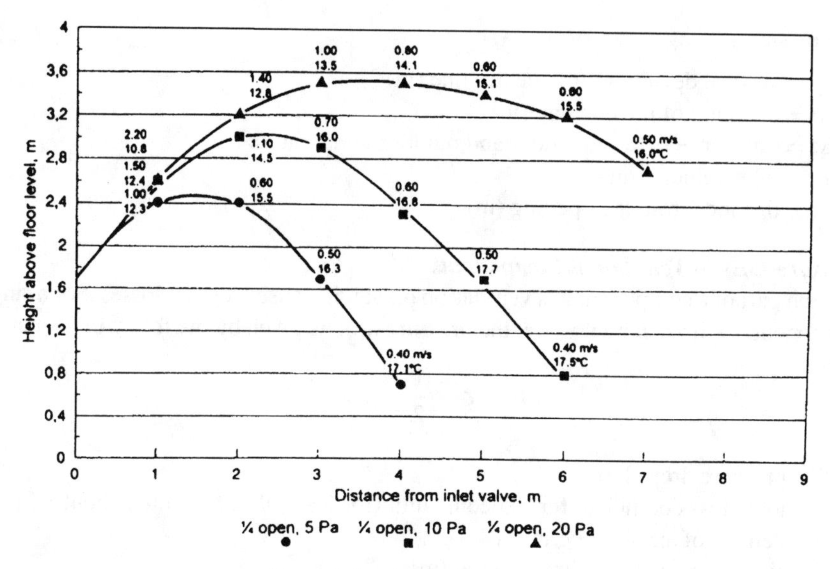

Figure 2.29. Nonisothermal jets.

progressively, but not in the second power as for a jet. Porous inlets differ from other types of inlets in not being adjustable by means of dampers according to the various needs of fresh air. Because the ratio of minimum to maximum capacity in a livestock building can be 1:10 or more, there can be design problems, because the range in pressure drop will also be 1:10 for materials with fine pores and up to 1:100 for materials with holes. Therefore, for a system based only on a porous ceiling, the pressure drop will either be rather low at minimum capacity or very high at maximum capacity, thereby causing problems for the fans.

Nonisothermal Jets

For nonisothermal conditions, the jet will deviate from the initial central line [4]. If the temperature of the jet is higher than the room temperature, the jet will rise, and if the temperature in the jet is lower, the jet will fall. The deviation can be calculated by Eq. (2.57). Figure 2.29 shows the results of a test [5] with a nonisothermal jet from a 0.63 × 0.27 m air inlet, with a one-quarter open damper and a temperature difference of 20°C at pressure drops of 5, 10, and 20 Pa, respectively. The turning of the curve is greatest for the lowest pressure drop.

$$\frac{y}{d} = 0.0022\frac{\Delta t d}{v_o^2}\left(\frac{x}{d}\right)^3 \tag{2.57}$$

or

$$y = 0.0022\frac{\Delta t}{v_o^2}\frac{x^3}{d} \tag{2.58}$$

where

y = vertical deviation from center line (m)
d = diameter of the opening (m)
Δt = difference between indoor and outdoor temperature (°C)
v_o = inlet velocity (m/s)
x = distance from the opening (m)

Pressure Loss in Ventilation Components

Each part of a component in a ventilation plant will cause pressure losses. For a single component, such as an inlet valve, the pressure drop is given by the following equation:

$$P = \zeta \times \frac{1}{2} \times \rho \times v^2 \tag{2.59}$$

P = pressure drop (Pa)
ζ = local loss coefficient for a specific unit (for wall inlets ζ is often about 1.5)
ρ = density of air (1.22 kg/m^3 at 15°C)
v = average air velocity in the outlet (m/s)

For rough calculations Eq. (2.59) can be simplified to

$$P = \zeta \times v^2/1.6 \tag{2.60}$$

The equation shows that the pressure increases with the velocity in the second power. The same is the case for the outdoor wind pressure, taking into account the actual correction factors for different shapes of the building. There is a positive pressure on the windward side and normally also a negative pressure on the leeward side. Therefore, it can be perceived that the velocity in side wall inlets can be affected very much by outdoor wind. How much depends on the local loss coefficient for the used side-wall inlet. The greater the loss coefficient, the greater the pressure drop for a given flow, and the smaller the wind effect on the inlet velocity. On the other hand, the energy consumption increases with an increased loss coefficient. Another way to prevent this effect is to use windbreakers.

Windbreakers

By utilization of the ejector principle, different hoods and windbreakers can be used for reduction of the wind effect on side wall inlets. Figure 2.30 shows an example of such a construction [6]. The investigation is carried out in the laboratory with a 1:10 scale model as well as in a full-scale barn.

The effect of the windbreaker in the laboratory with a fixed inlet opening and constant windspeed perpendicular to the side-wall inlet is shown in Fig. 2.31, in which the inlet velocity ratio V_w/V_o is defined as the inlet velocity with wind, V_w, divided by the inlet velocity with no wind, V_o. The wind ratio W/V_o is defined as the wind velocity W divided by the inlet velocity at no wind. The figure clearly shows that the potential maximum effect at high wind velocities is considerable. The results from the full-scale experiments have shown an effect as well, although smaller, among other reasons because the wind

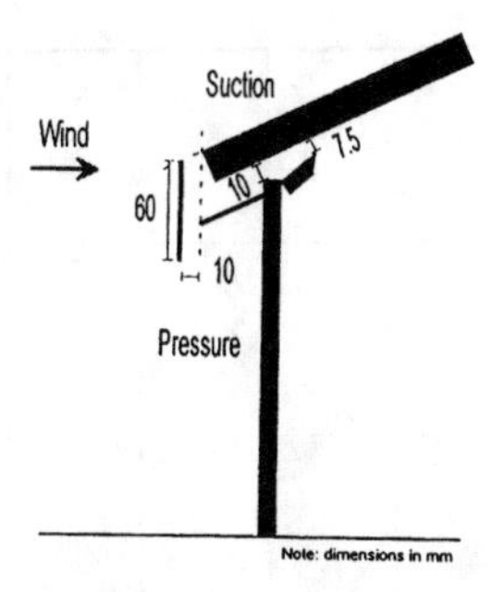

Figure 2.30. Side wall inlet with windbreaker.

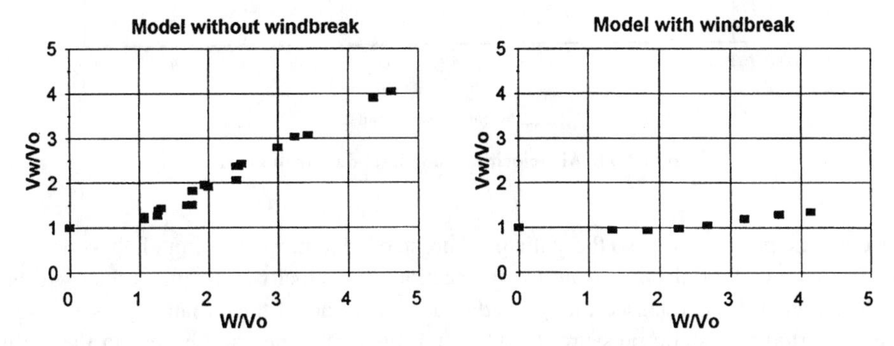

Figure 2.31. Velocity ratio for side wall inlets with and without windbreakers.

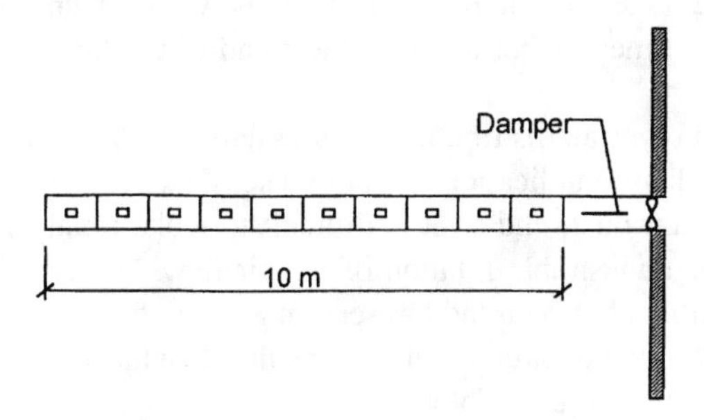

Figure 2.32. Inlet duct with nozzles.

is only partly perpendicular to the wall, and the opening area of the inlet valve varies in dependence of the animal density and the outdoor air temperature.

Inlet Ducts with Nozzles

In some cases the air is distributed by ducts with inlet slots or nozzles (Fig. 2.32). For such a duct the problem is to get an even air distribution through all openings along the duct, because the flow is normally highest at the farthest end of the fan section. The reason for this is that the total pressure in the duct closest to the fan section mainly occurs

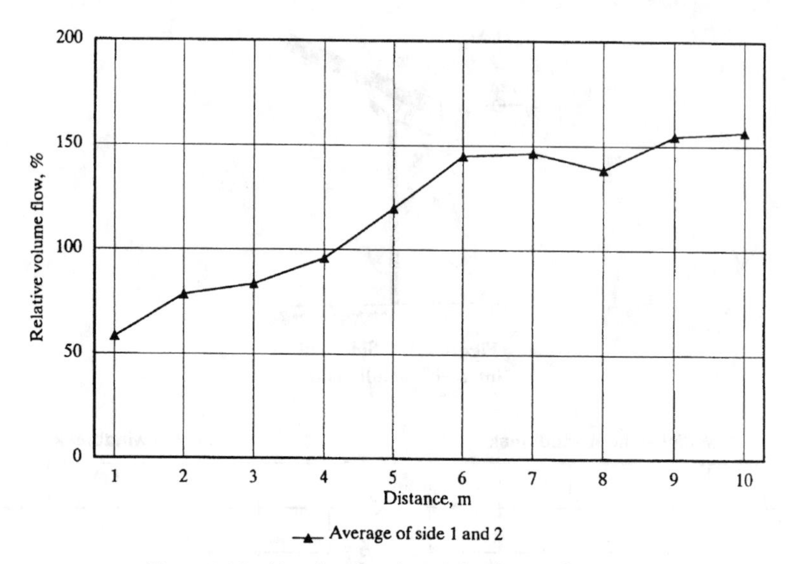

Figure 2.33. Air velocities along inlet duct with nozzles.

as dynamic pressure, and so the static pressure needed to move air through the side duct is low. At a certain distance from the fan section, the velocity in the main duct will be reduced when air is released through side ducts, and part of the dynamic pressure will be converted into static pressure. Even though there is some friction loss in the main channel, which reduces the static pressure, the airflow is normally highest in the farthest side ducts. Figure 2.33 shows the results from a test with a 10 m long sided duct with an air velocity three times higher at the farthest end of the fan section than at the fan section.

This means that an even air distribution requires dampers in each nozzle for adjustment of the air flow. Another complication is that by use of axial flow fans, the air normally rotates inside the main duct, and so it is difficult to make reliable calculations of the air distribution. The undesirable rotation of the air flow, however, can be reduced by insertion of air rectifiers between the fan section and the first side nozzle. As a rule of thumb, the total side-opening area should be smaller than the sectional area of the duct, if no dampers in the inlets are available.

Discharge of Jet Meeting Obstructions

When air is blown into a room, the jet will at some distance meet obstructions such as walls, poles, or light fittings, which will influence the jet and cause risks of draught. Therefore, it is very important to consider what the jet will meet on its way until it is discharged [7–9]. Figure 2.34 shows schematically how a jet is disturbed by a beam.

Figure 2.35 shows the discharge of an isothermal horizontal jet from a 0.1 m circular opening with a distance of 6 m to a wall and with velocities from 2 to 10 m/s. As shown, the discharge follows the rule of thumb closely. The velocity in a free jet at different distances from the opening corresponds to the velocity at a distance of 1 m divided by

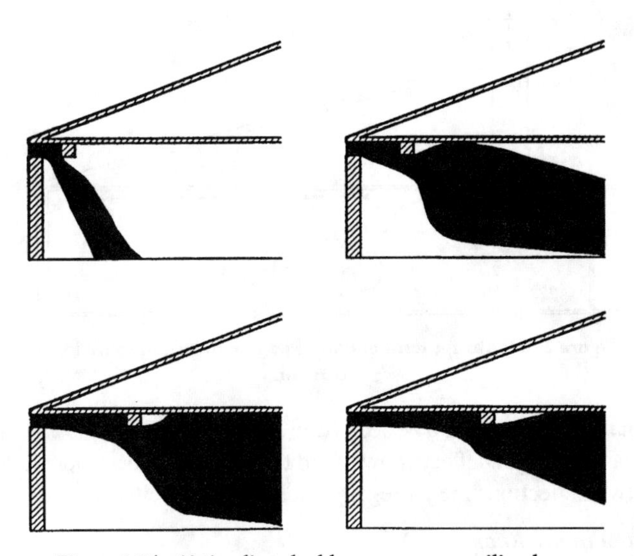

Figure 2.34. Air jet disturbed by a transverse ceiling beam.

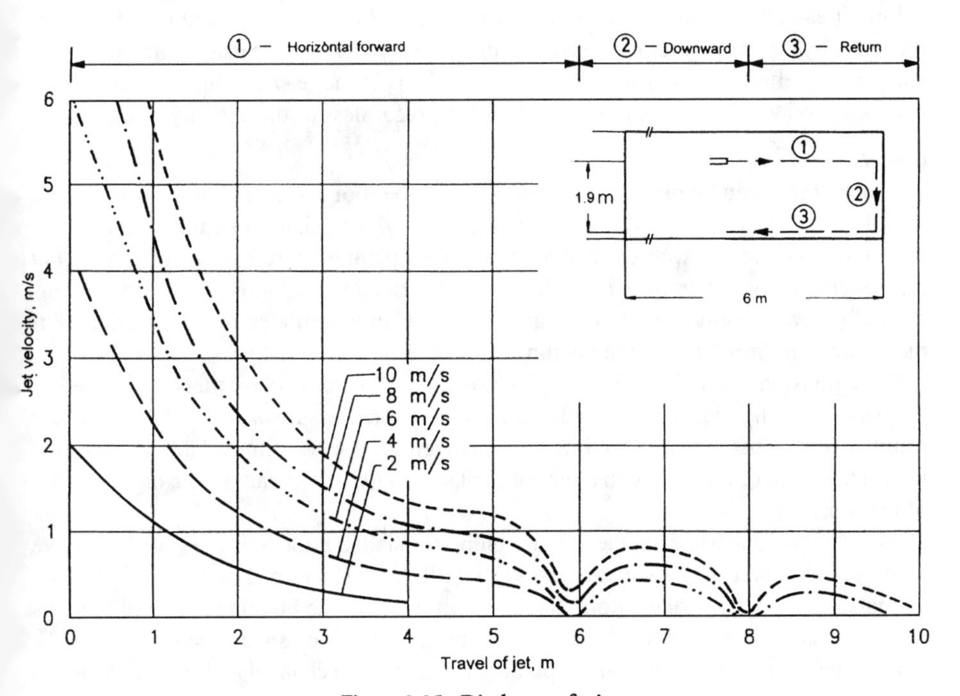

Figure 2.35. Discharge of a jet.

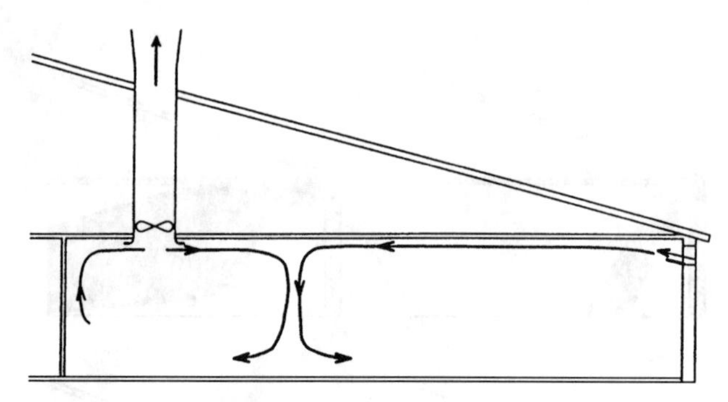

Figure 2.36. Abrupt termination of jet due to lack of space for the return air.

the distance in meters, as can also be seen from Eq. 2.54. Close to the wall the horizontal velocity decreases. The jet deflects downward towards the floor, where it deflects again. Between the two deflections the jet regains most of its velocity.

Air Distribution in the Room

Figure 2.36 shows another situation in which the inlets are placed one-sided in the outer wall, and the outlet is placed in the roof. Here, it is important to consider the necessary space for the return air. For a linear attached jet the expansion is approximately 0.3 m for each metre the jet moves downstream [10]. Taking into account that the center line velocity is relatively high, most of the air movement takes place inside a much smaller part, and the effective extension can be set to 0.15 m per metre. This means that it will be necessary to include a space of 0.3 m per metre, regardless of the velocity of discharge.

Outlets

Exhaust fans can be placed in the walls, down the roof surface, or at the ridge of the roof. Figure 2.37 [11] shows the air velocities around the exhaust opening of a fan.

Although there is a well-defined jet at an inlet opening, there is no directional jet at the outlet opening. Therefore, it can be concluded that the positions of the exhaust fans normally have a negligible influence on the air flow in a ventilated room at distances of more than 1 m from the suction opening.

In regions with small wind velocities it does not matter if the outlets are placed in the walls or at the roof, but in windy regions the wind pressure will disturb the airflow stability of exhaust fans if placed in walls. Therefore, it is recommendable to place the fans at the ridge of the roof with their outlet placed at least 0.5 m above the highest point of the roof.

By means of suction from the manure pit below slatted floors it is possible to remove a considerable part of the volatile gases (e.g., NH_3, H_2S) and bad odors released from animals, manure, and urine before they get whirled up in the breathing zone of animals and humans. Figure 2.38 shows the layout of an underfloor suction system [12, 13], which consists of a main air channel parallel to the manure channel and some side ducts. The main problem is to obtain a homogeneous flow all over the slatted floor. In theory,

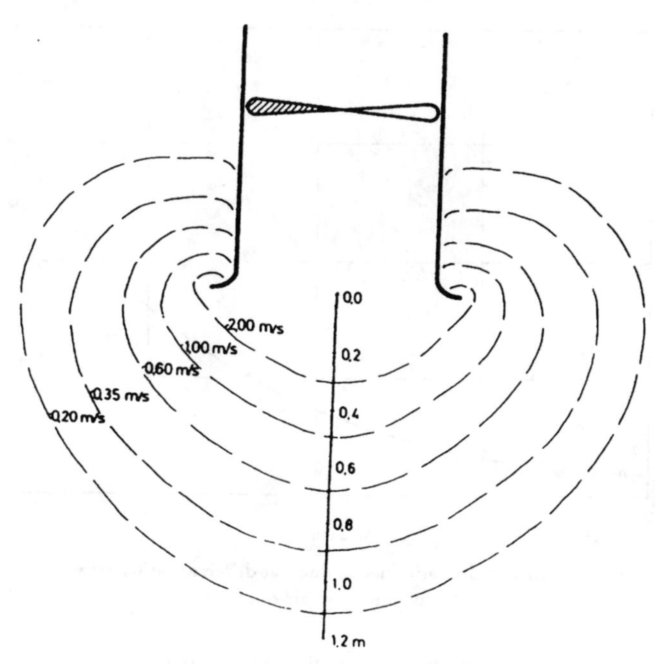

Figure 2.37. Air flow near the exhaust fan.

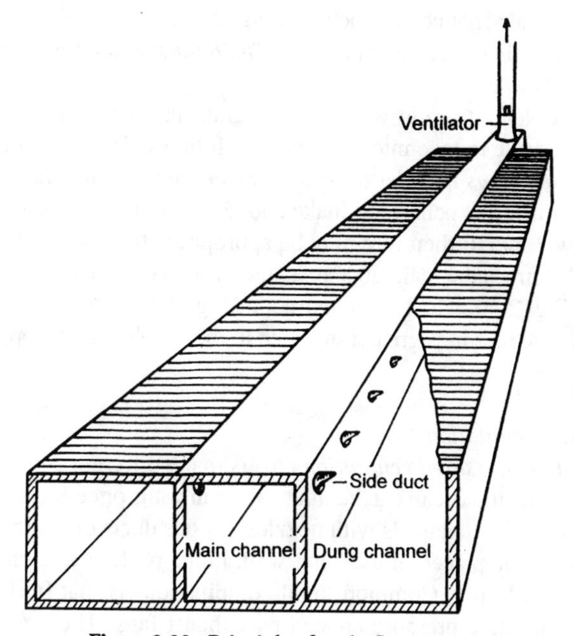

Figure 2.38. Principle of underfloor suction.

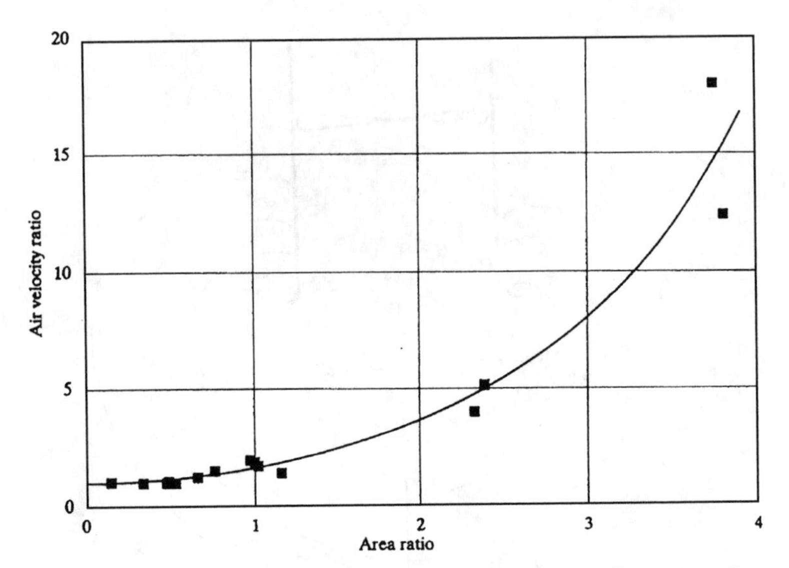

Figure 2.39. Air velocity ratio in side ducts at different ratios between total side-duct area and air-duct area.

the computation of the main air channel and the side ducts is time-consuming, because the suction is highest nearest to the fan and lowest furthest from the fan. The larger the air channel compared with the sum of the cross-area of the side ducts, the more equal the suction along the air channel, as shown in Fig. 2.39.

The following can therefore be concluded: *The total side duct area must be smaller than the main air channel.*

If that is so, the velocity ratio will not be higher than 1.6, which in most cases is acceptable. Another rule to take into account is as follows: *The air velocity in the main duct should be below 3 m/s in order to restrict the energy consumption.*

In some cases it is too expensive to make enough space for the main air duct to keep the velocity below 3 m/s. It therefore may be appropriate to increase the side duct area to, for example, 1.5 times the main duct area, and in such cases the velocity ratio will be 2.5 according to Fig. 2.40, which is an extract of Fig. 2.39. In that case, the side ducts should be placed at a 2.5 times greater distance to the fan than at the opposite end.

Ventilation Principles

Negative Pressured Ventilation

The term *negative pressured ventilation* covers many different combinations of inlet and exhaust types. The inlets can e.g. be inlet valves in walls or ceilings, inlet slots along the overhang, distribution channels with nozzles, porous ducts or ceilings, and so forth. The exhaust fans can be placed in the outer wall, in the roof, or as a part of an exhaust system under a slatted floor. Common to all combinations is that fresh air is supplied to the room by a negative pressure created by exhaust fans. The size of the negative

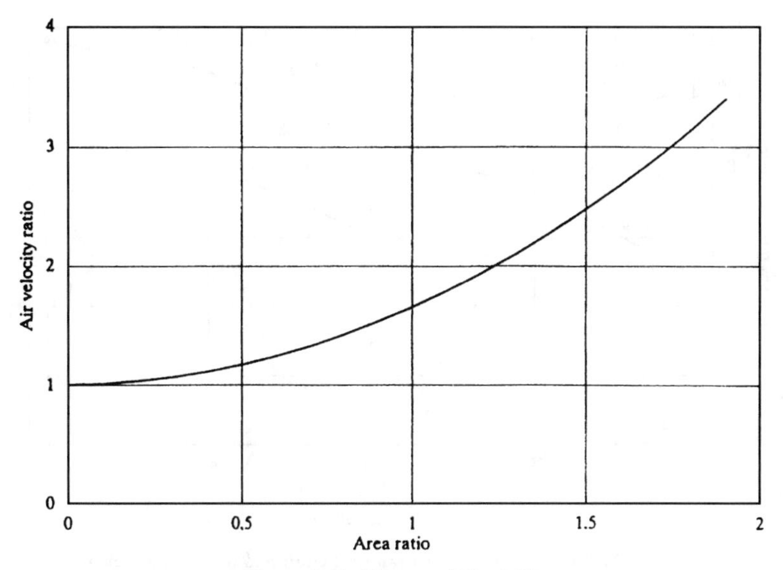

Figure 2.40. Extract of Fig. 2.39.

pressure depends on the used inlet velocity and the inlet air resistance. Normally, the indoor pressure is −5 to −15 Pa.

The advantages of negative pressured ventilation are the guarantee that humid indoor air will not pass walls and roof leakages and the fact that the energy consumption is low compared with equal pressured systems. The disadvantage is that the outdoor wind speed can influence the inlet jets, for example if wall inlets are used.

Equal Pressured Ventilation

Equal pressured ventilation means a system in which the capacities of the inlet and the exhaust fans at any time are the same. The inlet part normally consists of inlet ducts or heads with nozzles. It can eventually be combined with simple gable fans without any real distribution system. The exhaust fans can be of the same types as for a negative pressured system.

The advantages are that the system is only slightly dependent on wind, and that the inlet velocity can be kept at a certain level regardless of the building tightness. Because the ventilation flow rate is well controlled, no unnecessary energy loss of artificial heat occurs.

The disadvantages of the system are that it is more complicated and costly to maintain, and that the energy consumption is approximately twice as high as for negative pressured systems.

Positive Pressured Ventilation

In *Positive pressured ventilation* systems the ventilation flow rate is controlled by fans and the outlet air leaves the chimneys by means of low positive pressures of a few

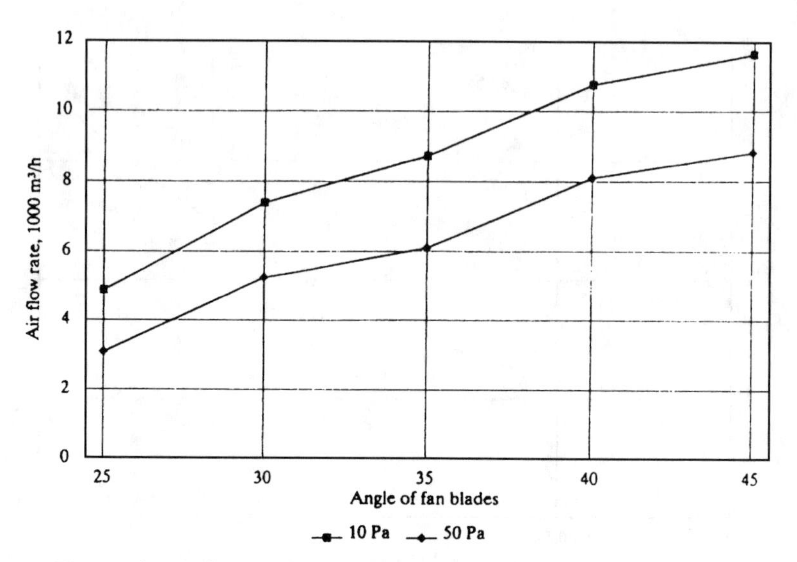

Figure 2.41. Airflow rate for exhaust fan unit equipped with fan wheels with different blade angles.

pascals. The same types of inlet systems as those used for equal pressured ventilation can be used.

The advantages are that the inlets are well controlled, as seen for equal pressured systems, and that the energy consumption is low. The disadvantage is the risk of pressing moist air into the building construction.

Fan Capacity

Because systems with forced ventilation in livestock buildings are normally categorized as low pressure systems, axial-type fans are normally used. Only in some special cases with pressure drops of 50 Pa or more centrifugal fans are used. The fan capacity depends on the dimension of the fan wheel, the number of blades, the blade angles, and the rotational speed. The more blades used and the larger the blade angles, the more air is moved. Figure 2.41 [14] shows how the blade angle affects the airflow rate, and Fig. 2.42 shows the energy consumption and the rotational speed for a basic unit with a diameter of 615 mm, with 10 blades and blade angles varying from 25 degrees to 45 degrees. The figures show that if the blade angle is increased from 25 degrees to 45 degrees, the ventilation flow rate at 10 Pa will more than double, and the energy consumption will increase by more than three times.

Another way to increase the capacity is to increase the number of blades. For the same basic unit [15] the number of blades was 4 and 8, respectively, and the air flow at a 20 Pa pressure drop was increased by about 23%. Likewise, the fan capacity at a high pressure drop was increased. In cases with high pressure losses, the most efficient way to increase the pressure is to increase the fan speed. Most fans use so-called four-pole

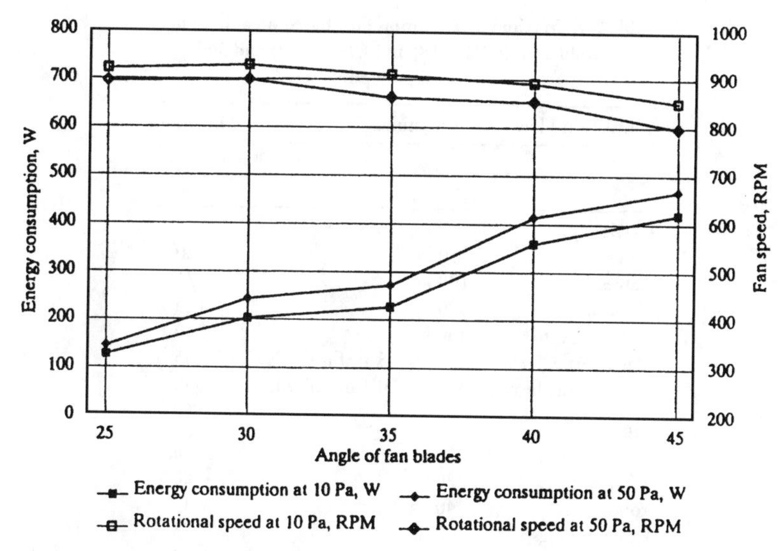

Figure 2.42. Power consumption and speed of the fans in Fig. 2.41.

motors, with a speed of about 920 rpm at 50 Hz (1100 rpm at 60 Hz). The influence of the fan speed on ventilation flow, pressure drop, and energy consumption can be calculated by the below fan laws:

$$\text{Air flow} = k_1 n \ (\text{m}^3/\text{s}) \tag{2.61}$$

$$\text{Pressure drop} = k_2 n^2 \ (\text{Pa}) \tag{2.62}$$

$$\text{Effect} = k_3 n^3 \ (\text{W}) \tag{2.63}$$

where k_1, k_2 and k_3 are constants, and n is fan velocity, s^{-1}.

The laws show that the flow rate increases proportionally with the speed, and the pressure increases with the speed in the second power. This means, for example, that for a flow rate of 6000 m^3/h with a pressure of 20 Pa at 920 rpm, the flow rate for a two-pole motor at 1450 rpm is 9500 m^3/h with a pressure drop of $(1450/920)^2 \times 20 = 50$ Pa. Because of the increased noise, it is normally not recommendable to increase the speed.

On the basis of a sample of fans, empirical equations for noise are given in a common fan-test procedure for Germany, the Netherlands, and Denmark [16]:

$$\text{Wall fans:} \quad N = 39.8 + 0.77 \times u \tag{2.64}$$

$$\text{Roof fans:} \quad N = 49.0 + 0.57 \times u \tag{2.65}$$

where

N = the noise at a distance of 2 m from the outlet opening and at an angle of 45 degrees, (dB(A))

u = the peripheral speed at a nominal voltage and a pressure drop of 0 Pa.

Table 2.9. Maximum ventilation flow based on animal heat production (CIGR [18] and CIGR [19]) and 300 m^3/h per hpu.

Ventilation Flow Rate per Animal		Maximum m^3/h per hpu
Cattle		
Calves,	50 kg	40
Calves,	100 kg	70
Heifers,	300 kg	175
Fattening cattle,	400 kg	240
Dairy cattle, Jersey,	400 kg, 15 1 of milk/day	265
Dairy cattle, heavy,	600 kg, 20 1 of milk/day	350
Pigs		
Weaned pigs,	10 kg	18
Weaned pigs,	20 kg	28
Young pigs,	40 kg	45
Fattening pigs	100 kg	75
Pregnant sows,	200 kg	80
Lactating sows,	200 kg (incl. Piglets)	220
Poultry		
Broilers,	1.5 kg	4.0
Layers, light race,	2.0 kg	3.5
Layers, heavy race,	3.5 kg	5.5

The reason for using different equations is that the wall fans normally have no accessories, as opposite to roof fans, which contain chimney dampers and so forth. For the previously mentioned example with an increase in speed from 920 to 1450 rpm, the noise increases by about 14 dB(A) for a wall fan.

Figure 2.43 shows the results of a test with a complete exhaust fan with a bell-shaped inlet and a diffusor as outlet [17]. The drawings are made according to a common test procedure in Germany, the Netherlands, and Denmark [16].

In regions with hot climate the ventilation flow rate often is valued relatively highly for creating a cooling effect at increased air velocity. Because the indoor temperature in such cases will only be 1°C or 2°C above the outdoor temperature at maximum design ventilation flow, an increased ventilation flowrate cannot reduce the indoor temperature noticeably, and so it is difficult to put forward recommendations for ventilation flow rates. In regions with lower temperatures, such as in Northern Europe, the desired indoor temperature normally can be maintained by selecting a ventilation flow rate within the minimum–maximum range. Maximum design ventilation flowrate can be based on animal heat production [18, 19] and a ventilation rate of 300 m^3/h/hpu, where hpu is a unit defined as an animal total heat production of 1000 W. Some examples are shown in Table 2.9. Experience from praxis shows that it often is recommendable to use higher flow rates than shown in Table 2.9, such as for calves, for the health of which it is appropriate to use high flow rates.

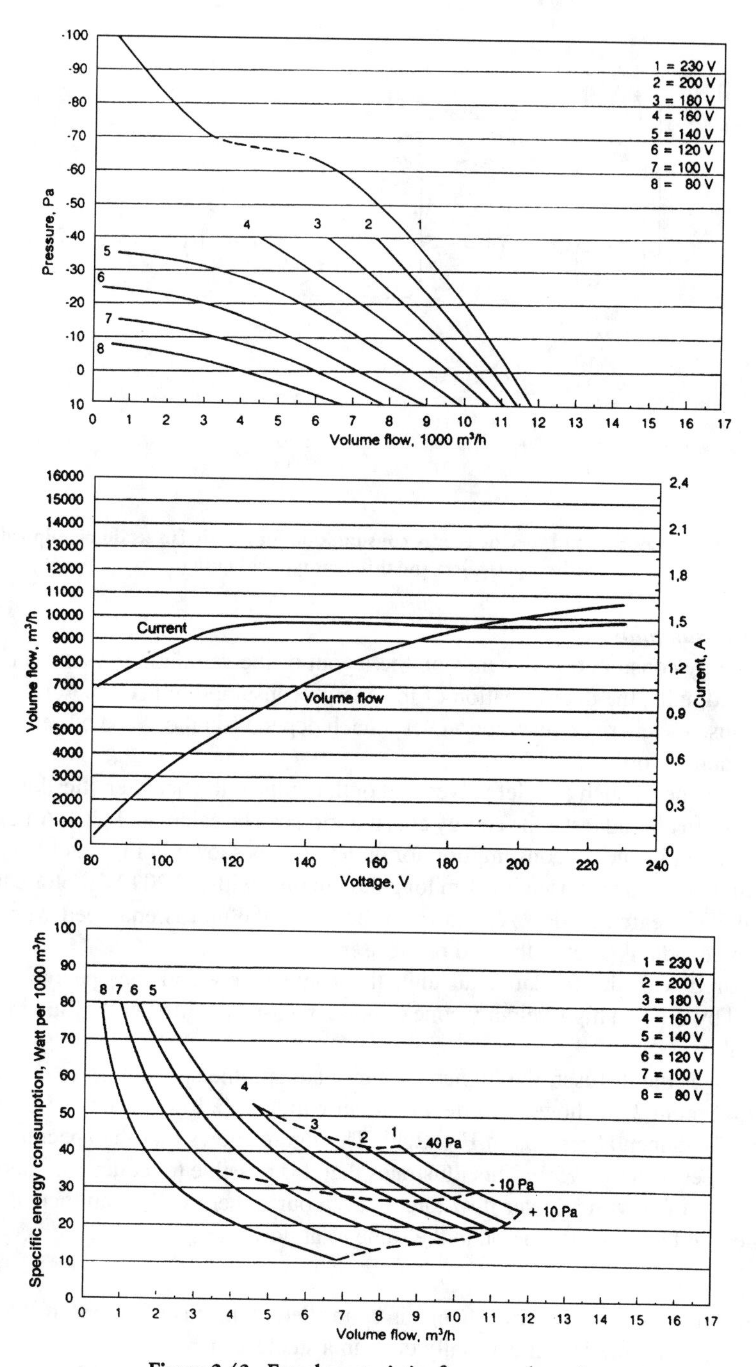

Figure 2.43. Fan characteristics for an outlet unit.

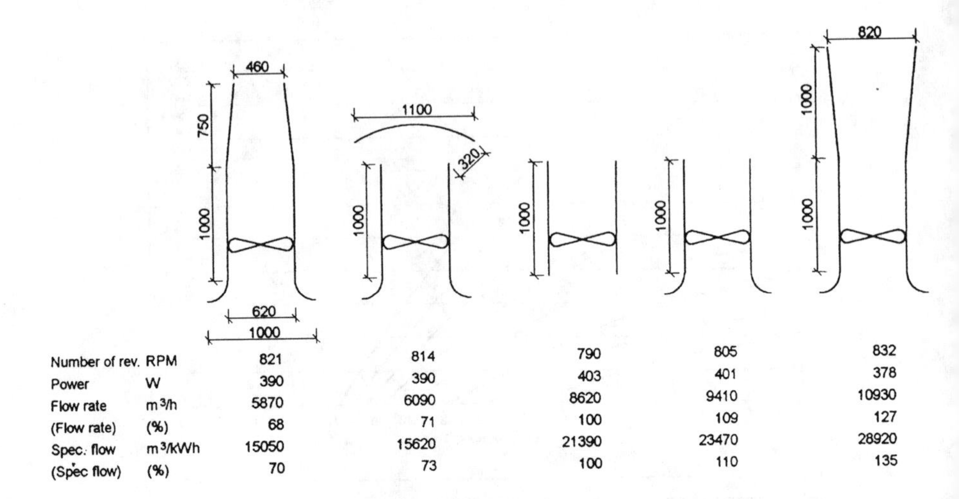

Number of rev.	RPM	821	814	790	805	832
Power	W	390	390	403	401	378
Flow rate	m^3/h	5870	6090	8620	9410	10930
(Flow rate)	(%)	68	71	100	109	127
Spec. flow	m^3/kWh	15050	15620	21390	23470	28920
(Spec flow)	(%)	70	73	100	110	135

Figure 2.44. Airflow rate and specific energy consumption for a basic fan section equipped with bell-shaped inlets and different types of outlet.

Energy Consumption

The energy consumption depends on the layout of the ventilating system. For more complicated units, the determination of the energy consumption needs careful calculation, because the energy consumption very much depends on the shape of bends, grids, dampers, and so forth.

For components such as inlet valves and outlet units with characteristics known, for example, from official tests, the energy consumption can be calculated [20]. An example of the variation in energy consumption for outlet units is shown in Fig. 2.44 [21].

The fundamental unit used is a 1 m long 0.64 m pipe with a 0.20 kW motor equipped with five 37.5-degree blades. As shown in the figure the unit is equipped with a bell-shaped inlet and an outlet with hood or diffuser.

Compared with the fundamental unit, the volume flow and the specific flow are decreased by use of a high-velocity cone or hood; by use of a diffuser they are increased considerably.

The longer the diffuser, the higher the flow rate and the specific flow, as shown in Fig. 2.45. Figure 2.46 shows the effect of a fan with a 30% lower flow rate inserted in the same fundamental pipe as in Fig. 2.45. The figure shows that the specific volume flow increases by about 75%. This illustrates that it is possible to reduce the energy cost by selecting a fan with a lower flow rate, but of course there will be an increase in the investment in fans in order to obtain the same total flow rate.

Operating Point

In the simplest way the air exchange is controlled by means of an on/off thermostat (Fig. 2.47a), but nowadays it is usually done in a stepless manner.

Up to the 1960s, most forced-ventilation systems were controlled by on/off thermostats. In the 1970s and 1980s it became common to vary the flowrate in steps, by means of special thermostats or by using stepless control. The latter could be achieved

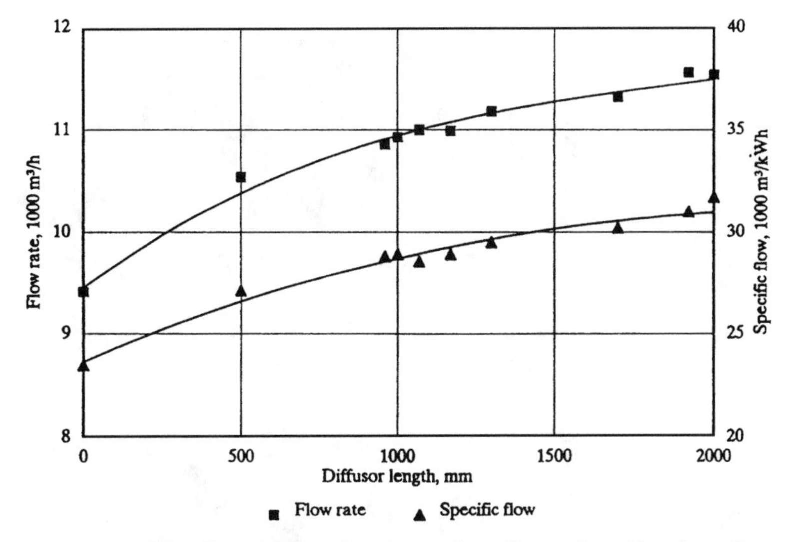

Figure 2.45. The effect of diffuser length on volume flow and specific volume flow.

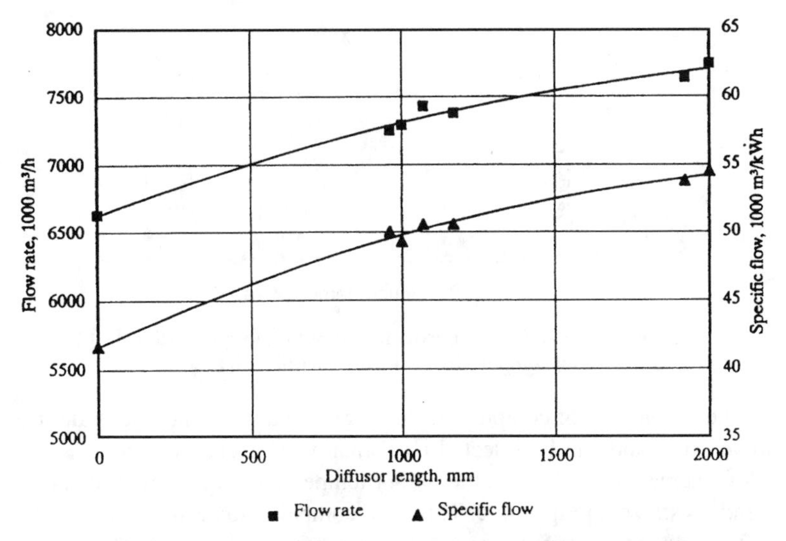

Figure 2.46. The effect of diffuser length on volume flow and specific volume flow with a fan with a smaller capacity than that shown in Fig. 2.45.

ɔy means of dampers or by varying the voltage by means of tyristors. At that time it was ɿormal to use controllers with a proportional band (xp band) of 4°C. This means that :he airflow increases proportionally to the increase in temperature. Such a principle is ;hown in Fig. 2.47b. In the 1990s the control units became more sophisticated due to the ɜasy access to computer technology. Because of a continuous fast evolution within this ırea, relevant information makes it necessary to make a close follow-up.

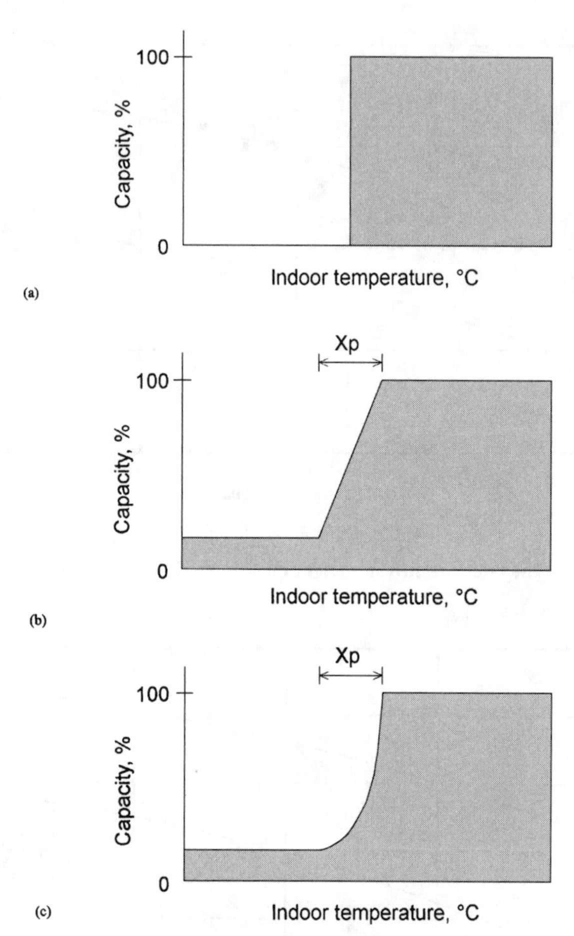

Figure 2.47. Characteristics for (a) on/off, (b) proportional, and (c) computerized control of ventilation flow rate.

One of the advantages of computer technology is that the ventilation rate inside the proportional (xp) band can be selected individually. This can be done in such a way that the flow increases slowly with the indoor temperature in the low end and faster in the high end, as shown in Fig. 2.47c. This way, compensation can be done for the great cooling effect, when the temperature difference between indoor and outdoor is high.

Because the different parts of ventilation systems can easily be controlled individually by climate computers, it can be determined how to correlate the increase of fan capacity with the opening.

References

1. ASHRAE. 1997. *ASHRAE Handbook, Fundamentals, SI Edition*. Atlanta, GA: American Society of Heating, Refrigerating and Air-conditioning Engineers.
2. Graee, T. 1974. Breathing Building Constructions. ASAE Paper 74-4057. St. Joseph, MI: ASAE.

3. Test Report No. 740. 1989. Diffust luftindtag (Porous air inlets). Horsens, Denmark: Research Centre Bygholm.
4. Koestel, A. 1955. Paths of horizontally projected heated and chilled air jets. *ASHVE Transactions*. 61:213–232.
5. Test Report No. 874. 1993. VSV vaegventil (VSV side wall inlet). Horsens, Denmark: Research Centre Bygholm.
6. Stroem, J. S., and Morsing, S. 1997. *Wind Protection of Side Wall Inlets: A New Windbreak Design*. Fifth International Livestock Environment Symposium, Volume II 851–858.
7. Nielsen, P. V. 1983. Air diffusion in rooms with ceiling-mounted obstacles and two-dimensional isothermal flow. 16th International Congress of Refrigeration, Commission EL, Paris.
8. Holmes, M. J., and Sachariewics, E. 1993. The effect of ceiling beams and light fittings on ventilating jets. Laboratory Report No.79, HVRA.
9. Nielsen, P. V., Evensen, L., Grabau, P., and Thulesen-Dahl, J. H. 1987. Air distribution in rooms with ceiling-mounted obstacles and three-dimensional isothermal flow. Room Vent 87, International Conference on Air Distribution in Ventilated Spaces, Stockholm.
10. Morsing, S., Stroem, J. S., and Zhang, G. Q. 1996. Make space for the return air in ventilated rooms. Paper 96B-057, AgEng, Madrid.
11. Pedersen, S. 1997. Velocities around an exhaust opening of a fan [unpublished]. Horsens, Denmark: Research Centre Bygholm.
12. Pedersen, S. 1978. Luftfordeling i gulvudsugningsanlæg til spaltegulvsstalde (Air distribution in livestock buildings with air exhaust through slatted floors). Ugeskrift for Agronomer, Hortonomer, Forstkandidater og Licentiater No. 1/2, Denmark.
13. Pedersen, S. 1997. Exhaust through slatted floors [unpublished]. Horsens, Denmark: Research Centre Bygholm.
14. Test Report No. 898. 1994. Rotor udsugningsenheder—Fantom (Rotor exhaust fans). Horsens, Denmark: Research Centre Bygholm.
15. Test Report No. 852, 1991. Echberg udsugningsenheder (Echberg exhaust fans) Horsens, Denmark: Research Centre Bygholm.
16. DLG/IMAG-DLO/SJF. 1993. Gebrauchswertprüfung von Stallventilatoren sowie von Zuluft-und Ablufteinheiten mit eingebauten Ventilatoren. DLG Prüfprogramm D/81. Deutsche Landwirt schafts-Gesellschaft e.V., Frankfurt am Main, Germany.
17. Test Report No. 908. 1996. Rotor udsugningsenheder (Rotor exhaust fans). Horsens, Denmark: Research Centre Bygholm.
18. CIGR. 1984. Report of the CIGR Working Group on Climatization of Animal Houses. Scottish Farm Building Investigation Unit, Aberdeen, Scotland.
19. CIGR. 1992. Report of the CIGR Working Group, Climatization of Animal Houses. State University of Ghent, Belgium.
20. Pedersen, S., and Stroem, J. S. 1995. Performance testing of complete ventilation units. *American Society of Agricultural Engineers* 11(1):131–136.
21. Pedersen, S. 1997. Exhaust fans with diffusor outlets [unpublished]. Horsens, Denmark: Research Centre Bygholm.

3 Livestock Housing

Vincenzo Mennella

3.1 Sheep Housing

In sheepfarming more than in other livestock sectors, buildings and related facilities are of a strategic importance in general breeding-management practice.

Buildings must therefore have a number of requisites, in order to satisfy the conditions of animal well-being in conformity with recent directives; allow an advanced level of mechanization and automation of production and management activities; have effect on the organization of the labor; permit management and equipment costs which are compatible with the livestock-breeding investments; and make the building compatible with the environment and the surrounding area in conformity with the new regulatory restrictions.

Thus, what are needed are not building volume units that can later be adapted according to the needs of the animal species being raised, but rather buildings that are characterized by and strictly correlated to the raising system utilized, the type of equipment used, and the degree to which it is to be automated.

A priority in the planning of building systems for sheep-raising centers is the study and analysis of topics that are fundamental for the innovation of planning procedures, such as the type of holding, production objectives, and farm reference models.

3.1.1 Types of Holdings

The most common types of sheep holdings are the following.

Semiopen Grazing

Characterized by direct grazing in the fields, where the flock spends almost all of its time. Fences mark the boundaries of the sheep's living space, which is also their principal feeding space. Feed supplements are provided in structures that have no special functional or constructional requirements. Protection from bad weather is provided, if at all, by simple shelters. Equipment and systems for feed supplements also may be more elaborate.

Extensive–Confined

The flock spends part of the year in pasture and is brought and kept inside the building for a certain period of the year (e.g., autumn–winter). This system is based on the possibility of dividing pastures into sectors and on the construction of buildings that give priority to economy over functionality, because they are not utilized year-round. The building functions include housing as well as feed preparation and distribution. A minimum degree of mechanization and labor organization is necessary for guaranteeing sufficient functionality and for limiting operating costs.

Extensive–Confined with Housing Overnight

The flock comes in every evening throughout the year and in the afternoon during the warmest periods. The building must allow the same functions as in the previous example, but for every day of the year.

Intensive–Confined

The flock stays permanently inside the building, which must permit all sheep-raising activities to be carried out: housing, feeding and controlling of the animals, removal of wastes, temperature control, and ventilation. The higher investment for buildings and equipment is compensated in this case by the higher revenues that can be obtained with more prolific and productive breeds.

3.1.2 Reference Background

Measures designed to promote the improving of livestock raising conditions and more efficient work organization have led today to the construction of large, specialized buildings.

For the defining of building models, the following parameters should be considered as references for a systematic approach: livestock husbandry (genetic improvement, diet, sanitation, healthcare and disease prevention); agronomics (crop organization and productivity, rational utilization of pastures); and management and economic aspects (building type, degree of mechanization, production process, profitability, compatibility with environment/regulations) [1].

The goal of sheepfarming lies in getting the highest performance in terms of production. Therefore the buildings for housing the animals must allow many activities to be organized and carried out in spaces defined as rationally as possible and also be compatible with the requirements of mechanization and automation.

3.1.3 Process and Product Characteristics

In defining a building system (number and type of buildings) for raising sheep, the first things to be considered and evaluated are the type of production and the number of animals in the holding.

Production types may be the following:

Sheep-Raising Specialized for Milk Production

Milk is the main product, and the main activities are centered around milk production. Necessary in this type of sheepfarming are milking parlor and cheesemaking and other facilities.

The production of lambs is a secondary part, because it entails different needs, labor, and organization that the holding is not set up for. In addition, unless bottlefeeding is done, the lambs drink the ewes' milk, which takes away from milk production and thus indirectly causes a lower yield.

Production of Both Milk and Meat

This farm is a complete-cycle type. The overall organization must take into consideration the needs of two production lines, and the breed of sheep must be suitable for both milk and meat production. The organization of the farm must take into consideration the specific dietary needs of the young animals, as they directly influence production and require special attention, especially in the first period of life.

Meat Production

This is essentially the production of lambs for fattening in confined, highly mechanized environments. In the Mediterranean area, small lambs weighing 13 to 16 kg are produced and consumed; there is increasing production of heavy lambs weighing 30 kg at 90 days after birth.

Wool Production

In general, with the exception of a few countries, the wool-production cycle is almost always combined on a secondary or equal level with milk production.

3.1.4 Criteria for the Defining of Building Systems

In each of the production types, animals that have a variety of needs, which differ among them, are made to live together, and the building in which they are housed must satisfy those needs. In order to define the quality, the number, and the characteristics of the individual buildings that form the overall system, it is necessary to identify the needs of each group of animals and the number of animals that it may contain, and to take into consideration organizational factors.

One design concept that may provide a suitable answer to the demands of the sheep-production world is that which is based on a systematic method, which makes it possible to identify specific reference parameters and guidelines by organizing functional, structural and installation aspects [2].

Although often involving building systems of considerable complexity in the functional organization of spaces, in most cases livestock building projects have simple, linear, single-level structures. The design procedure for defining the contents and size of the livestock buildings at the start provide the individuation of the fundamental spatial elements that constitute variants of each production type for each different species of livestock. The elementary activities linked with the subjects (person, animal, machine) involved in the production process constitute the input for the definition of the spatial elements.

Following is a list of the spatial elements that characterize the activities of sheep-holding:

- SE1 Feeding
- SE2 Covered sheds
- SE3 Open-air pens
- SE4 Corrals
- SE5 Lambing sheds E
- SE6 Alleys and passageways
- SE7 Milking
- SE8 Shearing areas
- SE9 Dipping vats
- SE10 Feed storage and preparation
- SE11 Product storage
- SE12 Fenced-in treatment areas
- SE13 Installations and control rooms
- SE14 Manure-carrying space (outside)
- SE15 Manure-storage area (outside)
- SE16 Liquid-waste storage
- SE17 Waste removal
- SE18 Equipment area (purification)
- SE19 Farm-machinery parking
- SE20 Farm-machinery moving (inside)
- SE21 Tool room (inside workshop)
- SE22 Equipment and material storage (inside workshop)
- SE23 Animal-feed storage
- SE24 Handling of animal feed
- SE25 Lunchroom
- SE26 Offices
- SE27 Pantry
- SE28 Kitchen
- SE29 Overnight quarters
- SE30 First aid–emergency
- SE31 Sitting room
- SE32 Lavatories/changing room
- SE33 Connection areas
- SE34 Roads
- SE35 Parking areas
- SE36 System- and installation-control rooms (outside)

The aggregation of coexisting elementary activities that are similar from a spatial–functional viewpoint and require a portion of space having the necessary characteristics for carrying out the activities in the correct manner creates *area units* (Fig. 3.1).

The area units can be repeated and aggregated by means of relational matrices and suitable models for the analysis of the flow of animals, people, material, and machinery in spatial systems of varying complexity that form *maximum aggregation units* or *functional areas*. These are the result of all the valid possibilities for the aggregating of area units and constitute autonomous functional layouts, therefore making the structural links of the buildings.

In this way integrated systems of area units take shape that are an answer to the carrying out of various activities, making possible the completing of a well-defined time cycle (meat production, feed preparation, etc.).

The final aggregation level, *functional line*, which is also identified by means of relational matrices among functional areas, constitutes the spatial models (Fig. 3.2).

3.1.5 Building Systems for Intensive Milk-Production Holdings

A diagram of the stages of the production process for obtaining milk for the primary production and lamb for the secondary production is given in Fig. 3.3. Consequently it is possible to define a production program that allows the identifying of the number of head in each single production stage, the total quantities of the end products, and the desired period in which they are to be obtained.

For this, the following factors must be taken into consideration: initial resources, number of sheep, the length of the individual production stages, the daily weight increase

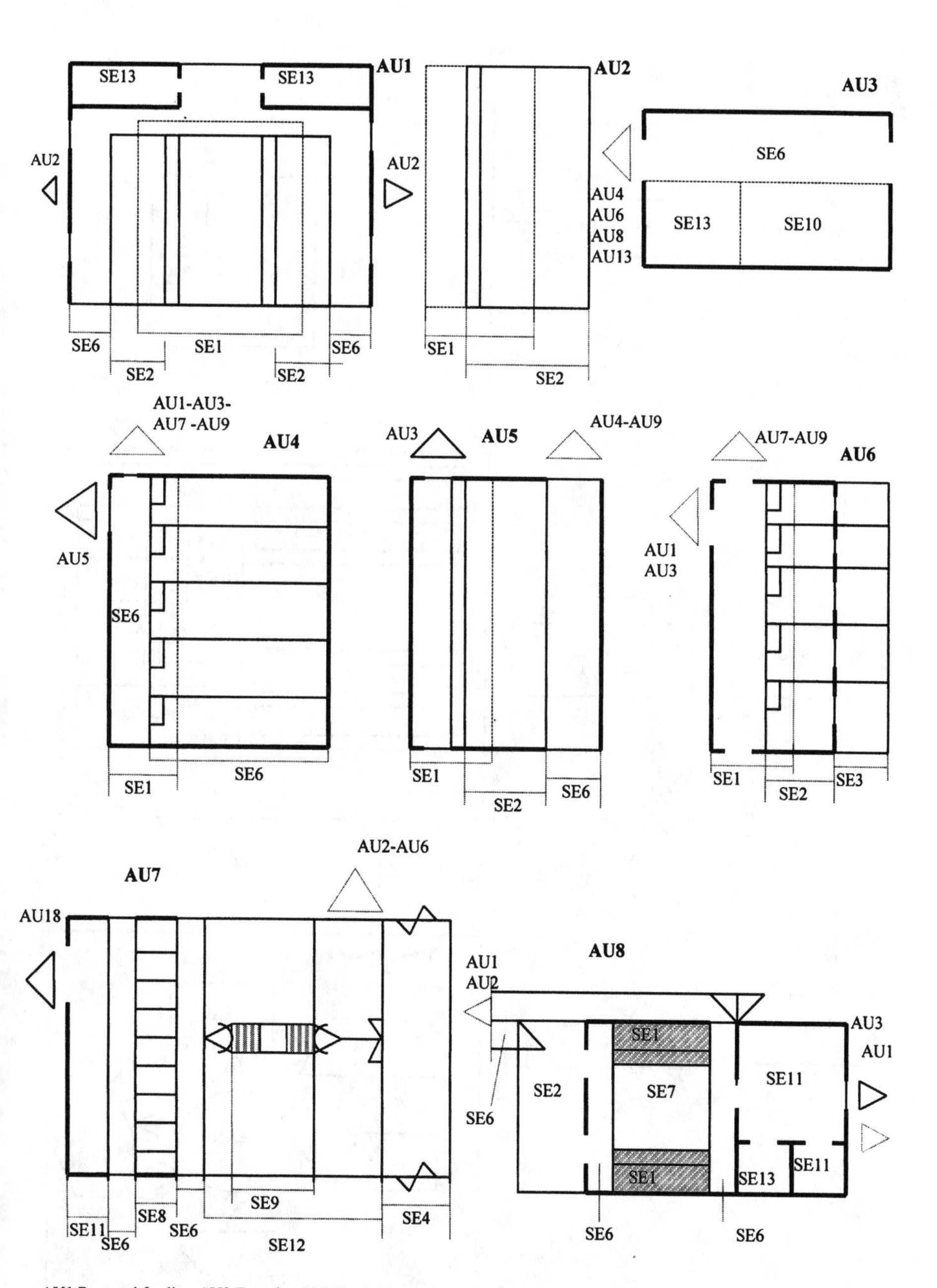

AU1 Rest and feeding AU2 Exercise AU3 Feed preparation AU4 Lambing
AU5 Weaning AU6 Rams AU7 Animal treatments AU8 Milking

Figure 3.1. The area units (AU) in the picture are the portion of space having the necessary characteristics for carrying out the coexisting elementary activities in the correct manner.

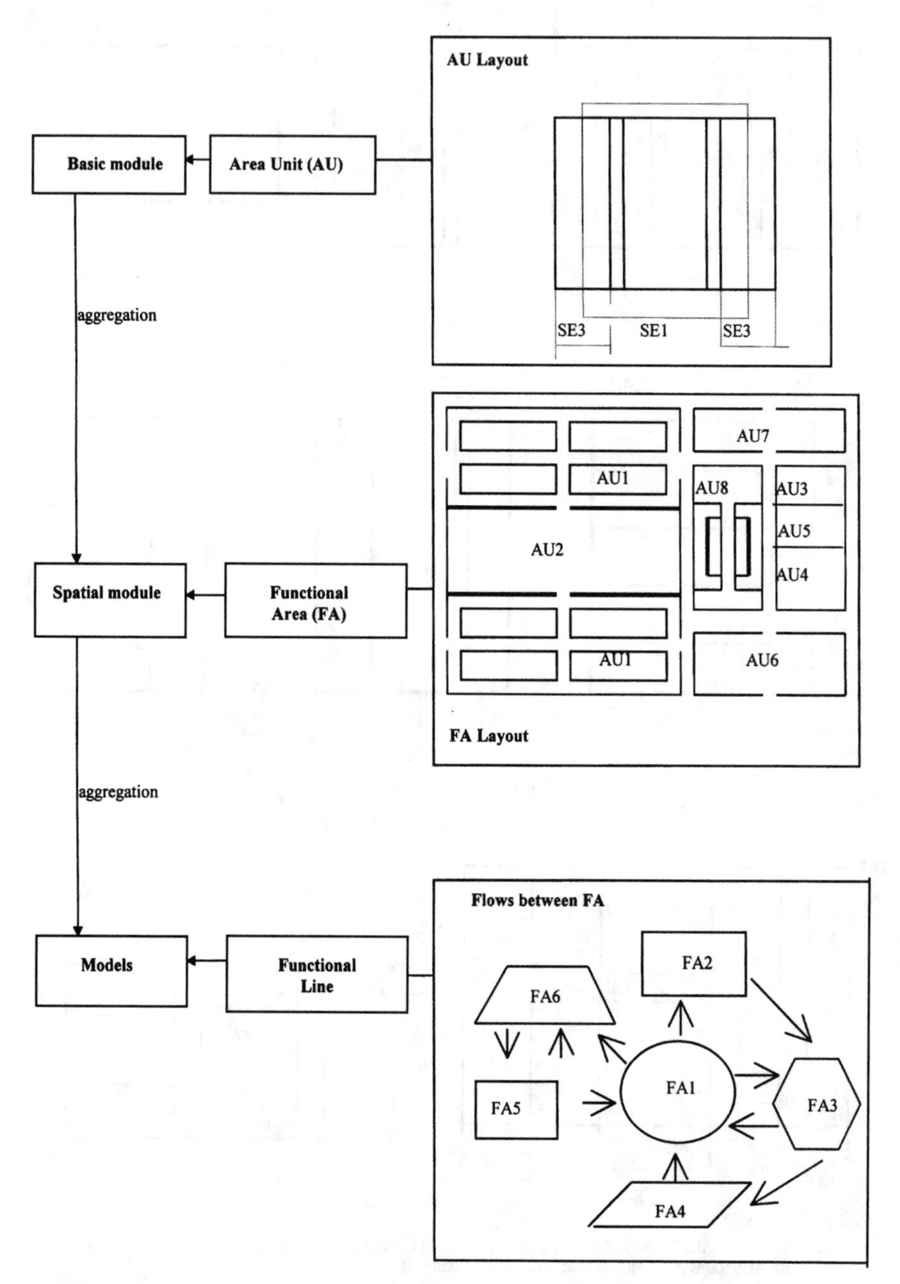

Figure 3.2. Outline for the defining of spatial models; constitutes the building system.

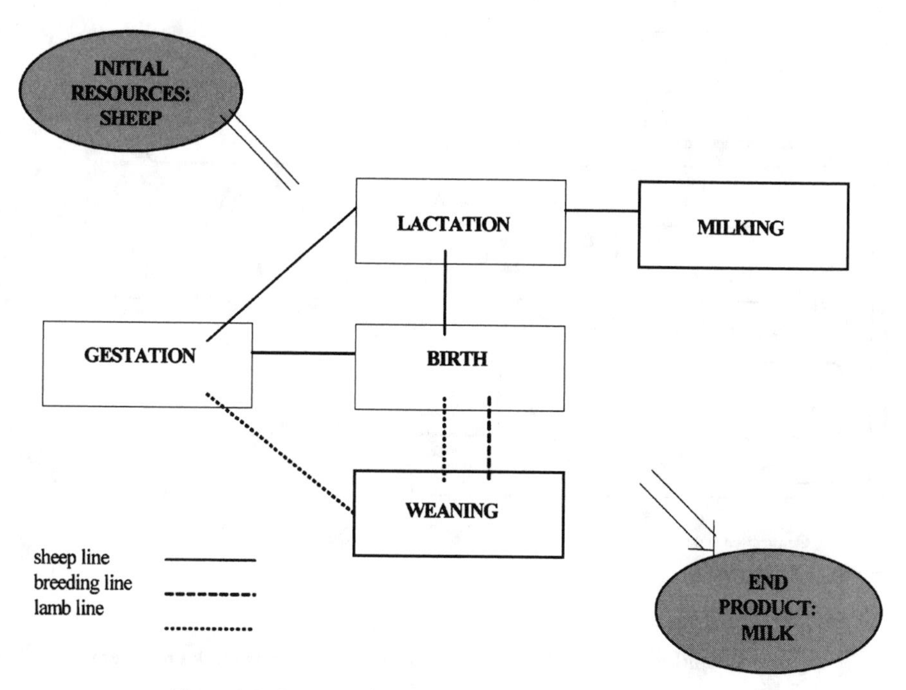

Figure 3.3. Stages in the process of intensive milk production.

(for lambs being weaned), diet and the feed-distribution method, the planning of lambing, and the end product [3].

It is thus possible to identify different "production lines" (e.g., sheep, lambs for breeding, lambs for finishing) and the different biological stages of the animals at different times in their productive careers (e.g., pregnant ewes, dry ewes, lactating ewes). Each production category has its own separate requirements, very different from those of other categories.

In order to facilitate work organization for holdings with more than 500 to 600 head, it is best to provide for separate buildings for each production category; for smaller holdings the different categories can be divided into areas or pens within a single building. In milk-production holdings, it is best to further divide the sheep into groups numerically the same as the number of places in the milking parlors, or into a multiple of this number. The choice of the type of production permits the identifying of the work categories necessary for normal operations, while the frequency with which they are carried out makes it possible to classify them.

Thus they can be divided into daily operations (distribution of feed, milking, and milk handling), regular shorter intervals (litter changing, health–sanitation controls), regular longer intervals (assistance during lambing, servicing of ewes, removal of animals, removal of waste, shearing, replacing animals, supplying of feed both at the farm and outside the farm), and those operations not repeated in the same cycle or the frequency of which cannot be determined (caring for sick animals).

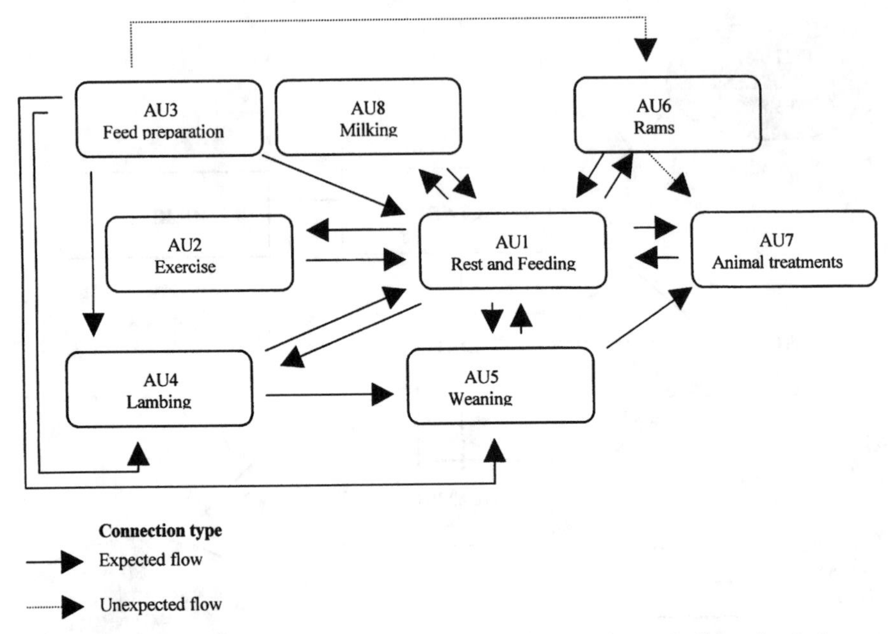

Figure 3.4. Outline of internal flows among the functional areas (milk production).

The *production operations* are made up of a number of activities carried out at the same time and sequentially in the area units. These aggregations of spatial elements are defined also by the type of environment and technological parameters that describe the carrying out of animal functions.

The defining of more complex functional spaces is done by the analysis and determining of flows, or of those values that express the number of movements of animals, workers, and machines in a given period of time. These emphasize the relationships of spatial–functional interdependency among the various area units in order to optimize and simplify all production operations and the use of labor, thus making the productive process economically valid.

The analysis of flows leads through different correlational levels (contiguity, vicinity, distance, lack of connection, exclusion) to the classification and spatial composition of the functional areas (Fig. 3.4). At the same time, the connections from the analysis of flows among the functional areas make it possible to plan the layout of the milk cycle (Fig. 3.5, 3.6), with synthesis of different possibilities for distributive–functional aggregations of the predefined functional areas.

3.1.6 Planning Parameters

The characterizing of the entire building system arises from the identifying of planning parameters, which provide a technological and dimensional definition of the spatial configurations. The possible combinations of various basic modules (area units) vary

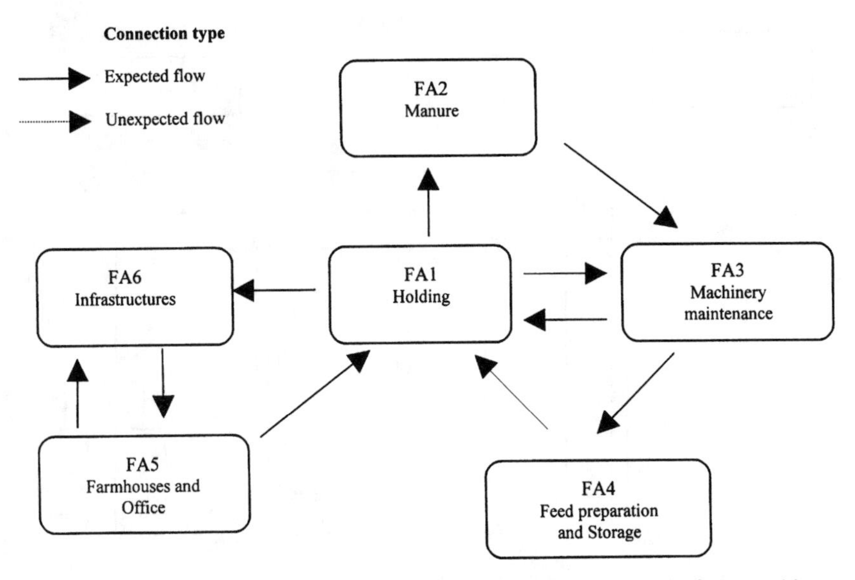

Figure 3.5. Outline of flows among the functional areas for defining the functional line.

Table 3.1. Standard sizes of area units

Area Unit	Covered Area (m^2/animal)	Feed-Trough Width (m/animal)	Feeding and Service Passage Width (m)
Rest/feeding	1.85–2.35	0.35–0.40	3.0/0.8
Exercise	2.00	—	2.5
Lambing	2.5–3.0	Individual troughs	2.00
Weaning	0.3	0.10	2/0.8
Rams	2.5–4.0	Individual troughs	2.00
Animal treatments	5.0		—
Milking	2.0	Individual troughs	0.8

according to planning choices, which must necessarily take different factors into consideration (production techniques, degree of automation, type of systems, etc.).

Tables 3.1 and 3.2 give the standard sizes of area units in buildings for sheep raised for milk production [4–8].

3.1.7 Building Systems for Meat-Production Holdings

The entire cycle is geared toward meat production and is characterized by five stages: housing of breeding ewes, dropping and natural weaning of lambs, fattening (feed with concentrates), breeding, carrying and storage of manure, as outlined in Fig. 3.7 [9].

Production-line requirements can be established in the same way as for sheep for milk production. Once the activities taking place in the area units are defined, they are aggregated according to function and correlation, making up the functional areas (Fig. 3.8).

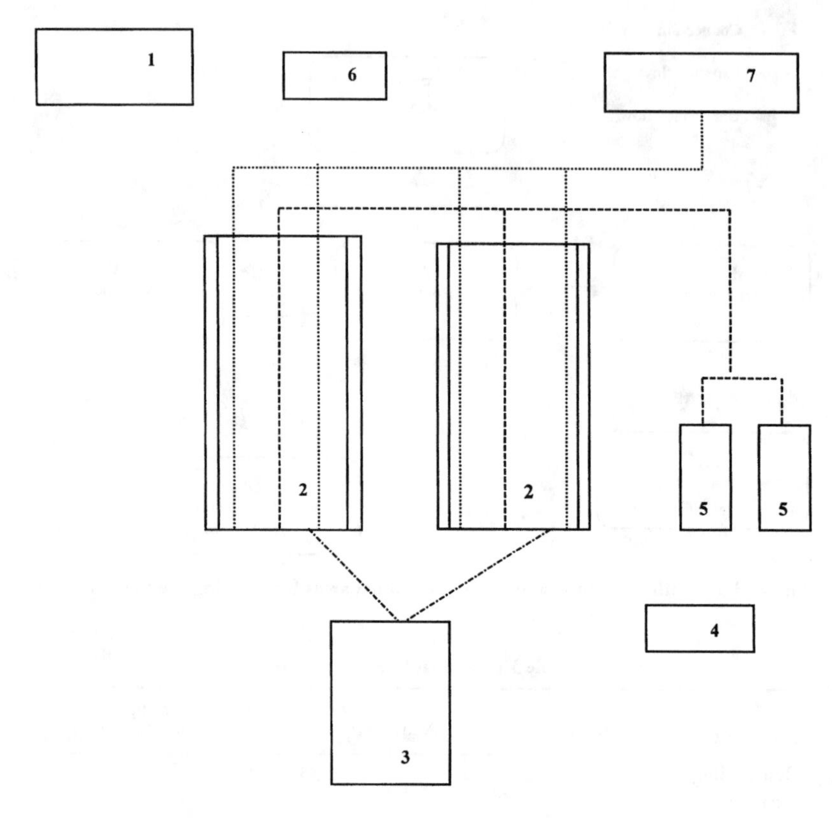

Legend
Basic operation and flows of movement
manure
feed
ewes to and from milking parlor

1. Farmhouses and offices **2**. Production buildings for sheep. Outside paddocks.
3. Milking parlor and milk storage **4.** Machinery and equipment storage
5. Hay barn, horizontal silos for feed **6.** Building for lambs
7. Platform for storing manure.

Figure 3.6. Building-system layout for sheep raised for milk production.

Table 3.2. Standards for temperature and relative humidity

Area Unit	Air Temperature (CE)	Relative humidity (%)
Rest/feeding	10–17	60–80
Lambing	10–17 (sheep) 20–22 (lambs)	60–80
Weaning	18–22	60–80
Rams	10–17	60–80
Milking	18–20	70–80

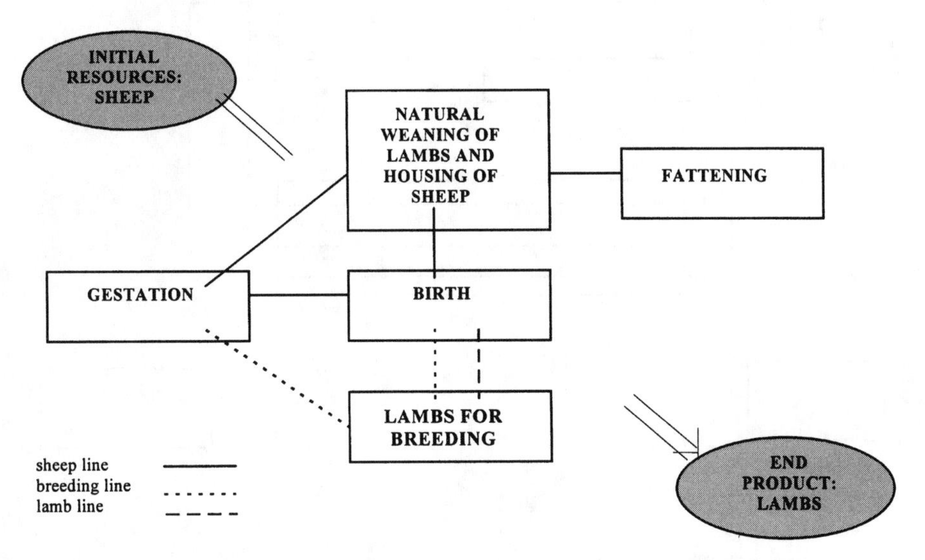

Figure 3.7. Stages of the production process for intensive sheep raising for meat.

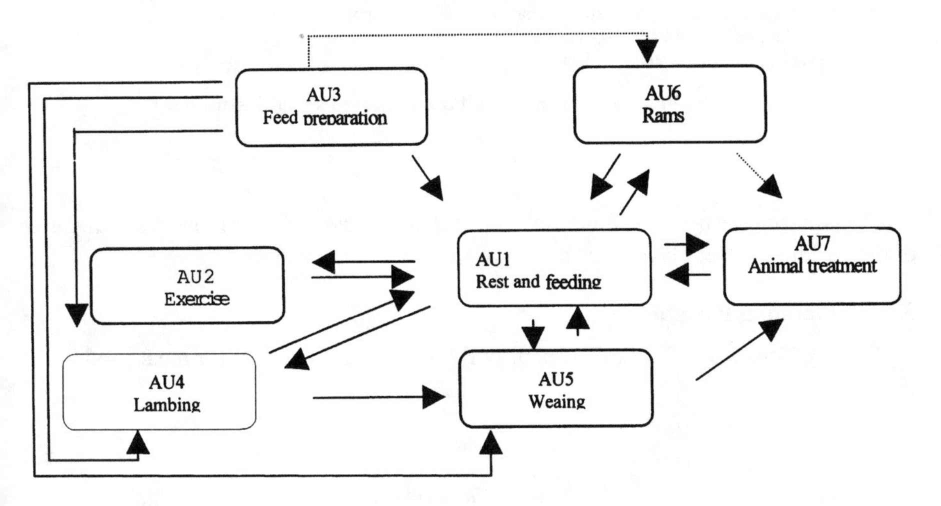

Figure 3.8. Outline of internal flows among the functional areas (meat production).

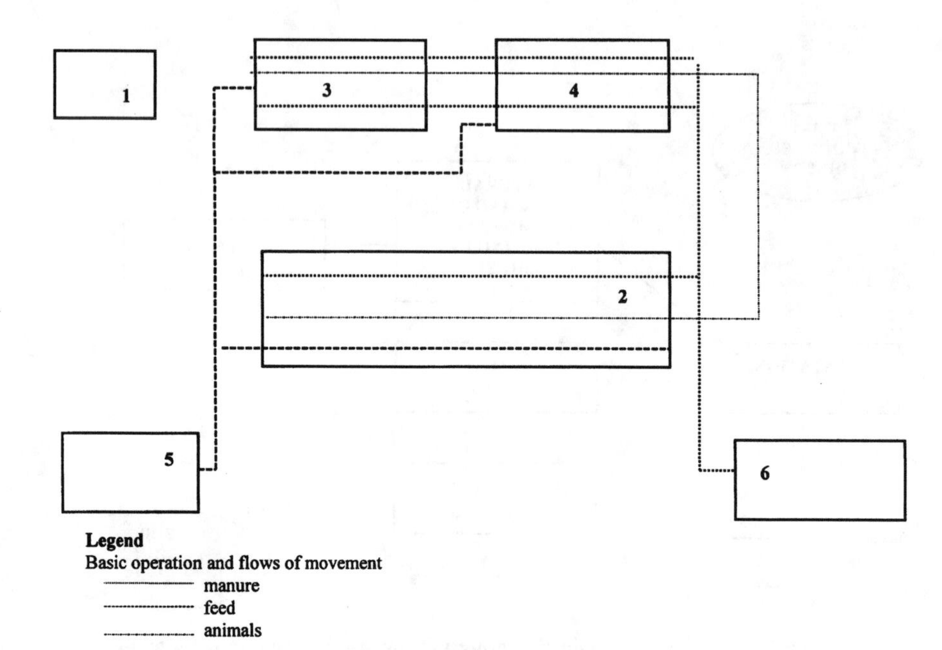

Figure 3.9. Building-system layout for intensive sheep raising for meat.

The connections between functional areas characterize and determine the boundaries of the building-system layout for intensive sheep raising for meat (Fig. 3.9).

3.1.8 Planning Parameters

Table 3.3 gives the standard sizes of buildings for sheep raised for meat production [9].

Table 3.3. Standard sizes

Area Unit	Covered Area (m^2/animal)	Feed Trough Width (m/animal)	Feeding and Service Passage Width (m)
Rest/feeding	1.30–1.70	0.35–0.45	3.80
Exercise	2.00	0.35	2.50
Lambing	2.5–3.0	Individual troughs	1.50–2.00
Weaning	0.3	0.2	2.80
Rams	2.5–4.0	Individual troughs	2.00
Fattening	0.6–0.8	0.2–0.3	3.80

3.1.9 Conclusion

In recent years livestock breeders have begun computerizing the management and operation of systems and installations (feeding, temperature control, waste removal, etc.) through the use of complete automated systems. They also have introduced new functions (automatic distribution of feed, installations management, milking machines, etc.) inside buildings. It has become necessary to consider new requirements starting in the building-planning stage.

In this setting of technological change, the necessity to integrate the distribution of space and technological elements is clearly a fundamental basis for the defining of any type of productive structure. The procedure outlined here satisfies requirements because it overcomes the stiffness of predetermined layouts.

References

1. American Society of Agricultural Engineers. 1981. *Agricultural Engineers Yearbook 1981–1982*. ASAE. St. Joseph, MI:
2. Chiappini, U., Fichera, C., and Mennella, V. (1994). *Metaprogettazione per l'edilizia zootecnica*. Milano, Italy: BE-MA Editrice.
3. Mennella, V., and Borghi, P. 1985. Criteri progettuali e modalità costruttive per la realizzazione di centri di allevamento ovino da latte. Sheep and goats journal. *Il Vergaro* 9/85.
4. Barnes, M., and Mander, C. 1986. *Farm Building Construction*. Ipswich, England: Farming Press.
5. Bryson, T. 1985. *The Sheep Housing Handbook*. Ipswich, England: Farming Press.
6. De Montis, S. 1983. *Edifici pre l'allevamento ovino*. Bologna, Italy: Edagricole.
7. Fearghal, O., and Farrel, B. 1985. *Sheep Housing and Handling Handbook*. Dublin, Ireland: Farm Structures and Environment Department Agriculture Institute.
8. Mennella, V., and Borghi, P. 1983. Moduli edilizi per l'allevamento intensivo della capra da latte. *Il Vergaro* 9/83.
9. Mennella, V., and Borghi, P. 1985. Criteri progettuali e modalità costruttive per la realizzazione di centri di allevamento ovino da carne. Sheep and goats journal Editrice Poligrafica Roma I, II, III, IV parte (5/85, 6/85, 7/85, 9/85). *Il Vergaro*.

3.2 Pig Housing

3.2.1 Reference Scenario

In almost all countries, pig breeding is shifting toward intensive breeding systems, with a high number of animals and a reduced utilization of human resources. This must necessarily lead to greater attention on the part of the breeder, who raises the animals using industrial management methods, to the inspiring principles of animal protection in order to guarantee conditions that provide the maximum respect for the physiological and behavioral needs of the species of animal being raised.

It is only by analyzing needs in ethological terms that breeding systems can be created that respect the "natural" lifecycle of animals without reducing productivity.

Furthermore, when designing buildings, especially those that are large, the form of the building and its relationship to the environment in which it is to be placed are aspects that cannot be neglected.

This does not mean that these aspects are more important than those regarding functionality and productivity; however, working with this approach makes it possible to find solutions that allow both of these types of requirements to be satisfied, without taking away from either.

The building system must, in its separate elements and as a whole, satisfy the requirements of the end user, understood as the explication of needs determined by means of parameters to which the building environment must be adapted. These parameters are the result of a complex procedure using a metadesign approach to translate the needs expressed by the user into requirements correlated to the areas of the building compound.

Along with these parameters that characterize each area unit, it is appropriate to provide indications concerning the relationships among the various units. These relationships are formulated by means of the analysis and quantification of flows (of persons, animals, vehicles, and materials) and contribute to the defining of the building layout.

The various models for the aggregating of area units must be able to satisfy the different needs, expressed in building terms as requirements (well-being, protection, etc.). Thus the performance characteristics regarding heating, cooling, humidity, airflow velocity, and noise-level requirements (in relation to the animal species and the particular stage of the production cycle) must be considered for the different area units and are also indispensable parameters for building design.

Once the various aggregation models are defined, it is also necessary to evaluate the various solutions through the determining of certain cost categories upon which one aggregation criterion might have an effect, while another might not. Examples of these are costs for land usage, construction, systems installation, operations, energy, and maintenance.

Along with the area system, the analysis requires that legal regulations, which are fundamental in the defining of a technological system, be taken into consideration. The parameters in this case are determined in relation to a given technological category, such as inner partitions, flooring, or roofing.

3.2.2 Types of Pig Holdings

The continual evolution of livestock technology and the achieving of production results once considered inconceivable require the availability of specific, appropriate buildings with advanced features, which naturally vary according to the type of pig holding. Production activities are divided into two main types, closed- or open-cycle, depending on whether the holding is for breeding or fattening. The former requires a wide range of units, for housing piglet production, growth, and fattening; the latter only a unit for fattening.

Thus the types of pig holdings can be identified as

- Sow breeding. This requires careful, expert management, as well as excellent organizational skills, for producing a large number of pigs at low cost.
- Fattening. This does not require any sophisticated management techniques, but

rather business skill in purchasing quality animals and feed at low cost and in selling mature animals at the right time and for the right price.

- Breeding and fattening. This requires both sets of abilities and capacities already described and especially good organizational skills to produce and breed pigs at a profit.

The area units follow the different life stages of the pig, which are birth; weaning; stimulation, waiting, heat, and covering of sows and the first stage of gestation, up until the diagnosing of pregnancy; the second stage of gestation; growth and fattening stage; and replacement.

3.2.3 Criteria for Defining the Building System

Because the needs of the animals to be housed are several and diverse according to the various living conditions, it is necessary to set up a design plan based on a systematic method that is capable of responding to the functional, structural, and installational needs.

The *area units*, which can be repeated and aggregated in simple or complex spatial systems, are the basic elements of the structural grid, or the "building skeleton." Although it involves highly complex organizational–functional systems, in most cases pig breeding requires a fairly simple, linear, single-floor structural grid.

There are two categories of project input: area and technological. The former include the dimensional characteristics of the basic invariants (area units) of the building system. In fact, these are represented by reference modules composed of boxes or pens having the same dimensions, installations, and equipment, in which an entire stage of the animal production cycle takes place (birth area unit, fattening area unit, etc.).

The technological input includes the level of technological complexity of the installations and equipment and the types of installations and systems; the organization of building areas and volumes must first of all take into consideration the spaces allotted for installation and system networks. For example, the type of ventilation system for pig breeding must also be chosen on the basis of other technological subsystems, such as the type of flooring (solid floor, grated flooring over drainage channels on the ground); whether or not there are other systems installed (such as for the distribution feed, the removal of wastes); and the type of roofing (the presence of false ceilings). Thus, if in the metadesign stage the structural grids of the area units or, rather, of their aggregation are defined on the basis of the area and technological requirements, the adaptations necessary both at the level of planning and of construction will be limited.

For livestock structures, the metadesign analysis of the area system provides standard area modules, which are of a size that remains constant according to the type of animal being raised and the specific stage of the production cycle. For example, the birth area unit for pigs makes use of modules composed of 8 to 10 boxes arranged in two rows along a central passageway. The limiting to only 8 to 10 boxes, which can be repeated modularly in one or two rows, derives from needs for better control of temperature, for ventilation, and above all to guarantee a peaceful environment for the sows and piglets during birthing and weaning.

In order to formulate the structural module, the size characteristics of the individual standard units are taken into consideration, and for each of these a definition is provided for the various uses, the temperature and humidity characteristics, the installing of systems, the space for the separate activities taking place there, and so forth. The aggregation of the Structural Modules into more complex forms for the various Units therefore is based on two separate stages: the organization of space distribution and the verifying of compatibility with system installations.

The organization of space leads to numerous possibilities for the aggregation of base units. These aggregations also are dependent upon the type of structure that may contain one or more area units, either identical or of different types.

Truss-type structures offer solutions with a lateral passageway for access to several area modules along a row. Structures with intermediate supports also allow for larger openings (16–20 m), with a layout of a double row of modules with a wide central passageway.

The stage of verifying compatibility with system installations requires suitable integration between the structural type chosen (the size of the structural span) and the housing of system networks. For intensive-breeding buildings, in which temperature, humidity, and sanitary and feeding conditions are the basic elements for obtaining satisfactory production cycles, the appropriate arrangement of ventilation, heating, food-distribution, and waste-removal systems is fundamental for the organization of the building structural grids.

The planning of installation networks must therefore be based on some basic informational data, which are provided for the most part by the metadesign charts:

- The type of system, according to the species being raised and the type of production.
- The analysis of technological requirements, referring to each subsystem and to each technological class, which must direct design choices towards the most satisfactory technical solution. For the animals, the requirements refer to the needs of well-being and protection; for operators, the requirements regard production (efficiency of service), energy saving, and the maintenance, integrability, reparability, and inspectionability of systems.
- The analysis of system-distribution organization, which requires the defining of technical volumes, that is, the planning of suitable housings in relation to the type of system, as well as the organizational needs of the spaces already determined (module aggregations).

In pig breeding, for example, the air-conditioning system often requires false ceilings and raised floors, which are necessary for the removal of waste that falls into the drainage channels below. False ceilings usually provide space for housing air ducts or for areas in which air is introduced under pressure and is then sent into the breeding boxes at various speeds and volumes.

For heating systems, the water is distributed in pipes that are sometimes placed under the floor, and if the floor is made from gratings over drainage channels, it is necessary to integrate these technological subsystems. The automatic distribution of feed is largely used in pig buildings in which feed is supplied in increasing amounts in close relation with the cycle of animal development during the growth stage. The positioning of this system also requires careful planning for the functional organization of space, size, technical volumes, and, in this specific case, also the measurement of the volume of the central feed units (silos for various types of feed, mixing tanks, etc.).

3.2.4 Systems Planning and Integration Parameters

The characterization of the entire livestock building structure must arise from the determining of planning parameters that are able to provide those guidelines necessary for spatial and technological planning.

In the previous section, reference was made to the ways of transferring the information furnished by the results of the metadesign process (charts, diagrams, matrixes, flowcharts) for the planning of a livestock-breeding center. In the following, the procedure is given for the formulating of a plan, applied to a specific case (closed-cycle pig-breeding center). The reference parameters can be taken from the results of metadesigning, starting with the area units. They are used, according to the different applications, either alone or in aggregates to form the base modules, which are dimensionally defined.

The possible combinations vary in relation to the planning choices, which must necessarily account for various factors (breeding techniques, degree of automation, type of systems, etc.). The aggregation of several similar or different base modules, put together according to functional, spatial, and organizational relationships, defines a spatial module. In the relational system, the spatial module corresponds to a functional area, where the latter is exclusively the place where a complex function is carried out following the defining of the spatial layout.

The models are obtained by aggregating spatial modules (functional areas), so as to obtain multifunctional spaces that define the closed-cycle pig-breeding center (functional sectors) (Fig. 3.10). The size variables of the different base modules and their combinations provide a first indication of the development of building systems, leaving open many possibilities for linear aggregation structures. The factors to take into consideration for aggregation are:

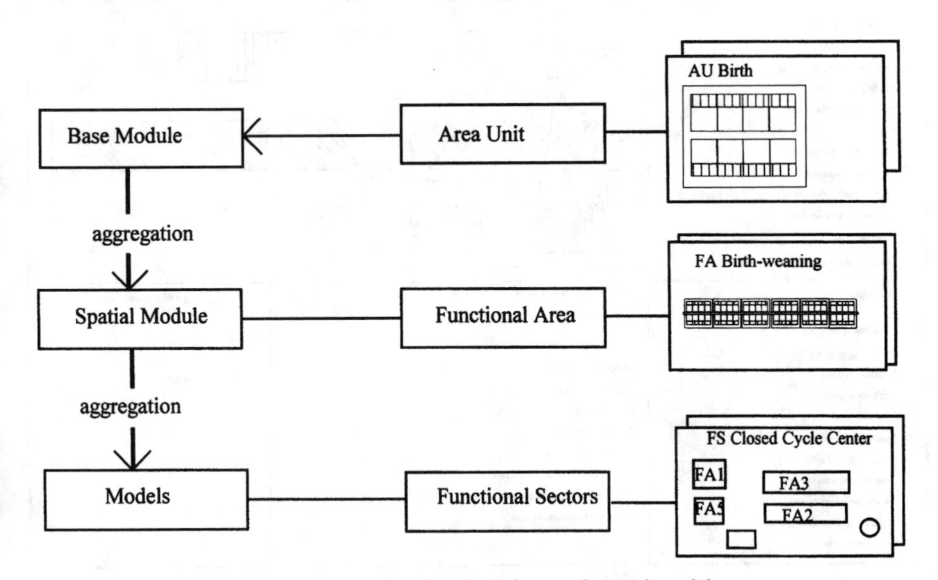

Figure 3.10. Outline for the defining of spatial models.

- Dimensional, organizational–functional, base-module relations and area parameters, for the defining of the layout of the spatial modules
- The utilizable construction characteristics (list of products for livestock buildings) with indications for the maximum dimensions of structures, which for livestock buildings are usually made out of reinforced concrete or metal. In this way, a structural grid compatible with the already defined and aggregated Spatial Modules is created
- Data on the equipping and type of systems for each area unit. To define these data, it is necessary to study the distribution of the system networks according to the flowcharts for materials (for example, the frequency of feed distribution), to the area unit organizational–functional charts, to the size of channeling, and to the system integration with other technological subsystems (and the structural grid in particular)

3.2.5 Closed-Cycle Pig-Breeding Center

Following is a plan for an integrated organizational–functional model corresponding to the closed cycle pig-breeding center functional sector created on a metadesign basis. A list of reference charts for a specific functional sector can be used as the preliminary instrument for outlining the size and contents of the building (Fig. 3.11). These charts indicate:

- The parameters for determining dimensions, divided into general parameters,

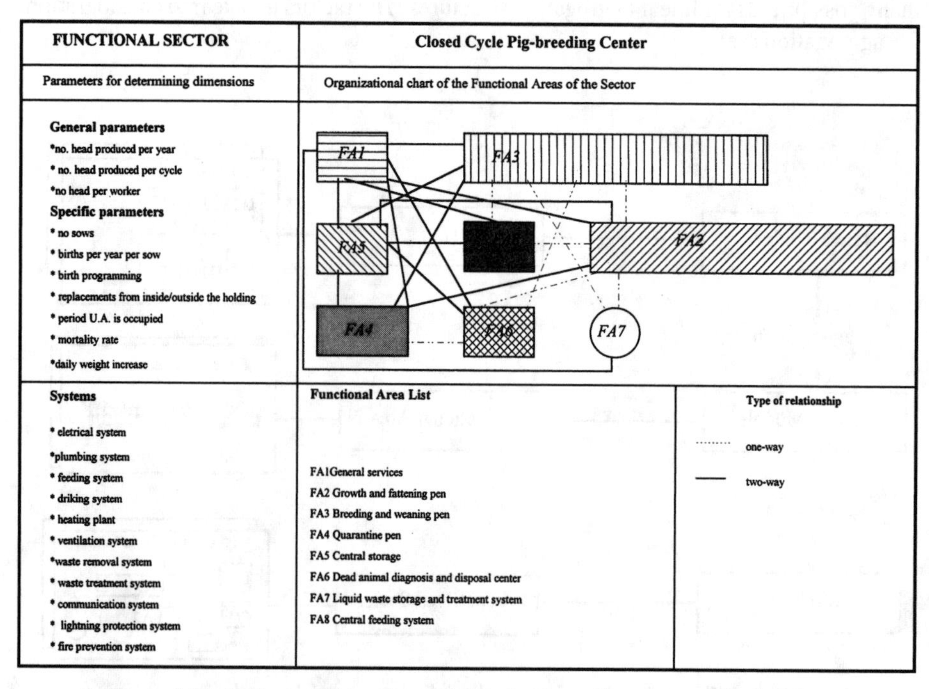

Figure 3.11. Reference charts for outlining the size and contents of the building.

referring to the overall building system and common to all functional areas, and specific parameters, which identify the type of production

- The organizational–functional diagrams of the functional sector, with indications as to the relations between the functional areas
- The systems and installations
- The main and secondary area units, the dimensions, and the number of animals and operators present.

For the specific case of the closed-cycle pig-breeding center, the data are those given in Table 3.4, which are used to determine the organizational diagram of the functional sector, starting with the area units and, generally, with the minimum characteristics that the area and technological systems may possess.

In order to transpose the metadesign results into the planning level, reference may be made to the diagram in Fig. 3.12; subsequently, it is necessary to individuate and define the production program in detail. To do this, the following factors may be taken into consideration: final product, the final weight of the pigs to be put on the market; initial product: the number of sows and type of replacement; the duration of the individual stages (e.g., very early/early/traditional weaning, etc.), from the daily body weight increase; type of feed; feed-distribution method; birth programming; empty stage—disinfecting.

For example, for a closed-cycle pig-breeding center, starting with 160 sows and carrying out programmed births and the completely empty/completely full technique, the holding will have 270 sucking piglets, 570 weaning piglets, 570 growing pigs, 660 fattening pigs, and 64 young replacement sows, and thus the center buildings must be made big enough to hold a total of 2300 pigs, corresponding to an annual production of 3440 pigs weighing 100 to 110 kg.

Thus one is able to determine the elements necessary for defining the number of area units for each base module (the base modules can be made up of a single or, in some cases, several area units), the size of the base module and the number of base modules that make up a spatial module corresponding to each functional area (Figs. 3.13–3.21).

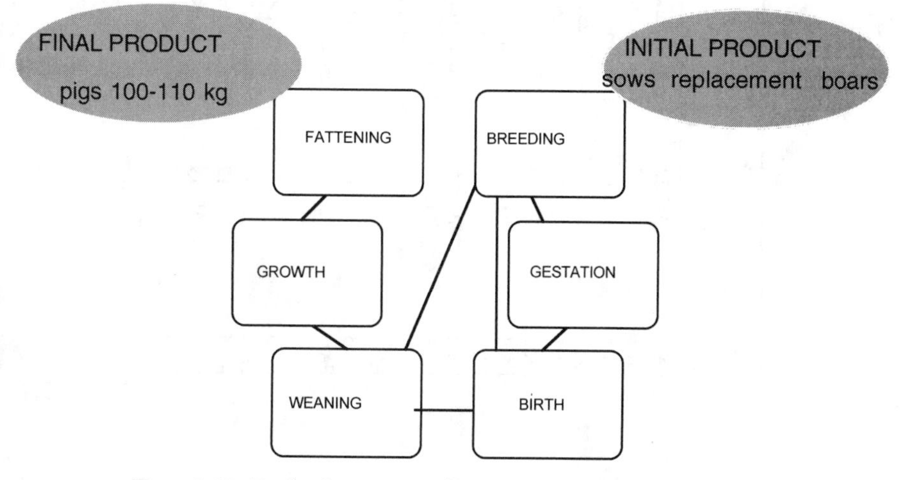

Figure 3.12. Production program for a closed-cycle pig-breeding center.

Table 3.4. Typical area-unit chart for closed-cycle pig-breeding center

Functional Area	Area Unit		Animals (head/m^2) Regular	Animals (head/m^2) Occasional	Operators (no.) Regular	Operators (no.) Occasional	Basic Module Dimensions (m)
FA 1	AU 22	Vehicle parking				1–4	min. 2.40 × 8.00
FA 1	AU 12	Changing rooms				1–2	min. 3.50 × 4.00
FA 1	AU 13	Lavatories				1	min. 1.80 × 3.00
FA 1	AU 17	Employee cafeteria				2–4	min. 2.50 × 3.70
FA 1	AU 14	Visitor check-in area				2	min. 2.00 × 3.00
FA 1	AU 10	Office			1–2		min. 3.00 × 3.50
FA 2	AU 5	Growth	1–3			1	2.10 × 5.50
FA 2	AU 6	Fattening	1–2			1	2.00 × 4.50
FA 2	AU 15	Systems-control room				1	min. 2.00 × 2.00
FA 3	AU 1	Fertilization				1–2	0.60 × 2.35
		stimulation	0.16–1.00				2.50 × 3.00
FA 3	AU 7	Replacement area	1–2			1	2.00 × 3.70
FA 3	AU 15	Systems-control room				1	min. 2.00 × 2.00
FA 3	AU 2	Gestation room	0.5–1			1–2	0.60 × 2.35
							3.00 × 3.00
FA 3	AU 3	Birthing room	0.18–2.50			1–2	2.20 × 2.60
FA 3	AU 4	Weaning area	5			1	2.00 × 2.50
FA 4	AU 12	Changing rooms				1–2	min. 3.50 × 4.00
FA 4	AU 14	Visitor check-in area				2	min. 2.00 × 3.00
FA 4	AU 20	Quarantine	1–5			1–2	2.00 × 3.70
FA 4	AU 15	Systems-control room				1	min. 2.00 × 2.00

FA 4	AU 21	Quarantine waste accumulation			min. r = 2.50
FA 5	AU 22	Vehicle parking		1–4	min. 2.40 × 8.00
FA 5	AU 16	Equipment storage		1–2	min. 2.75 × 3.50
FA 5	AU 12	Changing rooms		1–2	min. 3.50 × 4.00
FA 5	AU 13	Lavatories		1	min. 1.80 × 3.00
FA 5	AU 8	Feed preparation		1–2	min. 1.80 × 2.70
FA 5	AU 9	Feed storage		1–2	min. 2.00 × 3.00
FA 6	AU 14	Visitor check-in area		2	min. 2.00 × 3.00
FA 6	AU 12	Changing rooms		1–2	min. 3.50 × 4.00
FA 6	AU 13	Lavatories		1	min. 1.80 × 3.00
FA 6	AU 11	Laboratory	1	1	min. 1.80 × 2.80
FA 6	AU 18	Postmortem examination room	1	1	min. 2.80 × 4.00
FA 6	AU 19	Incinerator		1–2	min. 2.25 × 2.50
FA 7	AU 23	Liquid wastes treatment storage tank			min r = 5.30
FA 7	AU 15	Systems-control room		1	min. 2.00 × 2.00
FA 8	AU 8	Feed preparation		1–2	min. 1.80 × 2.70
FA 8	AU 9	Feed storage		1–2	min. 2.00 × 3.00
FA 8	AU 15	Systems-control room		1	min. 2.00 × 2.00

Note. min., minimum.

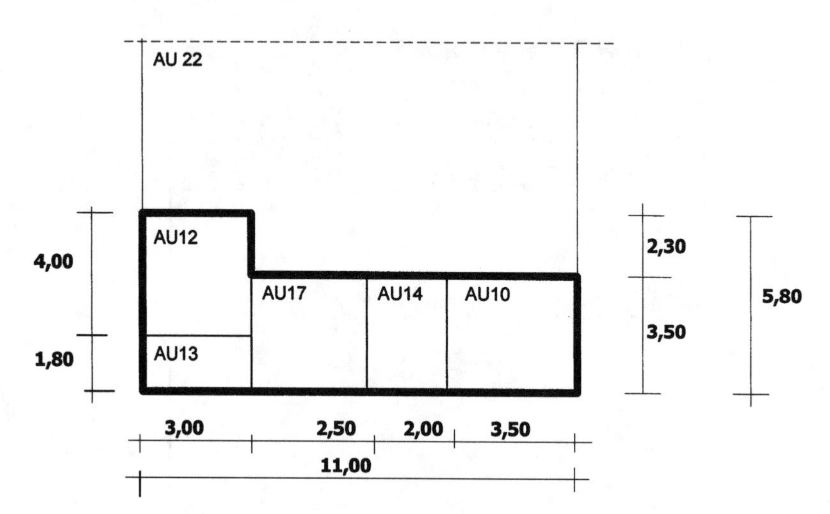

Figure 3.13. Spatial module 1 (general services) defined by six base modules, each composed of one AU.

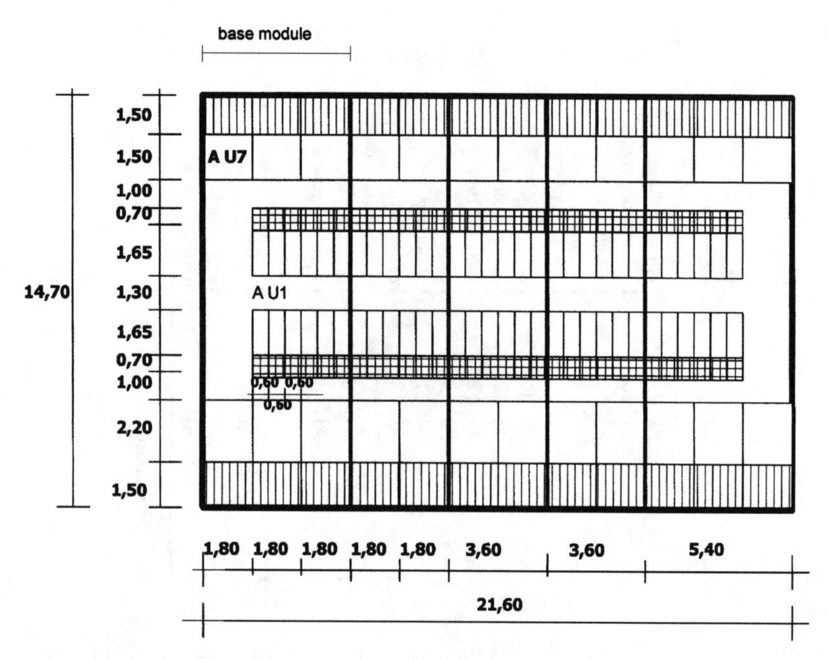

Figure 3.14. Spatial module 2 (stimulation–fertilization) defined by five identical base modules, each composed of AU1s and AU7s for a total of 60 sows, 16 boars, and 64 replacement sows.

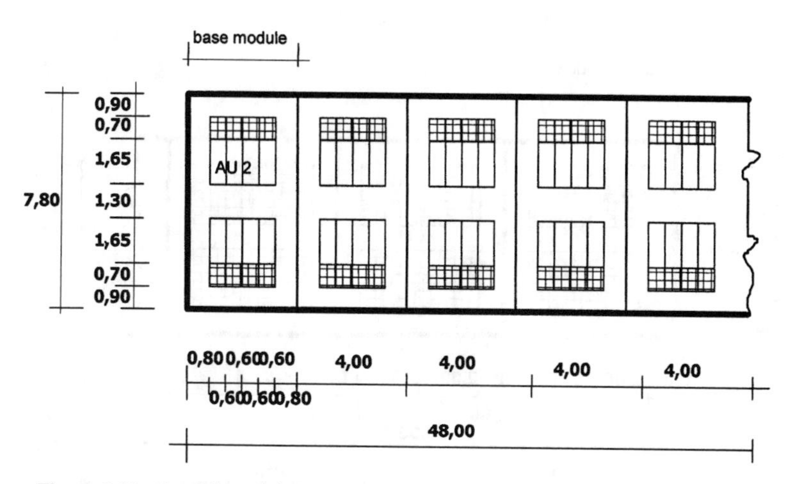

Figure 3.15. Spatial module 2A (gestation) defined by 12 identical base modules, each composed of eight AU2s, for a total of 96 sows.

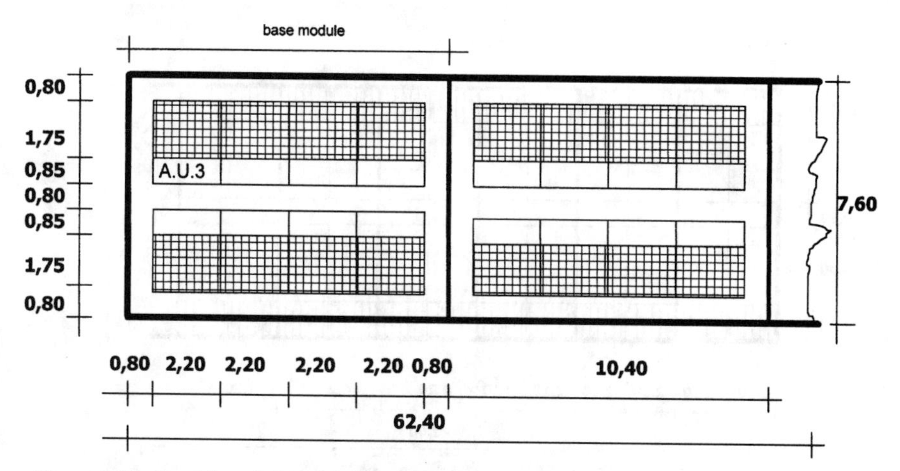

Figure 3.16. Spatial module 2B (birth) defined by six identical base modules, each composed of eight AU3s, for a total of 48 sows.

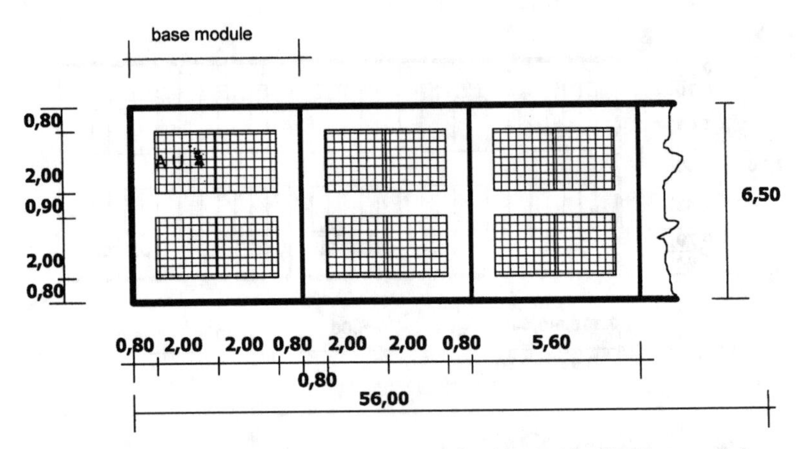

Figure 3.17. Spatial module 2C (weaning) defined by 10 identical base modules, each composed of four AU4s, for a total of 650 piglets up to 30 kg live weight.

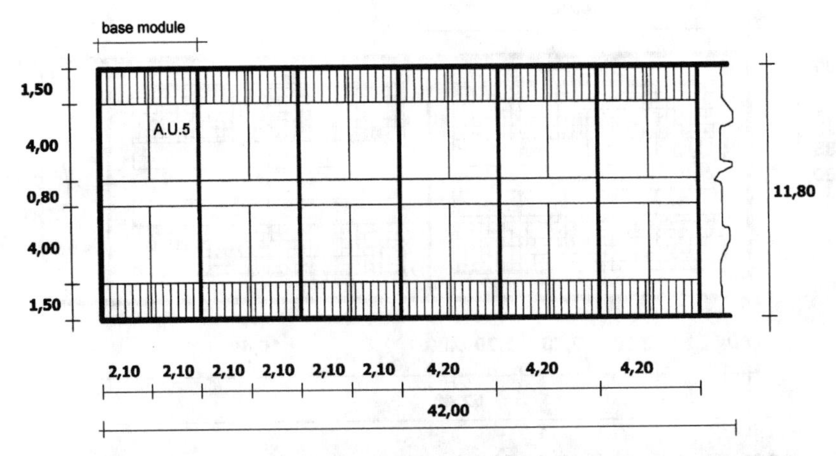

Figure 3.18. Spatial module 3 (growth) defined by 12 identical base modules, each composed of four eight AU5s, for a total of 680 animals up to 60 kg live weight.

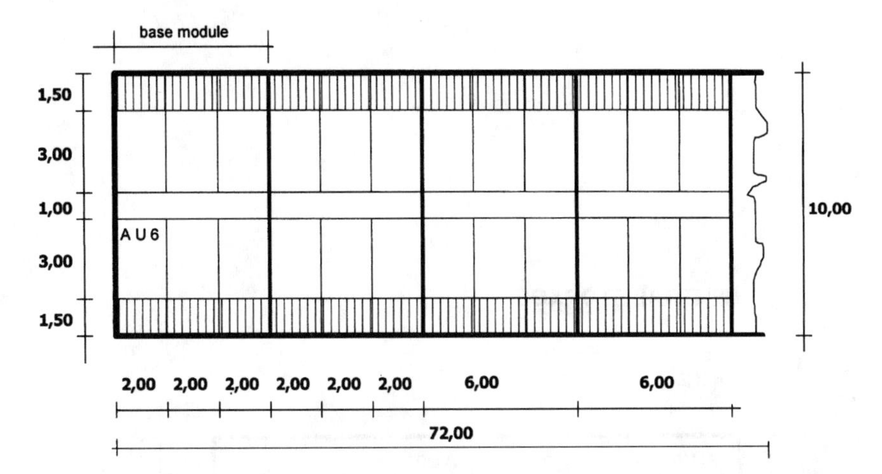

Figure 3.19. Spatial module 3A (fattening) defined by 12 identical base modules, each composed of six AU6s, for a total of 720 animals from 60 to 110 kg live weight.

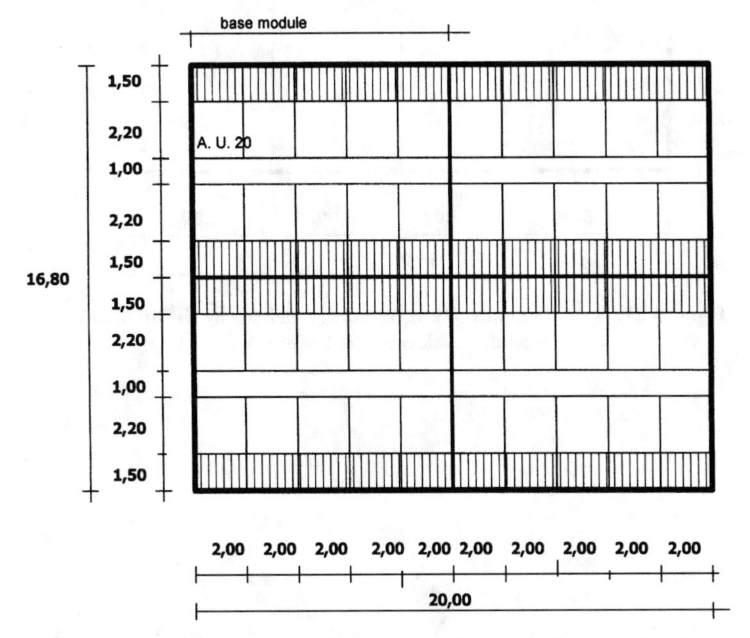

Figure 3.20. Spatial module 4 (quarantine) defined by four identical base modules, each composed of 10 AU20.

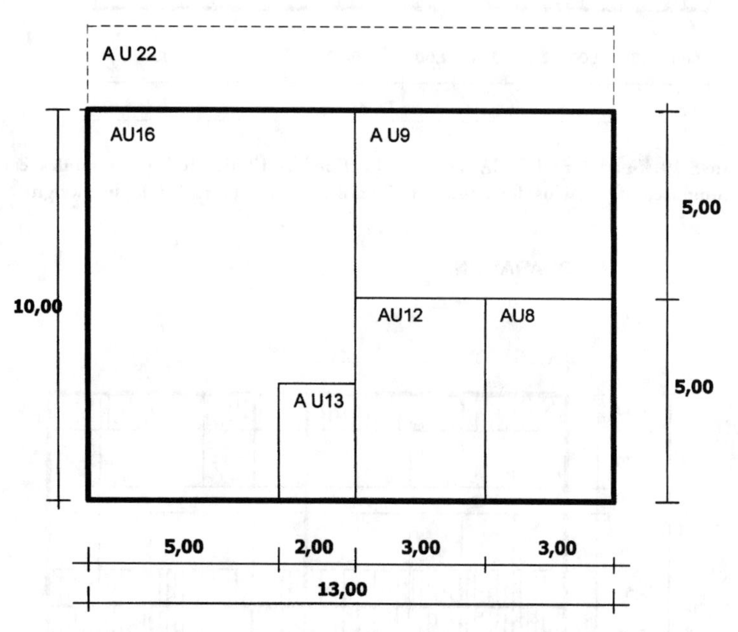

Figure 3.21. Spatial module 5 (central storage) defined by six different base modules, each composed of one AU.

4 Equipment and Control

Michel Tillie

4.1 Feed and Supply Distribution

The means of feed and drink supply are specific to each species. They should be adapted to the height of the animals so that they can satisfy the vital needs of food and drink. Regarding working conditions the means should also meet the requirements of the person who takes care of the herd, and, finally, they should meet the requirements of the economic situation of the farm.

Equipment may be quite different according to the breeding system. Intensive systems generally are high in investment expenses; extensive systems are generally lower.

4.1.1 Intensive Conditions

Cattle Breeding [1]

Adult cattle spend about 5 hours per day eating. In order to avoid competition, frustration, and aggression, there should be sufficient feeding places for all animals to feed at the same time, particularly with dairy cows. Any restriction in the number of places may, depending on quality and quantity of the feed, result in low-ranking animals receiving insufficient feed, and as a consequence they will produce less milk.

Feeding barriers and mangers for cattle should provide access to a large volume of feed, prevent bullying, prevent feed wastage, and be noninjurious to the animals. Configurations should not cause discomfort or injury, while offering maximum forward reach.

Feeding Alley

The alley must be of some minimum width that allows for unrestricted circulation and agonistic social behavior. Passage sizing is considered in terms of passage width. The minimum width required depends on the body size plus a minimum clearance for circulation. The minimum dimension (Wp) of a feeding alley is given by the following equation:

$$Wp = L + 2.7 \cdot W$$

where L is body length and W is withers width of animal in meters (Fig. 4.1).

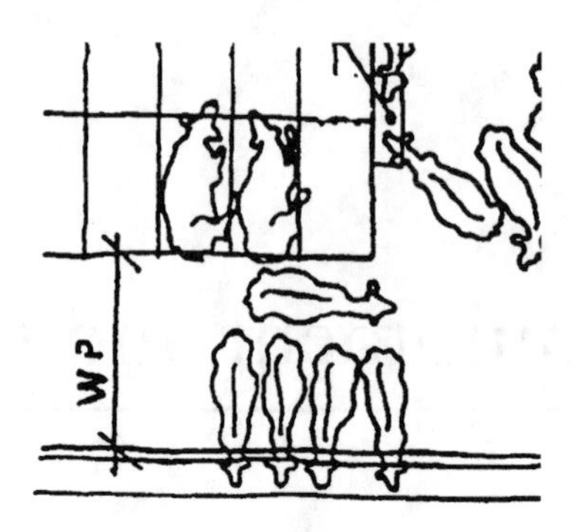

Figure 4.1. Width of passage.

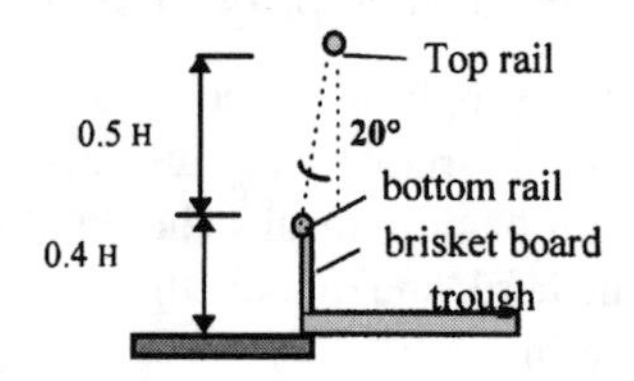

Figure 4.2. Post-and-rail barrier dimensions.

Troughs

It is important to allow all animals to feed. The minimum length (FL) of trough per animal is:

Ad-lib self feed: $\text{FL} = 0.5 \cdot W$

Semi restricted feed: $\text{FL} = W$

Restricted feed: $\text{FL} = 1.15 \cdot W$

For pregnant female: $\text{FL} = 1.25 \cdot W$

Feed Barriers

Research has shown that inclining the barrier by 20 degrees towards the manger increases the volume of food within reach without causing injuries (see Fig. 4.2). The withers of cattle are particularly sensitive and prone to injury, which means that positioning of the equipment is crucial. The spacing between cattle places (CS) is given by the following equation:

$$\text{CS} = k \cdot W$$

where $k = 1.15$. However, with female animals, if a large percentage of cows in a group are likely to be pregnant at the same time, then $k = 1.25$.

A *post-and-rail barrier* is very simple. It is composed of horizontal bar fixed to posts and a board close to ground. The dimensions are given in Fig. 4.2.

A *diagonal barrier* forces cattle to angle their heads to gain access. The angle of the bars prevents food wastage. Sideways movement is limited to some extent. For maximum reach and minimum feed wastage the angle should be 65 degrees. Dimensions are given in Fig. 4.3 (H is the height at the withers).

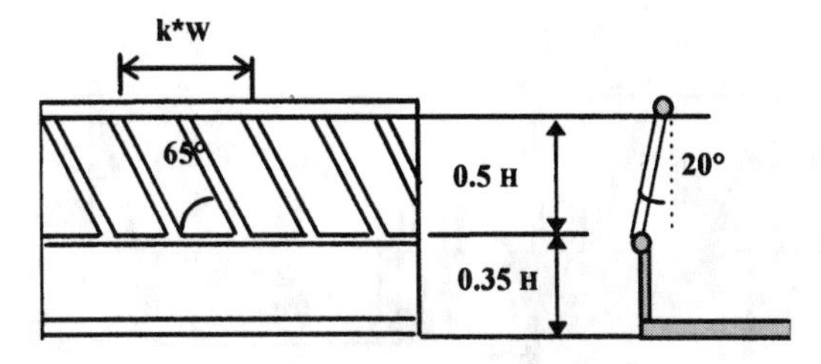

Figure 4.3. Diagonal barrier.

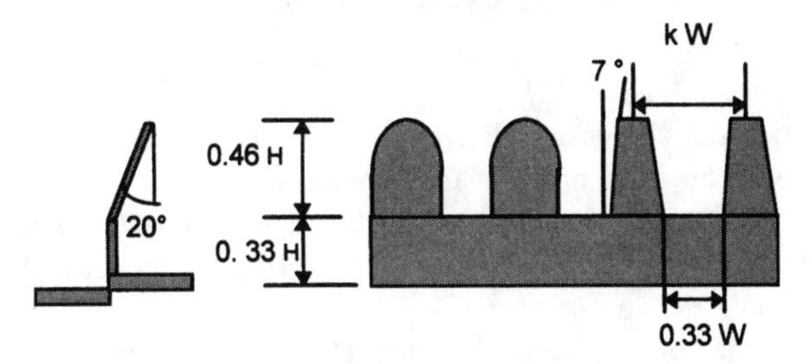

Figure 4.4. Tombstone barrier.

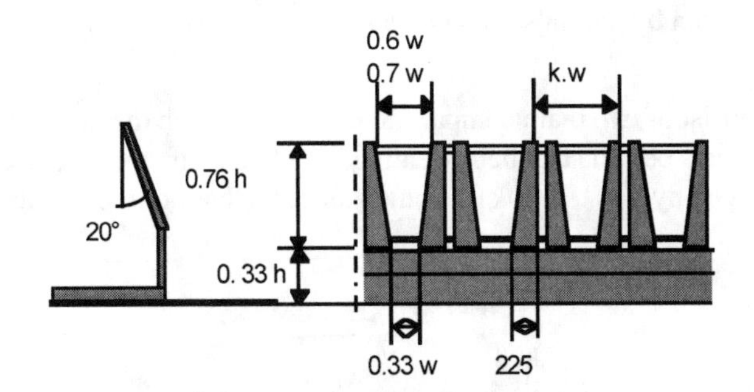

Figure 4.5. Dovetail barrier.

The *tombstone barrier* limits sideways reach. Optimun sideways reach is achieved by splaying the tombstone at an angle of 7 degrees. The cantilever design of the barrier requires a heavy construction. As with other barriers, the tombstone should be angled by 20 degrees towards the manger, as shown in Fig. 4.4.

The *dovetail barrier* provides the best access to feed without causing injuries or feed wastage. The barrier may be constructed in wood on site at a relatively low cost or purchased in prefabricated steel sections. Dimensions are shown in Fig. 4.5.

Self-locking barriers allow cows to be constrained and offer the management advantages that cattle can be held for observation, veterinary care, and for females, artificial insemination; feed wastage is reduced; and bullying is reduced. The dimensions of a typical design are given in Fig. 4.6.

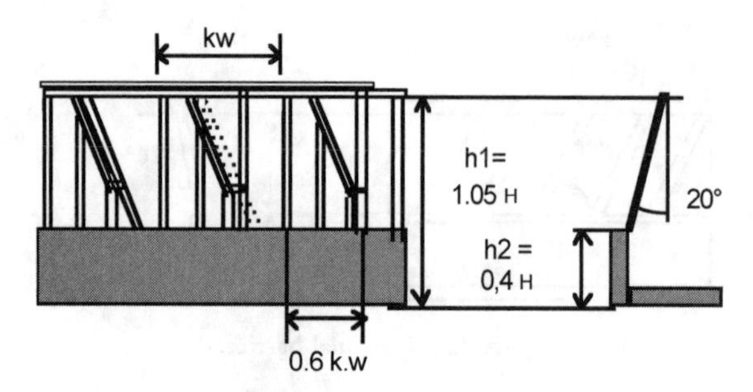

Figure 4.6. Typical self-locking barrier.

Automatic Concentrate Feeders [2]

A concentrate feeder should not deliver too-small portions, otherwise feeding motivation is reduced. The availability of feed should also be very predictable, to avoid unrewarded visits. If the animals know when they can get concentrates and when not, they will quickly learn the relationship between visiting the feeder and obtaining concentrates (Fig. 4.7).

In such systems the concentrate feeder should be located in the feeding area, near the feed barrier, in rows of cubicles, or in an (exterior) exercise yard rather than in the lying area. The advantages of the automatic feeder are the possibility to control the quantity of concentrates taken by animals each day and a decrease of labor for the cowman.

Hay Racks

Hay racks consist of two main compartments: one in which the hay is placed and one in which the animals pull the hay before eating. The two compartments are separated by vertical bars spaced by 140 to 160 cm. A manger at the base of the eating compartment

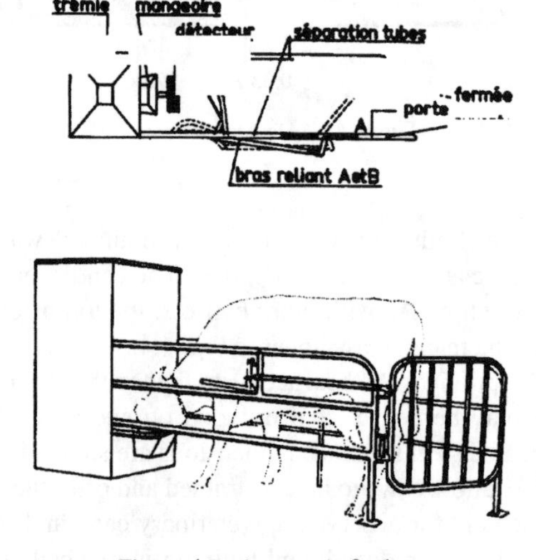

Figure 4.7. Automatic feeders.

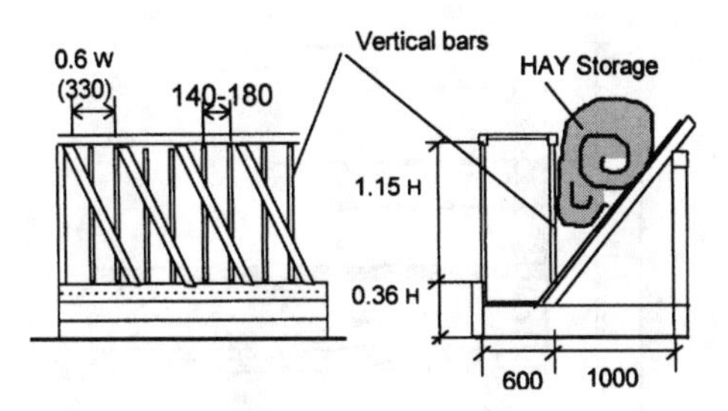

Figure 4.8. Hay rack.

collects the hay dropped by the animals and this is eaten subsequently. The cattle gain access to the feeding compartment through a feed barrier (Fig. 4.8).

Sheep Breeding [3]

The materials used for feeding equipment for sheep are very simple. Generally they are made of wood and are movable.

Technical Requirements

To allow each animal to feed according to its needs, depending on breeds, 30 to 45 cm per ewe are required, that is, five to six ewes for each 2-m element. But for pregnant ewes 8% to 10% more length is required. For lambs, 20 to 30 cm are necessary, that is, eight lambs for each 2-m element.

Feeding Alleys and Troughs

The trough and the feed-distribution alley can be one movable element (Fig. 4.9). Dimensions for ewes are shown in Fig. 4.10, and for lambs in Fig. 4.11.

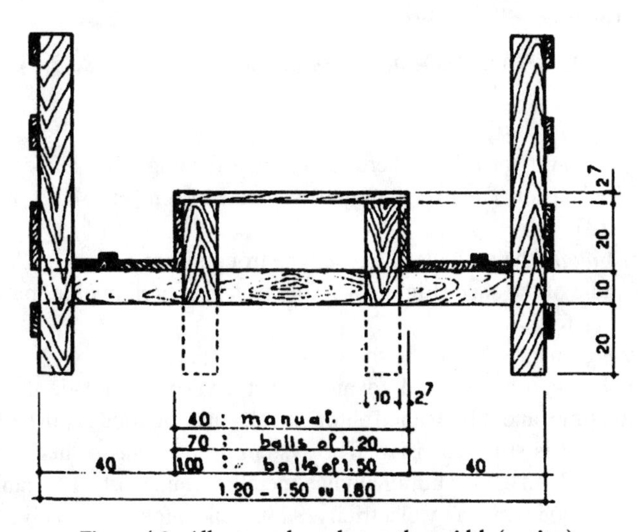

Figure 4.9. Alley trough made together width (section).

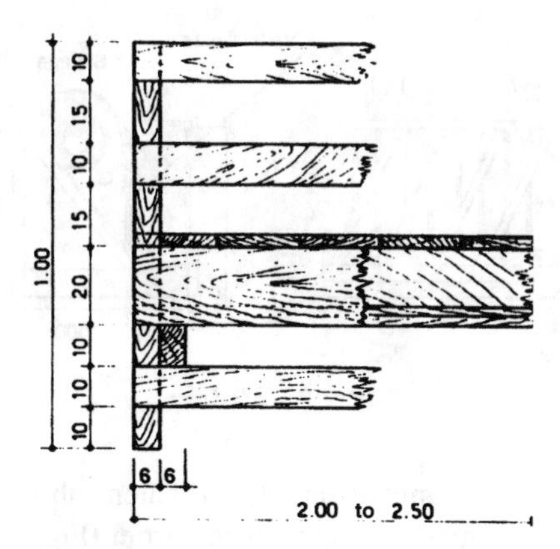

Figure 4.10. Dimensions of barrier and trough for ewes.

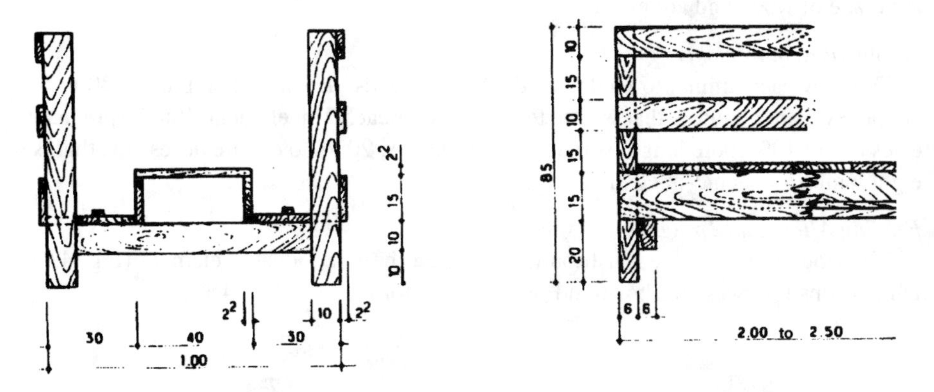

Figure 4.11. Dimensions of barrier and trough for lambs.

When a shed is built for a large herd, a concrete feeding alley makes mechanized distribution of feed easier. The trough may be separated from (Fig. 4.12) or included in the alley (Fig. 4.13).

With a *crate trough*, the animals can reach the trough from both sides. The trough is placed in the middle of the shed (Fig. 4.14). This system is used in older buildings or on permanent pastures for hay.

A *wooden trough with a barrier* is shown in Fig. 4.15.

A *mechanical trough* has several advantages: it saves room inside the building, it saves distribution time, and it is cleaned automatically and the food is pushed back when dropped. However, it is still expensive. The system used in sheep sheds consists of a belt conveyor drawn by a cable sliding along a metallic framework. The cable winds up around a pulley, which is driven by electricity. At one end a food bin receives the fodder. Concentrates can be distributed at the same place but separately (Fig. 4.16).

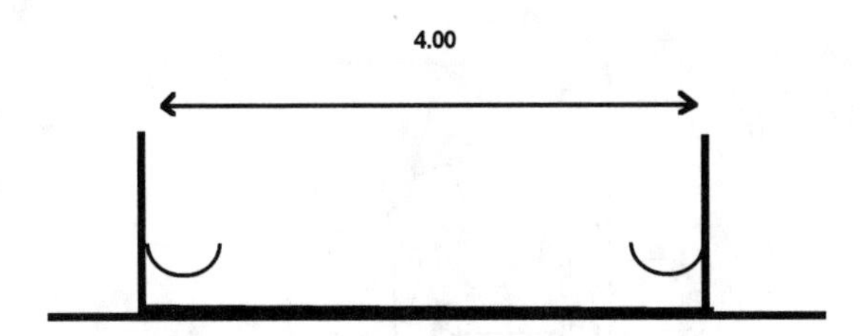

Figure 4.12. Alley with separate troughs.

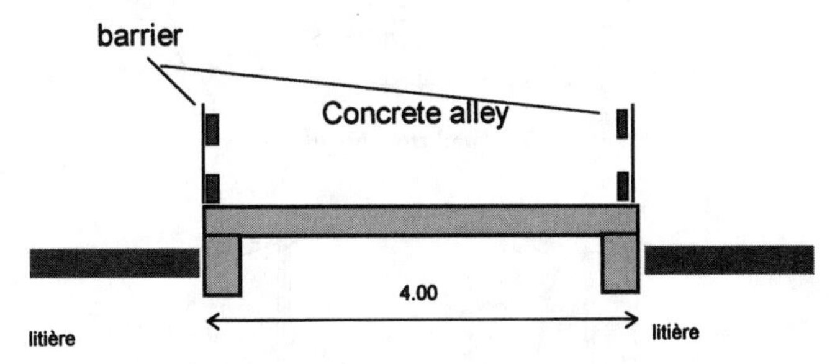

Figure 4.13. Concrete feeding alley.

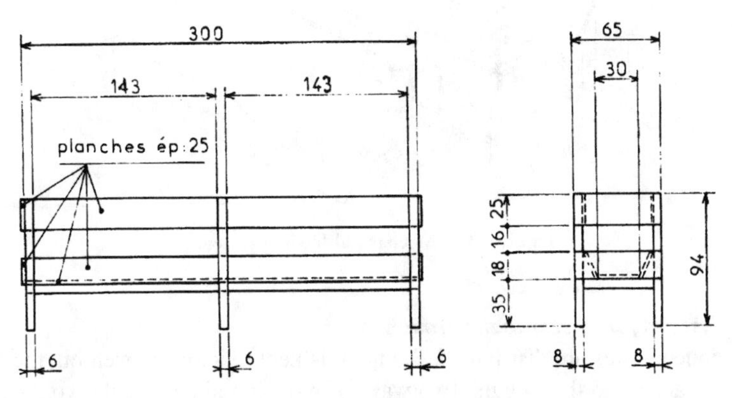

Figure 4.14. One-row wood trough.

Racks

Racks are necessary for hay. Their size must be adapted to the type of bales that are used.

A rack can be placed over the trough. It is used only for bulk hay or small bales (Fig. 4.17). In that case, metal is used as building material.

Using round bales implies large-size racks (Fig. 4.18). They are round or square in shape, and a fixed yoke is set at the base. The basic principle is the same as for cattle racks (see Fig. 4.8).

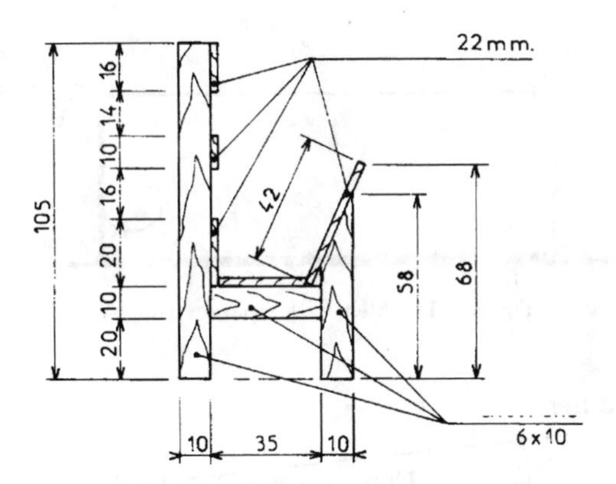

Figure 4.15. Wood trough with barrier.

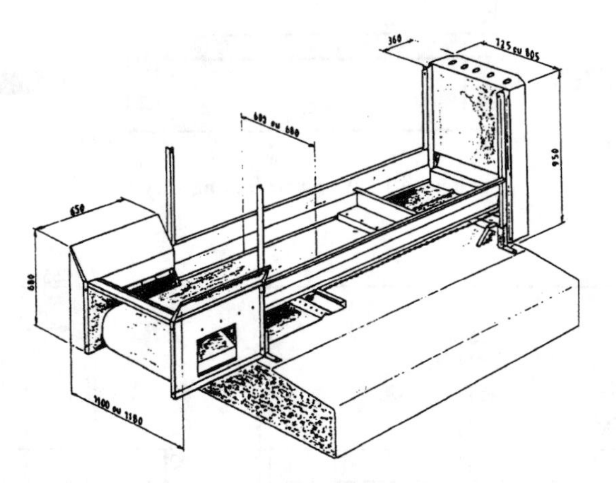

Figure 4.16. Mechanical feeding trough.

Blocking Systems for Access to the Trough

When concentrates are distributed, the flock is kept quiet by preventing the animals from getting access to the trough. Two ways of blocking are available: yokes with the same closing system as for cattle (Fig. 4.19), or a board that is placed between horizontal elements of the trough barrier. The operating principle and dimensions are shown in Fig. 4.20.

Goat Breeding

Equipment should be designed to maintain the animals inside their lots. The device depends on the managing system. Self-blocking yokes or individual closing systems are recommended (Figs. 4.21, 4.22).

Technical requirements for goat breeding are as follow: The feeding space per goat is 0.40 m, that is, six goats per 2.50-m yoke. The space between the closed bar of the yoke should be 0.09 m for the neck to be held. The trough is 0.40 m wide at its bottom and 0.60 m at brim level, so that no fodder is dropped over.

Figure 4.17. Trough with rack for little ball.

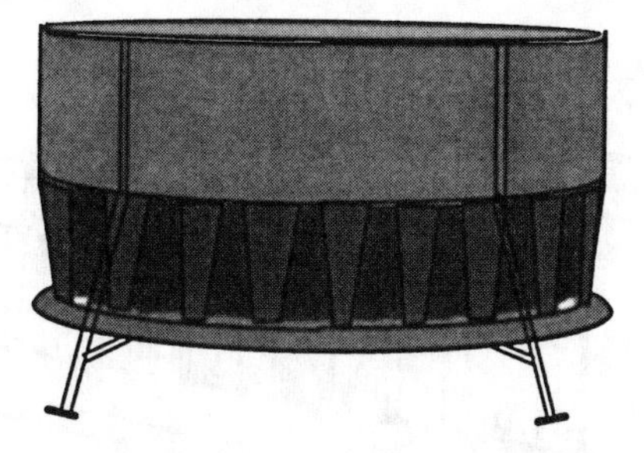

Figure 4.18. Round ball rack for sheep.

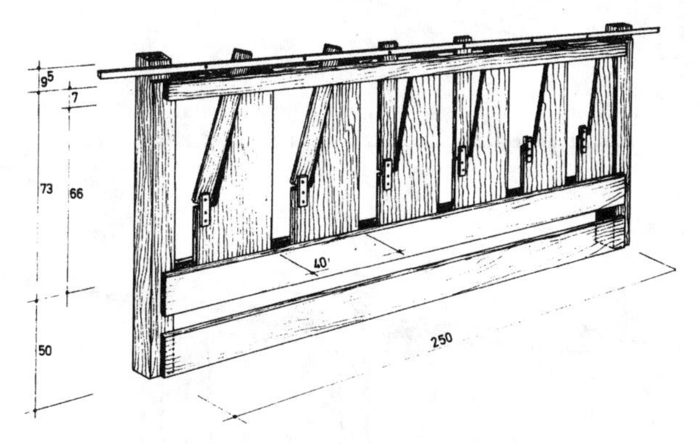

Figure 4.19. Locking yoke in wood.

The poles holding the yokes are fixed about 0.35 m deep into the ground along the passage. Three solutions are available (Fig. 4.23):

- A yoke with a trough opening 1.50 m wide on the passage if distribution is made with a wheelbarrow, 3 m wide if it is mechanized
- A yoke without a trough; the passage is used for feed distribution (3 m wide)
- A yoke with a belt conveyor or chain conveyor similar to those that are used for sheep (see Fig. 4.16).

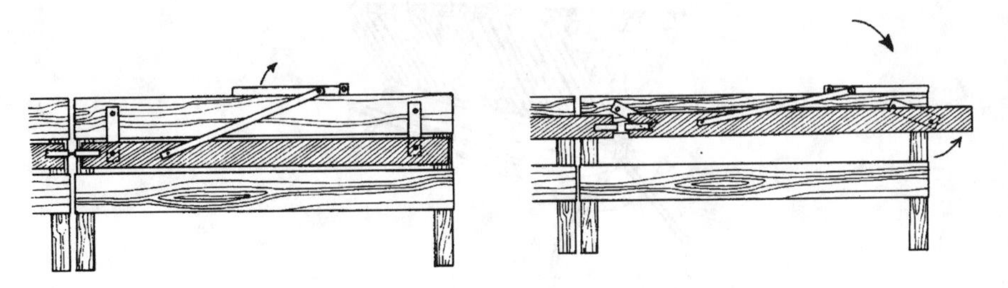

Figure 4.20. Closing system with a plank.

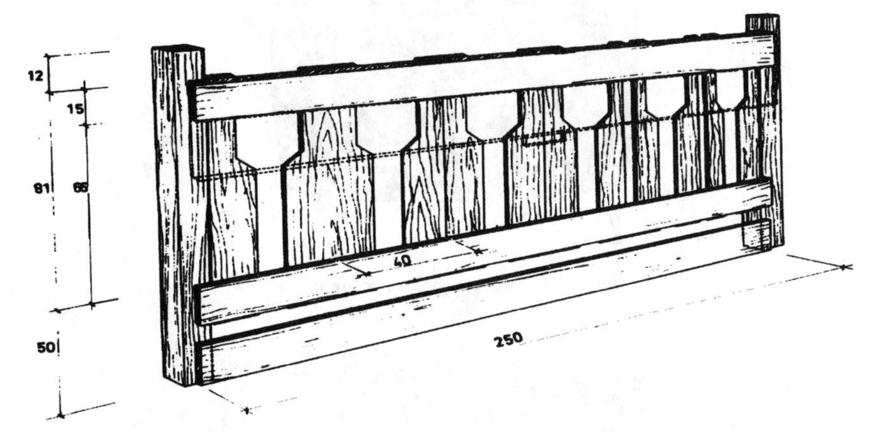

Figure 4.21. Closing system for fixed yoke.

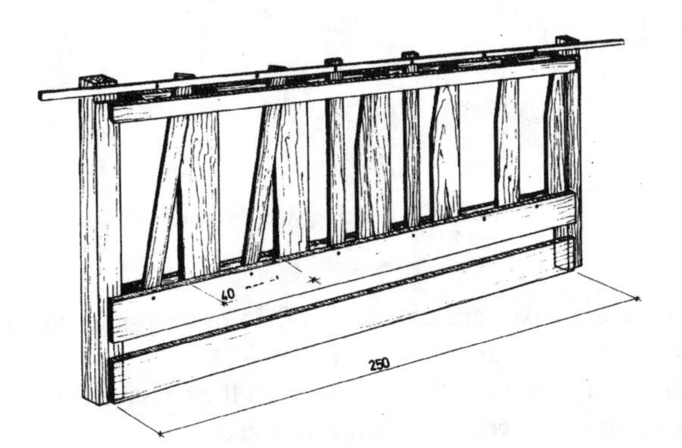

Figure 4.22. Locking yoke.

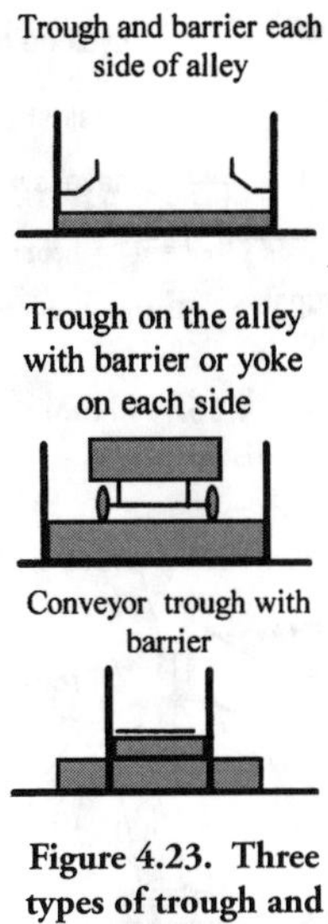

Figure 4.23. Three types of trough and alley.

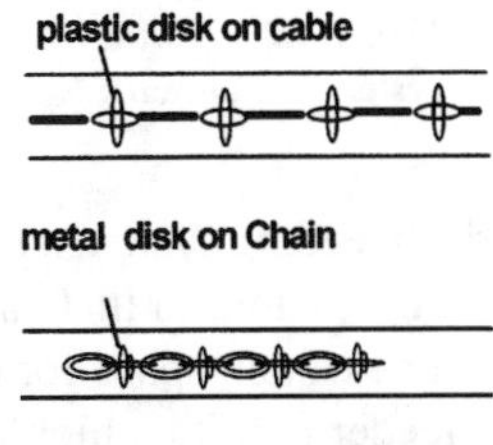

Figure 4.24. Tube cable and disk.

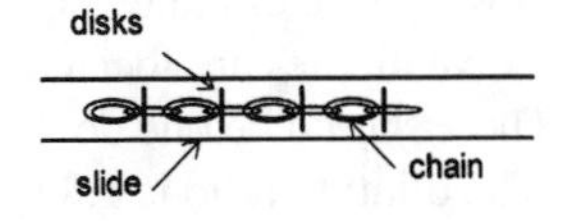

Figure 4.25. Gutter with chain and slide.

Pig Breeding [4]

Feed for pigs is available in five forms: pellets, crumbs, dry meal, moist meal, and soup. The system of distribution ranges from rationing with one trough place per animal to partial self-service or full self-service with one place for 3 to 15 animals.

Dry Feeding

The feed stored in a silo is conveyed to the delivery place by means of an auger-conveyor or a chain (Figs. 4.24, 4.25) or pneumatically (Fig. 4.26). The curves and slopes reduce the flow along the conveyors or chains.

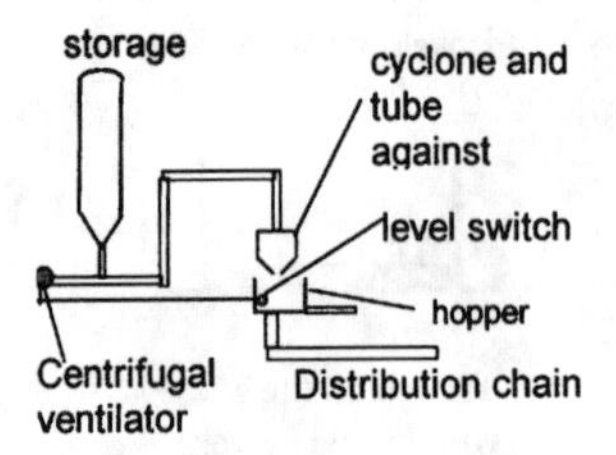

Figure 4.26. Pneumatic distribution.

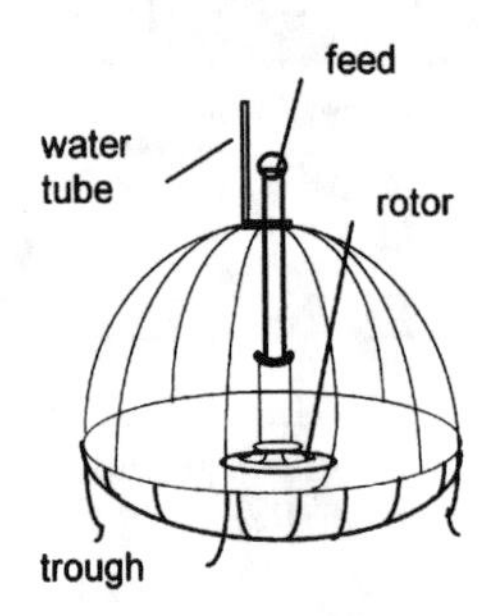

Figure 4.27. Rotor system (turbomat).

Feeders generally are used in ad-lib feeding systems. Rationing is possible by limiting the access time or by delivering a daily ration. In that case, it is necesssary to set apart males from females and to provide one trough space for two or three pigs.

A soup feeder is a single-place feeder fit with a nipple. The feed is diluted in a ratio of 2.0 to 2.2 L of water per kilogram of feed. It is used mostly in self-service systems, with one device for 12 to 15 piglets.

In a turbomat the feed runs down into a circular tray, 75 cm in diameter, when the pig acts upon a rotor (Fig. 4.27). Five nipples are located below the trough and provide drinking water simultaneously. There should be one device for 20 to 25 pigs, or three to five per batch of 60 to 100 pigs. The dilution ratio is 1.8 to 2.21 L of water per kilogram of feed, which reduces the volume of slurry production. Delivery is clock-programmed.

In the Apor System the feed is delivered in a double-sided auger by means of a rotor placed on a slotted tray (Fig. 4.28). The auger is fitted with push-button drinkers. There should be one device for 50 animals, or 25 pigs per pen, the auger being located across two cubicles.

Liquid System

In a liquid system the feed is conveyed to the delivery place by centrifugal, spiral, or piston pumps, by means of pipes.

With centrifugal pumps, the flow rate is sensitive to head losses and depends on the viscosity of the product. With volumetric spiral pumps, the flow rate is constant. They are suitable for long circuits but tend to deteriorate if there are any foreign bodies,

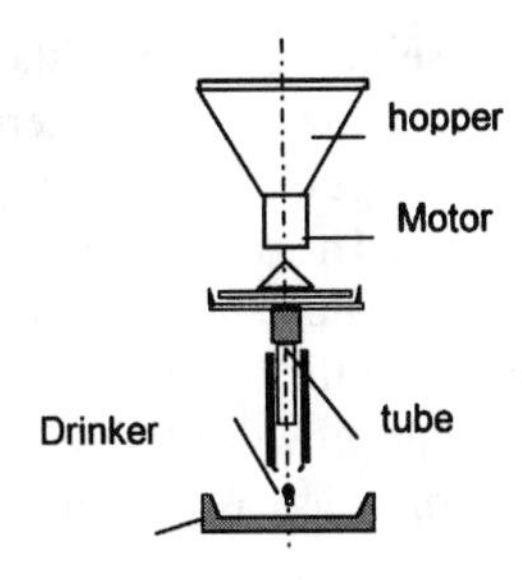

Figure 4.28. Rotative tray.

uch as stones. With piston pumps, the flow rate shows little sensitivity to head losses. 'heir operation is volumetrical and similar to that of a syringe, each stroke of the piston eleasing the desired volume of feed.

Delivery into the augers is operated through a shutter, either hand-worked or automatic vith compressed air. The feed quantities for each pen are controlled before conveying y load cells located under the mixing tank.

To reduce the quantities of nitrogen and phosphorus outputs in slurry, multiphase ystems are used. Two diets are mixed in varying proportions corresponding to the needs f pigs and conveyed in the trough.

oultry [5]

aying Hens

In ground-level breeding every 1000 hens should have trays in the starting period and 0 m of chain, that is, 8 cm of auger length per fowl, or 40 trays. Three systems are used: spiral, a flat chain, or a trolley.

In a spiral system, the auger and the conveying device are one. The spiral lies at the ottom of the auger (Fig. 4.29). It is operated by a clock.

The flat chain moves along the bottom of the auger. The forwarding speed is 12 m/min.

There are two types of feeding trolleys: a trolley with only one feeding box placed bove the cages, in which a conveyor pushes the feed towards the downward pipes; and side trolley with a feed box for each level on each side. The trolley moves along rails. he rails, placed on the lower part of the batteries, reduce the strain laid on the cages. ccording to the type, it is possible to adjust the amount of feed in the augers.

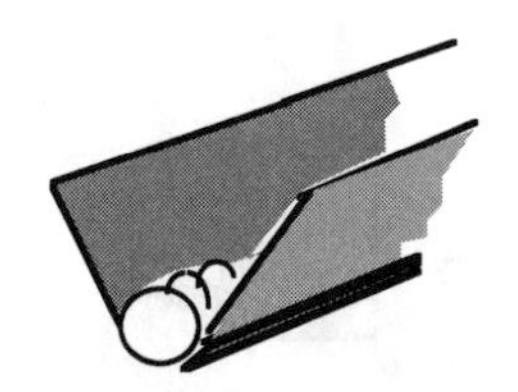

Figure 4.29. Spiral in auger.

Amount control is ensured by a weighing system either on the silo or automatically. Electronic weighing at the cage level allows one to determine the daily intake through permanent control over a sample of cages. This system gives a precision of 99% and provides a very significant daily weight. The size of the sample is determined statistically.

Computer calculation allows one to calculate the intake according to several parameters: weight of the fowls, temperature of the building, laying performances, and feather condition. Usually the same computer provides the technical management of the unit and the monitoring of each batch of hens, as well as the calculation of the various ratios.

Broilers

In the starting period for broilers, one feeding plate or tray for 100 chicks should be set. In the breeding period, one plate for 70 fowls and access along 50 m for 1000 fowls is needed.

Aerial conveyors are not used. A tubular ground chain fitted with a tray is better. It is composed of a 125- to 175-L feed box (storage) with an agitator and a monitoring system in order to avoid loadless running. The dispensing line consists of a spiral inside a pipe (diameter: 45 mm). The trays are distributed according to the width of the building the number of lines. Synthetic materials are preferred to metallic ones. The diameter of the trays ranges from 32 to 38 cm. The height above ground is 6 to 8 cm.

Turkeys [6]

One tray to 40 fowls is to be supplied. The mangers should be placed above the backs of the fowls.

For the starting period there should be feeding trays for the fattening period, the mangers are set in lines and fitted with first-age plates 0.30 to 0.40 m up to 6 weeks, then with adult plates 0.50 m.

4.1.2 Extensive Conditions

Cattle Breeding

For outdoor breeding or an extensive system, equipment consists only of the devices necessary for feed addition, generally hay and concentrates. Hay is distributed in racks. Concentrates are available in augers or by means of conveyors.

Racks

Hay racks are similar to the one in Fig. 4.18 but are under shelter or have their own built-in roof (Fig. 4.30). For dimensions, see Fig. 4.8.

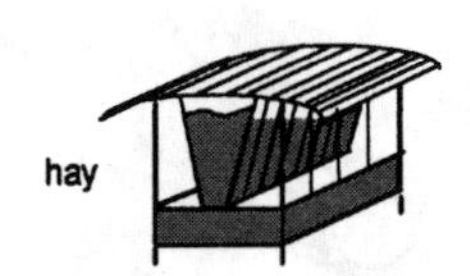

Figure 4.30. Covered rack.

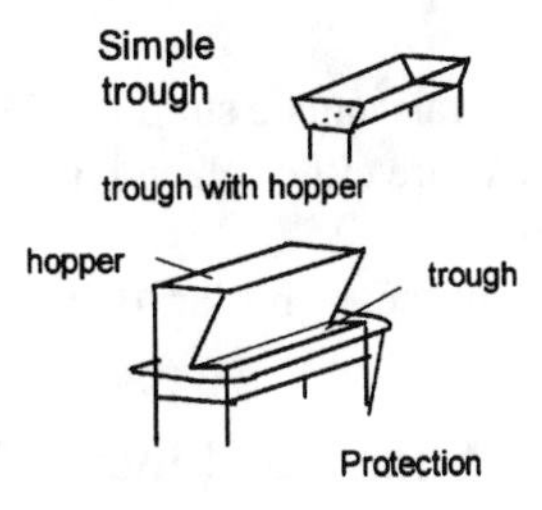

Figure 4.31. Pellet feeder.

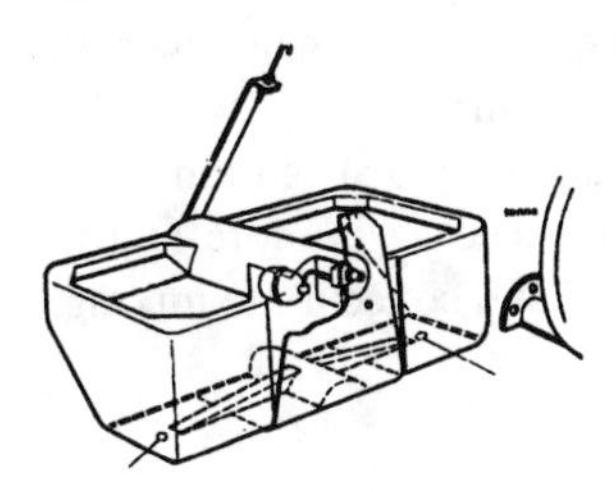

Figure 4.32. Drinker for tank trailer.

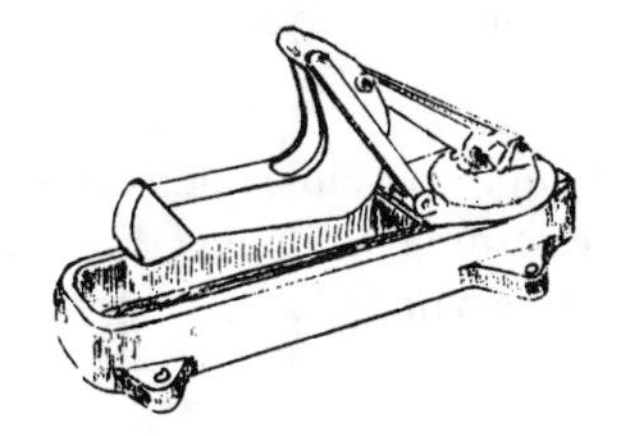

Figure 4.33. Pump drinker for pasture.

Troughs

Troughs are merely basins placed up on legs, or they include equipment for feed storage (Fig. 4.31).

Watering Equipments

Drinking supply is vital if there is no natural water source available. Two types of drinking systems are suitable: a cistern with drinking troughs designed to be fixed on it (Fig. 4.32) or a pump worked by the animal itself and linked to a water stock or a river (Fig. 4.33).

Pig Breeding [7]

In extensive breeding systems the animals are kept in the open on equipped areas. A semicircular shelter and a series of augers are placed on the ground. In many buildings pellets are spread on the ground.

Sheep and Goat Breeding

It is important that salt and minerals can be supplied. Gutter troughs are used for bulk feed and licking blocks. The blocks are either placed in basins on the ground or attached to a pole, depending on the shape.

Supplementary hay is brought in racks similar to those used in sheep breeding.

4.2 Feed Mixers and Intake Control

4.2.1 Cattle Breeding

Feed distribution is generally an important time and labor factor in breeding activities. Even though self-service from the silo is possible in small dairy herds, in most cases the feed is distributed mechanically in throughs.

The breeder seeks efficient and easy distribution solutions that require as little labor time as possible. Complete diet feeding is more and more widely used but requires technical means regarding silage and concentrate mixing.

Feeding Trailers

Trailers

The distribution trailer consists of a container for the feed whose bottom is mobile, and a screw or a side conveyor that drops the fodder into the trough (Fig. 4.34). With this type of equipment the feed cannot be mixed.

Mounted Silo Unloaders

A simple version of this system is available: a distributing silo unloader (Fig. 4.35). This machine can take the fodder from the silo and distribute it into the troughs. At the bottom of the machine a chain conveyor and slats throw the fodder sideways.

Figure 4.34. Feeding trailer.

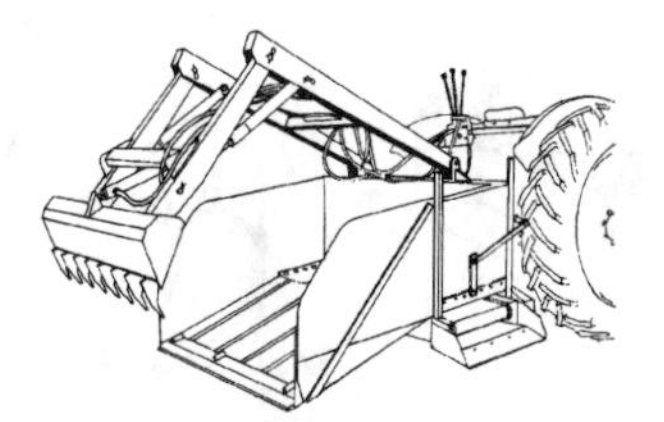

Figure 4.35. Fodder loading and spreading.

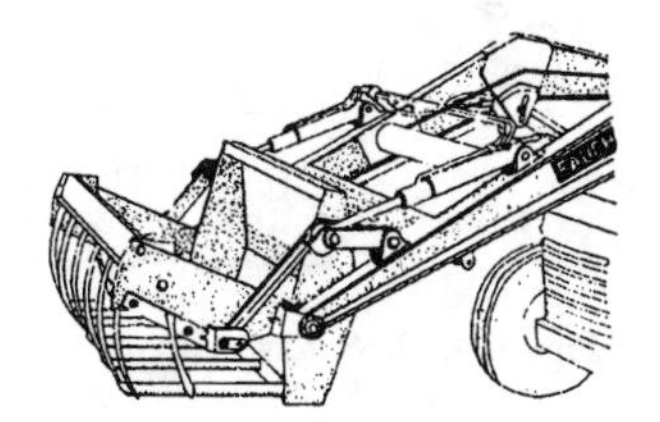

Figure 4.36. Hydraulic fork.

Figure 4.37. Mixer trailers.

Trailers and silo unloaders can include a system for distribution on both sides. The hay is either cut or torn from the silo. These machines generally can work at heights between 2 and 3 m.

Hydraulic Fork

The hydraulic fork is the simplest means for taking and distributing silage. However, it is impossible to spread the feed in the trough. It has to be piled first and spread later (Fig. 4.36).

Mixer Trailers (Complete Diet)

Mixing distributing trailers have a system for mixing the feed so that a complete diet can be prepared (Fig. 4.37). A weighing system with a gauge constraint allows for measurement of the quantities of feed put into the trailer. The mixing of feed and concentrates is achieved by one or several screws that move the fodder.

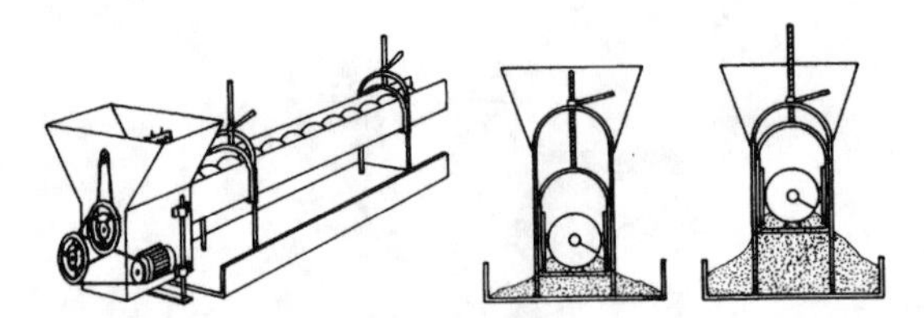

Figure 4.38. Screw in auger.

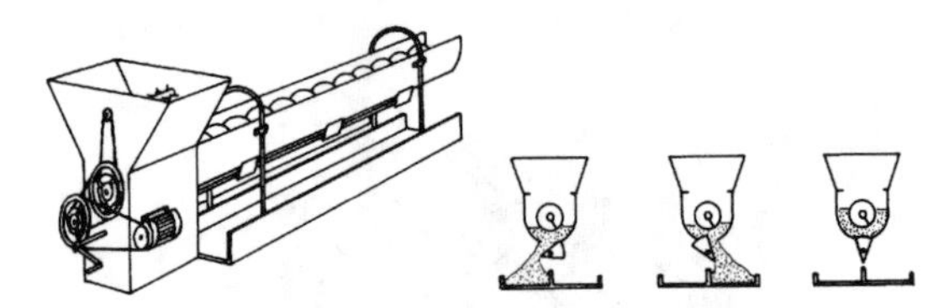

Figure 4.39. Screw in auger.

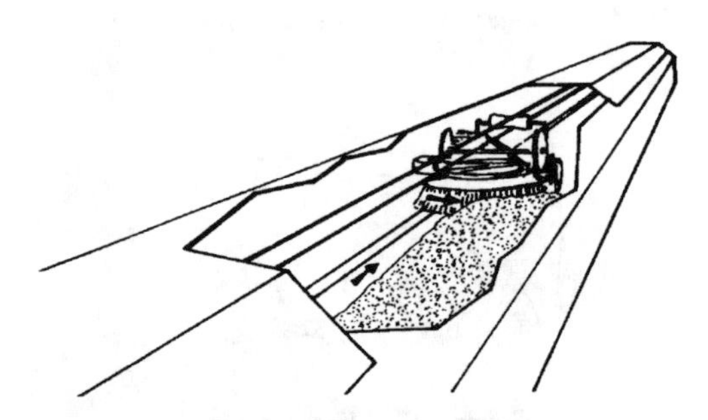

Figure 4.40. Mechanical auger with rotary brush.

Fixed Feeding Systems

In large breeding units feed distribution can be mechanized with a conveyor. This is a fixed piece of equipment that conveys and distributes the feed into a trough. Distribution is best with short minced feed. Generally these pieces of equipment are limited in length, especially those with screws.

Screws

There are two types of screw systems: Either the trough is fixed, in which case a distributor is placed below and throws the fodder to the right or left of the trough (Fig. 4.38), or the trough is mobile vertically and the feed falls into a prism shape into the trough (Fig. 4.39).

Conveyors

In this system the feed is carried along on a conveyor and a brush pushes it into the trough (Fig. 4.40).

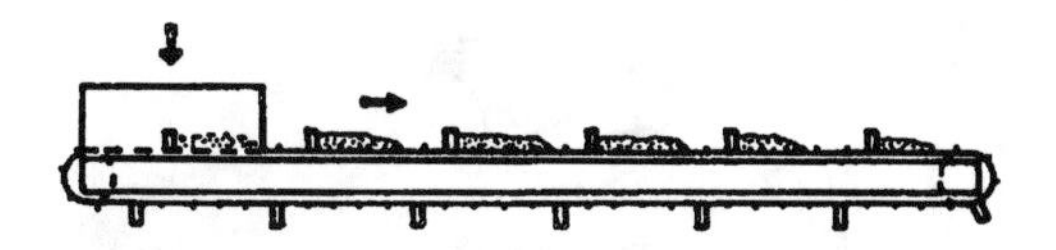

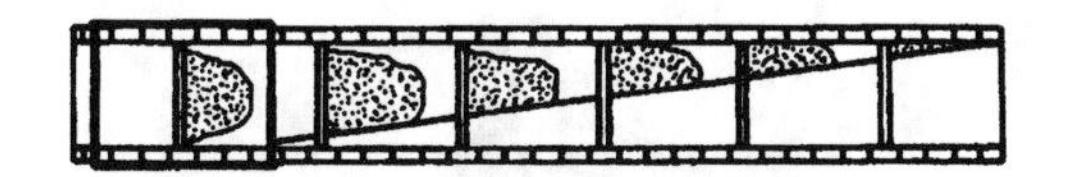

Figure 4.41. Mechanical auger with slats.

Chains with Slats

The feed is pushed forward by means of paddles, which are moved along a conveyor by chains (Fig. 4.41).

4.2.2 Sheep Production

Generally, the type of equipment used for sheep is the same as for cattle (see material on feeding alleys and troughs in the subsections on sheep and goat breeeding).

4.3 Watering Equipment

Water is very important for animals, and lack of water significantly affects production. Consequently, good-quality water supplied through adequate and accessible drinking facilities must be provided.

4.3.1 Cattle Breeding [8]

Drinking-Water Requirements

The quantities of water absorbed depend on various factors: production, dry-matter content of the feed, stage of lactation, and ambient temperature. The following equation [9] predicts water requirements:

$$y = 2.53x_1 + 0.45x_2 - 15.3\,(\pm 8.31)$$

where

y = water requirement (kg/d)
x_1 = milk production (kg/d)
x_2 = dry matter content of feed (%).

Drinking-Water Equipment

Water Bowl

Bowl-type drinkers may be used, provided they have a minimum surface area of 0.06 m^2 (Figs. 4.42, 4.43) and a water inlet flow rate of 0.16 L/s. The height of the bowl above the floor is 0.55 times the height of the animal withers (Fig. 4.44).

Figure 4.42. Constant-level water bowl.

Figure 4.43. Water bowl with stop valve.

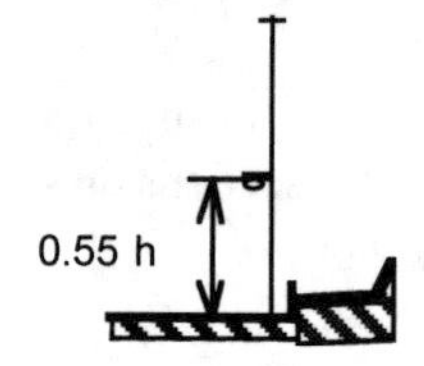

Figure 4.44. Bowl in tie stall.

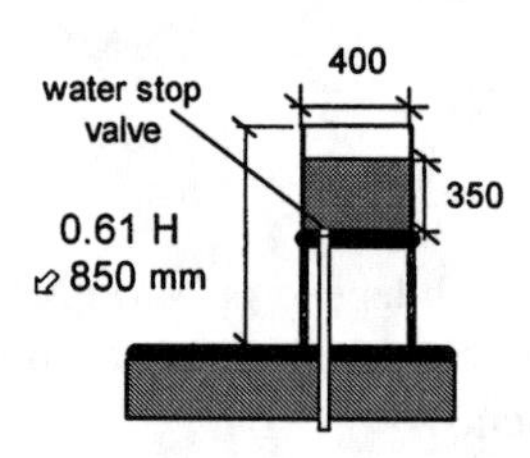

Figure 4.45. Water trough.

Water Trough

Water troughs are recommended for loose housing, with a capacity of about 200 L. The size of the trough allows cows that may take in 0.2 to 0.33 L/s to drink regulary. A trough 2 m length, 400 mm wide, and 400 mm deep is sufficient. Troughs should present a water surface to the cows at a height of 0.61 times the height of the animal withers (about 850 mm) from the floor (Fig. 4.45).

Number of Bowls

The number of water bowls provided should be equivalent to 15% of the number of animals. The number of water troughs should be equivalent to one per 25 cows.

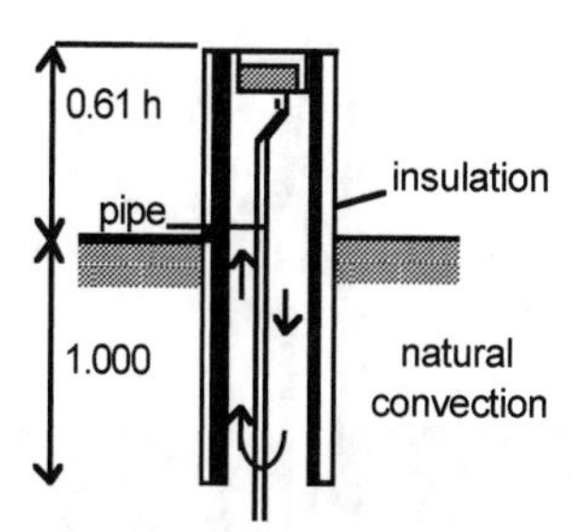

Figure 4.46. Antifrost water trough, passive system.

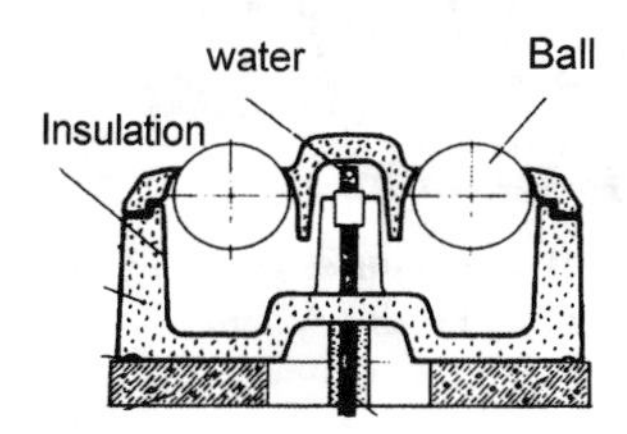

Figure 4.47. Insulated watering facility.

Location of Drinkers

Troughs should be located in passages at least 3 m and preferably 3.5 m wide. Several animals may be drinking at the same time, and sufficient space is necessary for others animals to walk behind.

Freezing-Control Systems, Heating Systems

The water-supply system and drinkers should be protected from frost. This may be done in various ways. Passive systems, using ground heat, might be appropriate in some climates (Figs. 4.46, 4.47). In other climates, heating systems with electricity, gas or some other fuel input may be necessary.

Among heating systems that may be used to prevent freezing are:

- Low-voltage electrical direct heating of the water in the drinker
- Low-voltage electrical wrapping around the supply pipes, valves, and the body of the drinker (Fig. 4.48)
- continuous circulation of heated water through the drinkers (particularly for water bowl) (Fig. 4.49).

4.3.2 Goats and Sheep

The general requirements for drinking facilities are similar for goats and sheep.

Drinking-Water Requirements

According to its physiological state a ewe can drink 3 to 5 L of water per kilogram of ingested dry matter. But it is very important to allow animals to drink easily. Thus

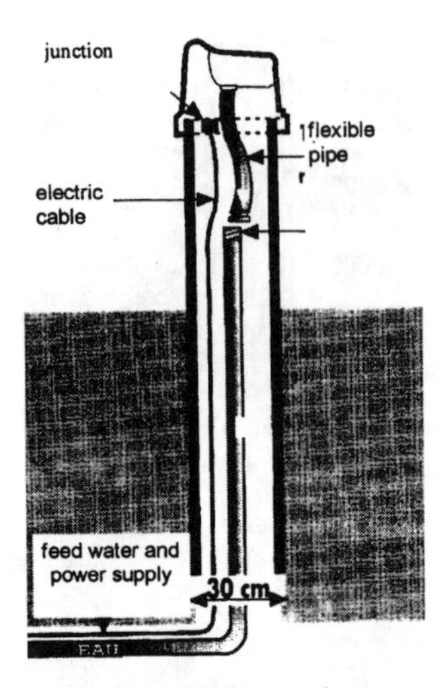

Figure 4.48. Low-voltage heating system.

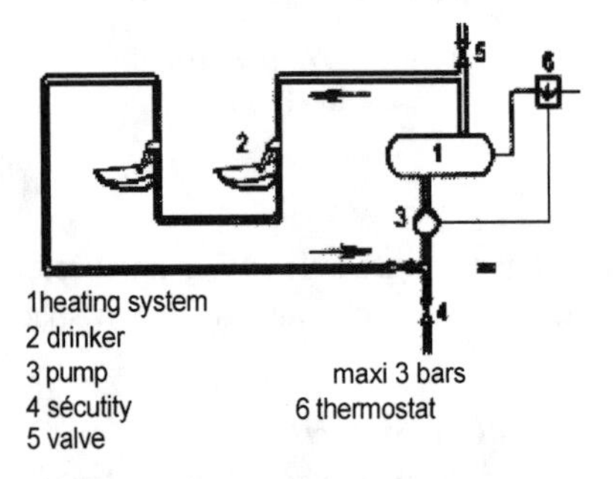

Figure 4.49. Continuous-circulation water.

three conditions must be respected: the shape of the drinking facility, its location, and its height. For these animals a water bowl is generally used (Fig. 4.50). Cleanliness of the water is very important. Sheep are reluctant to drink soiled water, and the drinkers should be cleaned very often.

Number of Bowls

The number of water bowls provided should be equivalent to 2.5% of the number of animals.

Location of Bowls

The bowls should be located on the opposite side of feeding trough. The height of the bowl should be at least 70 cm from the floor (Fig. 4.51). According to the thickness of the litter it may be necessary to adjust the position of the watering facility. A flexible pipe should be provided between the bowl and the water installation (Fig. 4.52).

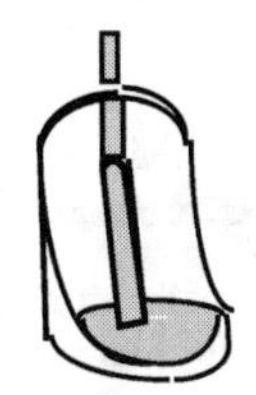

Figure 4.50. Water bowl for sheep.

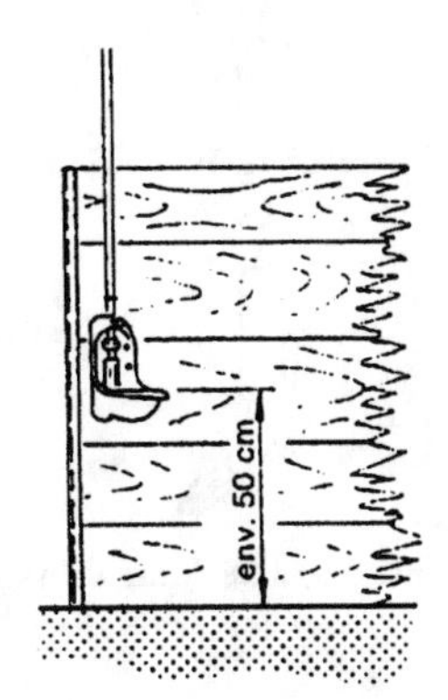

Figure 4.51. Drinker height position.

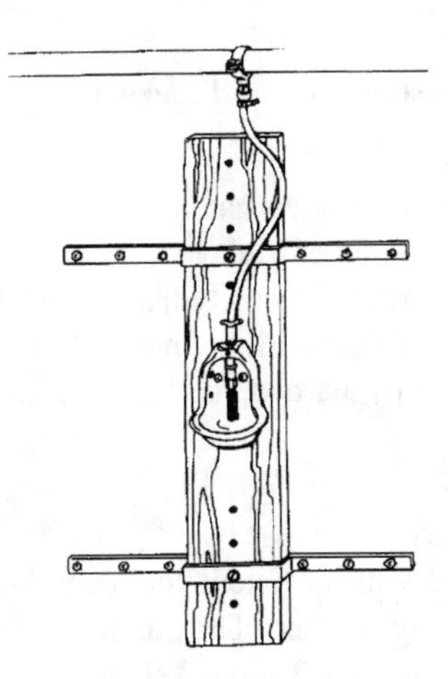

Figure 4.52. Flexible pipe adjusting height of the bowl.

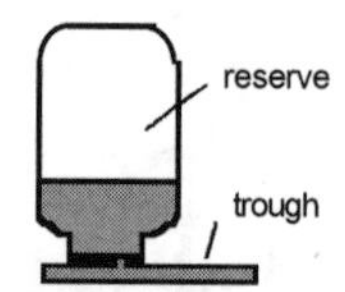

Figure 4.53. Siphoning drinker for piglets.

Figure 4.54. Tube drinker.

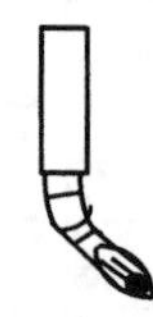

Figure 4.55. Suckling drinker.

4.3.3 Pig Breeding [9]

In pig breeding the water is supplied by drinking facilities in different ways: water bowl, water-tube system, water-suckling system, or directly with the feed.

Water Requirements

The recommendations are shown in the Table 4.1.

Drinking-Water Equipment

A siphoning drinking bowl for piglets can also deliver artificial milk. The water is at room temperature (Fig. 4.53). A tube drinker is easy to use, has no wastage, and stays clean (Fig. 4.54). A suckling drinker is easily learned by the animals, but a lot of water is wasted (Fig. 4.55). A constant-level drinker has little wastage, but the water gets dirty (Fig. 4.56). A finger drinker easily learned, has little wastage, and stays clean (Fig. 4.57).

4.3.4 Poultry [6]

Drinking Water Requirements

Poultry should be provided with sufficient cool drinking water without wastage. The drinking trough should be easily reached. The quality of the water should meet the local standards for drinking water. Tables 4.2 and 4.3 show the number and the type of devices needed.

Table 4.1. Recommendations of flow rate and position of drinking systems

Animal and Physiology Stages	Type of Drinking System	Flow Rate (l/60 s)	Height of Drinking System from the Floor (cm)			Maximum Number of Animals per Unit
			Without Step	With Step		
				Total Height	Step Height	
Piglet	Bowl	0.5–1.0	8	15	5	—
	Suckling system	0.5	20	30	5	—
Weaning pigs	Bowl	1.0–1.2	12	20	10	18
	Suckling system	0.5–1.0	30	35	10	10
Fattening pigs	Bowl	0.8–1.2	20	35	15	18
	Suckling system	0.5–1.0	50	60	15	10
Sows in collective box	Bowl	3.0	30	45	20	10
	Suckling system	1.5	70	80	20	5
Sows in individual box	Suckling system	>0.30	5–10	—	—	—

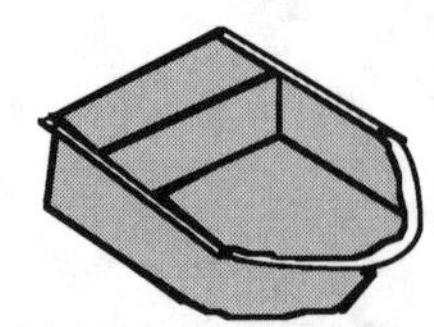

Figure 4.56. Constant-level drinker.

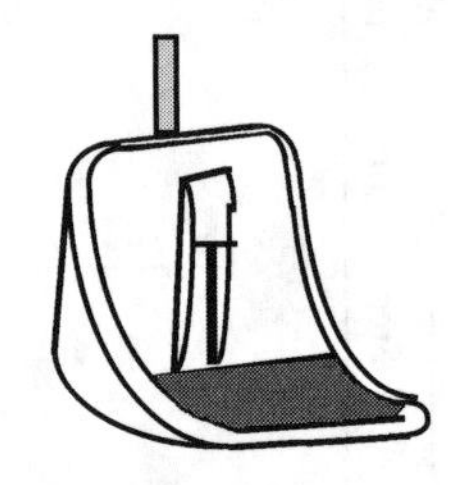

Figure 4.57. Finger drinker.

Water-Supply Systems

Laying Hens

In cage-breeding systems the dripping system is used (Fig. 4.58) or the suckling drinker into which the water is led by gravity from constant-level tanks located at the end of each battery.

In the cage two drippers should be available to each hen. A device for water recovery (cup or gutter) is fitted below each drinker (Fig. 4.59). This contributes to obtaining dry droppings.

Broilers

Dripping systems (pipette) fixed on a feeding pipe and hanging inside the building, supplied through gravity, or round drinkers can be used. Height above floor level depends on the size of the fowls and on the period (either starting or breeding).

Table 4.2. Number of drinking points for starting poultry by drinker type

Breeding Type	Dripping	Minidrinker	Little Bell	Dripping
Battery	1 for 10	1 for 50	—	—
Floor	—	—	10	80

Table 4.3. Number of drinking points for poultry by drinker type in various climates

	Number of Points		
Climate	Bell	Linear	Dripping
Mild	10	8	80
Warm	13	10	100

Note. Data for floor breeding in the period from 1 to 17 weeks.

Figure 4.58. Drop drinking

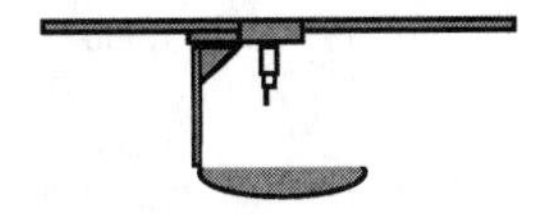

Figure 4.59. Cup for water recovery.

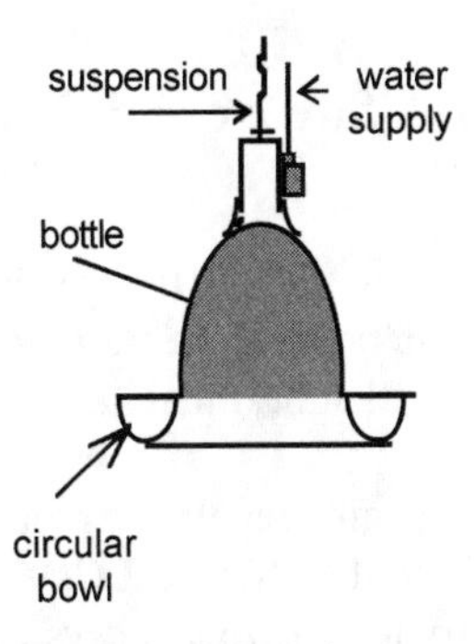

Figure 4.60. Bell drinker.

Turkeys

Automatic round drinking bowls (Fig. 4.60) can be used. A flap system regulates the water level and prevents overrunning, thus keeping the litter dry. The number of drinking bowls depends on the age of the fowls (Table 4.4). The drinkers are either hanging or set on the floor. For a 1000-m^2 building, 96 turkey-type drinking bowls are provided.

Table 4.4. Number and type of drinkers according to age of the fowls

	Age	
	0 to 6 Weeks	More than 6 Weeks
Type of drinkers	Little bell	Bells
Number of drinkers	1 per 80	1 per 120 to 150

Table 4.5. Influence of initial contamination and storage temperature of milk on its bacterial quality

Initial Contamination (germs/mL)	Storage Temperature (°C)	After 24 Hours Multiplication Rate	After 24 Hours Germs/mL	After 48 Hours Multiplication Rate	After 48 Hours Germs/mL
4,300	4.5	1	4,100	1	4,500
	10	3	14,000	30	12,800
	16	372	1,600,000	8000	33,000,000
40,000	4.5	2	88,000	3	127,000
	10	5	180,000	21	830,000
	16	113	4,500,000	2500	100,000,000
140,000	4.5	2	280,000	4	540,000
	10	8	1,200,000	100	14,000,000
	16	180	25,000,000	4300	600,000,000

Source: [12].

4.4 Milk Storage and Control [11]

4.4.1 General Points

Refrigeration of milk is necessary to avoid the development of microorganisms in the milk while it is stored on the farm, before collection. Refrigeration will not improve the bacteriologic quality of milk, it will at best maintain it, provided the milk cooler works properly and the tank is clean.

Within the European Union, specific regulations for hygiene define storage and cooling conditions for milk (see 85/397/EC, 89/362/EC, 92/46/EC).[1] These texts have been adapted by each member country to its own legal system, so it is recommended to refer to local regulations that indicate among other things the temperature for milk storage according to the intervals between collecting, and to the norm ISO 5708/1983[2] concerning bulk milk coolers.

4.4.2 Influence of Refrigeration on the Microbial Flora

Milk naturally contains antibacterial substances apt to stop contaminating flora from evolving if it is stored at ambient temperature. But this protection is effective only for a few hours and does not last enough for the storage time on the farm. Refrigeration is used for stopping or slowing down the activity of microorganisms. This bacteriostatic effect depends on three factors:

- Level of initial contamination
- Cooling speed
- Storage temperature

The influence of these factors is shown in Table 4.5.

The effect of refrigeration is not the same on all the germs contained in the milk: psychotropic bacteria, and especially *Pseudomonas fluorescens*, can develop below 4°C.

[1]EC: European Community.
[2]International Standard Organization.

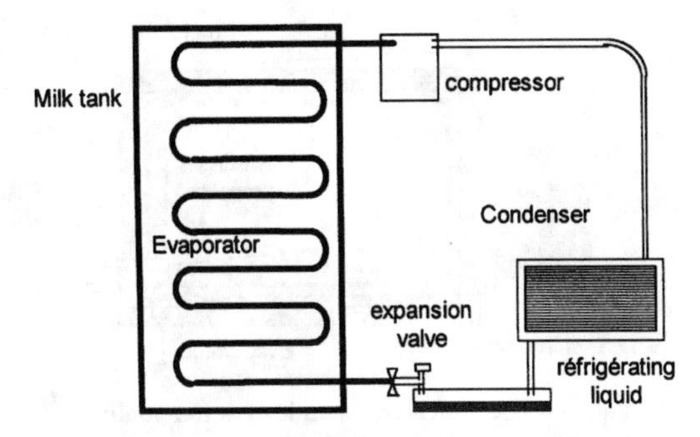

Figure 4.61. Simplified sketch of refreshing milk tank (S.B. Spencer).[1]

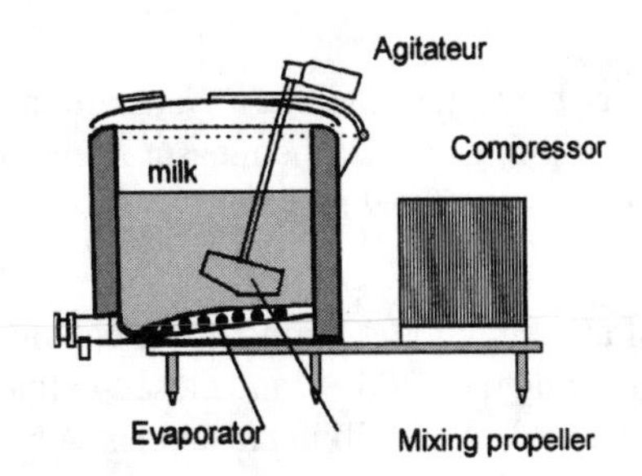

Figure 4.62. Refrigerator unit by expansion system.

4.4.3 Refrigeration Systems

A milk cooler works on the same principle as a home refrigerator. A cold-generating fluid in liquid form is distended into an evaporator and produces cold (Fig. 4.61). The heat created by the compression–condensation system is evacuated by means of a ventilator. The cooling level is regulated at the beginning by the mass of milk and then by a thermostat that stops the compressor when the milk has reached the desired temperature.

Direct-Distension Coolers

In a direct-distension cooler, the evaporator is in direct contact with the sides of the tank in which the milk is stored (Fig. 4.62).

Ice-Accumulation Coolers

The evaporator in an ice-accumulation cooler is in contact with a water tank in which ice is produced and accumulated. This water tank is around the tank that contains the milk; icy water is pumped to the tank (Fig. 4.63).

An agitator ensures uniform cooling of the milk and homogeinization so that the fat does not all come up to the surface. This system includes a blade set on an axis moved by a speed-reducer -; it should turn slowly so as to avoid churning and risks of fat desintegration.

[1] S. B. Spencer Prof. College of Agri. Sciences Coop. Extension. Dairy and Animal Science. Pennsylvania State-University.

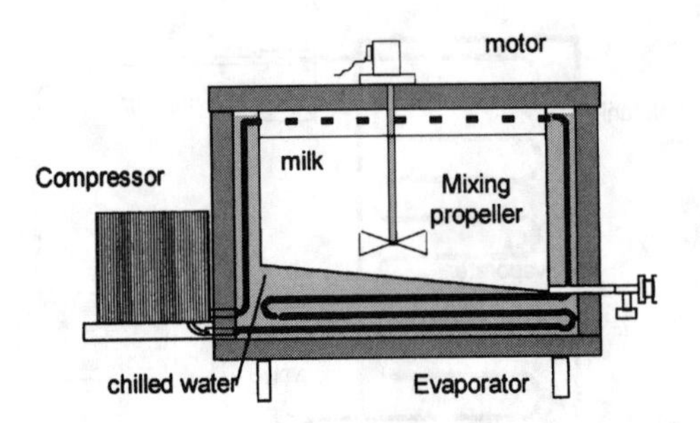

Figure 4.63. Refrigerator unit by water cooling.

4.4.4 Tank Equipment

Milk Measurement

In some countries, storage tanks must include a system for milk measurement, usually a graduated ruler or an electronic gauge. If the measurement system is used for trading transactions, it is controlled by an official service.

Other Equipment

A thermometer tells of anything going wrong in the equipment. A cleaning automat is necessary for tanks containing over 2000 L and closed-cylinder tanks, the principle is similar to that of a cleaning automat for milking machines. A heat exchanger recuperates the calories produced by the condenser–compressor system, which are used to warm up water in a tank.

4.4.5 Control of Coolers

Technical controls of the coolers are conducted to make sure that the equipment meets the demands of established norms. The points to be controlled are the maximum duration of milk cooling, conservation (average temperature of the milk between cooling periods), isothermy (average rise of temperature when the compressor and the agitator are not functioning), deep freezing of the milk (no ice should appear on the surface of the milk if the tank contains between 10% and 100% of its total capacity).

References

1. CIGR Section II.1, Working Group No. 14. 1994. *Cattle Housing* 1994. ADAS Bridgets Dairy Research Center FBRT.
2. Ministère de l'Agriculture Paris La stabulation libre pour vaches laitières. Série Construction rurales.
3. Institut de l'Elevage Paris Bergerie Moderne. 1985. Plans type FNGEDA–ITOVIC 1980.
4. Manuel de l'éleveur de porc ITP 1995.
5. La Pondeuse ITAVI Re-issue in progress1998.
6. L'Elevage des volailles ITAVI 1997.
7. Manuel de l'éleveur de porc ITP 1995.

8. CIGR Section II 1995 Recommandations of the Cattle Group.
9. Castle, M. E., and Thomas, T. P. 1975. The water intake of British Friesian cows on rations containing various forage. *Animal Production* 20:181–189.
10. Manuel de l'éleveur de porc ITP. 1995.
11. Institut de l'Elevage Paris. 1995. Références techniques pour l'hygiéne en production laitière bovine.
12. Davis and Killmeier. 1959. New method for determining the quality of bulk tank milk. *Milk Review* 21:37.

5 Storing Forages and Forage Products

Shahab S. Sokhansanj

5.1 Introduction

Losses in processed alfalfa from harvest to the time it is used can be mechanical, physical, or nutritional. This chapter focuses on losses during the storage of alfalfa and its products and considers only the losses due to biotic conditions. Harvest losses due to mechanical shattering of the leaves are discussed elsewhere [1, 2].

The chemical composition of alfalfa includes water, nitrogenous compounds, carbohydrates and pectin, lignin, lipids, organic acids, pigments, vitamins, and minerals. The growing plant contains 70% to 90% water, and often the proportion of chemicals in plant material is expressed on a dry-matter basis. Proteins are found in a soluble form in the cytoplasm and in a granular form in the chloroplasts. Carbohydrates in the plant are divided into structural and nonstructural. Structural carbohydrates consist of the cell walls and other tissues that give rigidity to the plant. Cellulose, hemicellulose, and lignin are structural carbohydrates. Nonstructural carbohydrates include digestible sugars and starches. Chlorophyll and carotenoids constitute the main pigments in alfalfa. Green chlorophyll that has no direct nutritional value often is associated with high protein and high carotene content. The beta carotene in alfalfa is the precursor to vitamin A. Carotene is a very unstable compound whose oxidative destruction commences as soon as the plant is cut.

The analysis of alfalfa as a feed starts with the determination of the moisture content, which is expressed in percent of the original weight of the material. Other compounds found then are expressed on a dry-matter basis (original weight minus water content). Crude protein content is the proportion of nitrogen multiplied by 6.25. Crude protein does not indicate the nutritional quality of protein and the value of the essential amino acids. To determine the fiber content of the plant material the neutral-detergent and acid-detergent fiber procedures are used. The neutral-detergent fiber provides the contents of the cell wall, which include cellulose, hemicellulose, and lignin. The acid-detergent fiber separates hemicellulose from the fiber, and thus the value does not contain the component of hemicellulose. The quality of forages also are given in the form of energy intake. A portion of this energy is digestible (often more than 50%), out of which about half of it is the metabolized energy.

Canadian data on production of hay in general and that of alfalfa in specific are difficult to compile. Table 5.1 lists the published statistical data on the production of cultivated hay in Canada, which includes alfalfa and other forages. Canada's annual production of hay is about 33 million tons, covering an area of about 8.5 million ha. About half of the balers are deployed in Western Canada, while the majority of forage harvesters are in Eastern Canada. In Saskatchewan about 10% of the hay produced is processed into cubes and pellets mainly for export. The percentage of the forage processed in Alberta is smaller than in Saskatchewan, as most of the hay is fed directly to the livestock.

5.2 Losses in Swath

The quality of field-cured hay depends upon such factors as maturity at the time of cut, method of handling, moisture content, and weather conditions during harvest. In addition to mechanical losses (loss of leaves), enzymatic and oxidative losses also occur during field drying. Dry-matter losses from respiration (enzyme activity) have been reported to be up to 15% of the plant weight. With the slow field drying, as much as 80% of the vitamin A (beta carotene) in the alfalfa plant is lost due to exposure to sun [3]. Sun bleaching of cut forages generally increases the loss of vitamin A. Vitamin D content of the cut plant, however, increases as a result of the synthesis of the vitamin from plant sterols in the presence of sunlight and ultraviolet rays.

Bruhn and Oliver [4] tested hay stored outside for tocopherols (vitamin E) and carotene (vitamin A) contents. Tocopherol is known to protect lipids from oxidation in foods and feeds. Swathed and baled hay samples were kept outside over a period of 18 weeks. The hay was tested for dry matter, tocopherol, and carotene content in California (presumably around Davis). A linear fit to the test data was obtained showing that the dry matter increased from 88% to about 92%, the γ-tocopherol decreased from 90 to 80 μg/g, the α-tocopherol decreased from 84 to 50 μg/g and the carotene decreased from 58 to 18 μg/g during the 18 weeks of outside storage. Because the climatic conditions during storage were not reported, we may assume that the losses probably were due to sun curing.

Losses in protein occur during field curing as the activity of the enzyme protease decreases N-protein (estimated protein based on the total nitrogen content) but increases the amount of soluble proteins, peptides, amino acids, amides, and other compounds. However, changes in N-protein are small and inconsequential with respect to the feeding quality of hay.

5.3 Losses During Storage of Square-Baled Hay

During field drying (curing), considerable losses in dry matter and nutritional value occur due to respiration, leaching, bleaching, microbial activity, and mechanical handling [5]. Stable storage of hay in rectangular bales requires that the hay have less than 18% moisture content. Buckmaster et al. [5] investigated the storage stability of baled hay stored in a hay shed in Michigan. They monitored the dry-matter loss, ash content, fiber content, crude protein, and development of insoluble proteins in the bales of alfalfa that varied in moisture content and density.

Table 5.1. Statistics on forage production, yield, area under cultivation, number of balers and forage harvesters in Canada (1990) and production of processed forgages (1992–1993)

Provinces	Production (t × 1000)	Yield (t/ha)	Area (ha × 1000)	No. Balers	No. Forage Harvesters	Pellets (t × 1000)	Cubes (t × 1000)	Compact Bales (t × 1000)
New Foundland	24	4.63		154	30			
Prince Edward Island	290	5.15		1844	339			
Nove Scotia	435	6.32		2379	416			
New Brunswick	417	5.86		2151	382			
Quebec	7000	7.06		26,247	8402			
Ontario	7439	7.21		37,282	14,513			
Eastern Canada	15,605	6.06		70,057	24,082	20		10[b]
Manitoba	3084	4.62	384	15,137	978	19[a]		
Saskatchewan	2722	3.20	5949	30,640	1560	233		5[b]
Alberta	9525	5.06	2104	37,578	3449	192		40[b]
British Columbia	2177	6.18	85	7484	1863	—		
Canada total	33,113	5.54	8522	161,076	31,936	464	286	55[b]

[a] Includes British Columbia.
[b] Estimated for 1993–1994 crop year.

Table 5.2. Storage stability of baled hay

Moisture content range (%)	10–15		15–20		20–25		25–30	
Stack number*	1	2	5	3	3	2	1	2
Moisture content (%)	11.5	14.4	17.7	16.8	24.0	22.4	27.7	26.4
Density (kg/m^3)	111	173	87	188	101	227	106	230
dry matter loss (%)	2.1	2.5	3.1	3.0	4.4	4.4	5.8	9.4
Maximum temp. (°C)	26	25	26	37	29	43	27	47
Average temp. (°C)	17	18	21	25	18	29	21	34
Ash (%)	7.2		8.3		8.9		8.2	
Retention ratio	1.04		0.99		1.02		1.00	
ADF (%)	42.5		34.3		31.2		37.5	
Retention ratio	0.92		0.99		1.06		0.95	
CP (%)	13.9		17.0		20.1		15.3	
Retention ratio	1.07		0.99		0.97		1.05	
ADIP (%)	1.13		1.01		1.07		1.12	
Retention ratio	0.99		1.02		1.13		1.20	

* Stack number is the location of the bale in a stack of bales of 5 rows.

The results obtained by Buckmaster et al. [5] cover a moisture range for the baled alfalfa from 11.5% to 48.0%. Because most alfalfa is baled when moisture content is less than 20%, Table 5.2 lists part of the results published by Buckmaster et al. Each moisture range includes at least two moisture contents. The bales were stacked in five rows and three columns; columns were separated by styrofoam insulation board. The location of the tested bale in the stack is identified by the stack number.

Buckmaster et al. found that dry-matter loss was directly correlated with moisture content of the bales but could be considered independent of the bale density. The dense bales, however, showed a higher temperature during storage than the loose hay (refer to Table 5.2 and compare the maximum or average temperature of the dense bale to that of the loose bale). Increased baling moisture content resulted in an increase in the concentration of crude protein, ash, and acid-detergent fiber when removed from storage. Quality retention ratios indicated that ash and acid-detergent fiber did not change, but the amount of insoluble protein increased with moisture content.

5.4 Losses During Storage of Round-Baled Hay

Rider et al. [6] classified different portions of a round bale according to susceptibility of each portion to deterioration. Figure 5.1 depicts the divisions in a round bale. As a round bale settles, approximately one third of the circumference contacts the ground; a substantial amount of moisture can be absorbed through the bottom of the bale resulting in spoilage as far up as 30 cm. If the weather affects only the outer 15 cm of the round bale not in contact with the ground plus an additional 15 cm at the bottom, 42% of the bale volume can be affected. Assuming a uniform bale density, the outer 15 cm of a round bale accounts for more than 20% of the total bale mass.

Rider et al. [6] investigated six types of storage of round bales: inside a barn; unsheltered, with a black polyethylene wrap around the circumference of each bale; unsheltered,

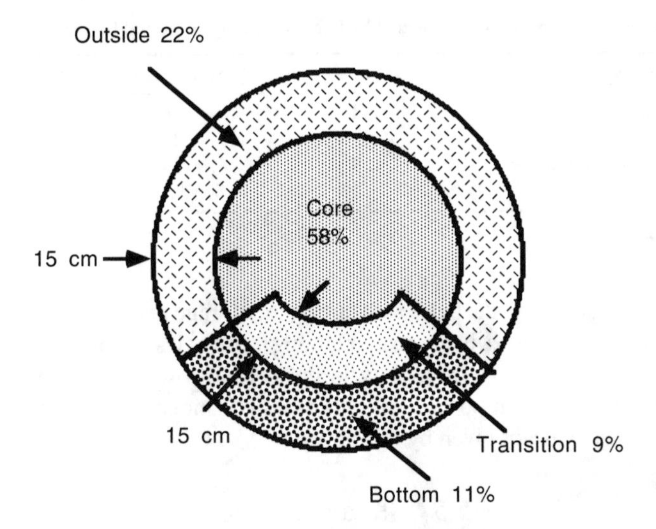

Figure 5.1. Division of a round bale.

with direct ground contact; unsheltered, with black polyethylene under the bale contact area; unsheltered, with black polyethylene under the projected bale contact area; and unsheltered, with a black polyethylene ground cover on the entire storage area. The results of the tests are summarized in Table 5.3. In Western Canada the losses in dry matter in round bales as a result of outdoor storage were reported to be 4% to 8% [7]. Schodt [8] investigated the losses in wrapped and unwrapped round bales in Manitoba. Hay, consisting of 60% alfalfa and remainder timothy and brome, was baled in July. The hard-core bales were stored using five storage treatments. Two were wrapped with polyethylene with ends left open (*wrap* in Table 5.4); three unwrapped bales included bales stored horizontally, end-to-end in a single row; bales stored vertically with the bottom bale on its end and the top bale on its side; and bales protected from weather inside an enclosed building (barn). Outside bales were stored in a well-drained area. Bales were monitored for 16 months, during which 423 mm of rain was recorded. Table 5.4 lists a summary of the results.

Dry-matter loss was calculated based on weighing the bales before and after storage, and spoilage was judged as not being fit for consumption by livestock. Table 5.4 shows

Table 5.3. Changes in alfalfa bale quality for the six storage types

Storage Type	Percent Change in Crude Protein from Initial 24.3%	Percent Change in Digestible Dry Matter from Initial 64.4%
Barn	−17.9	−4.2
Wrap	−21.4	−12.2
Ground	−29.2	−23.4
Contact area cover	−29.3	−20.9
Projected area cover	−25.6	−24.0
Total ground cover	−27.7	−22.5

Table 5.4. Results of tests conducted by PAMI on storage of round bales in Manitoba

Storage Type	Dry-Matter Loss	Spoilage Loss	Total Loss
In barn	Nil	Nil	Nil
Wrap	3.7	3.8	7.8
Single row unprotected	0.8	5.6	6.4
Vertical	9.8	1.0	10.8

Source: [8]

that storing in the barn kept the bales safe during the storage period. Not much difference was noticed in the total losses in dry matter among those bales wrapped and not wrapped. The vertical storage losess were mainly in the bottom row, and spoilage losses were centered around the bottom portion of each bale. The moisture content of the bales at the time of storage was not given by the author.

5.5 Storage of Cubes

Chaplin and Tetlow [13] studied the storage stability of both whole and ground samples of several grasses including alfalfa. The tests were conducted in a laboratory at a temperature of 21°C in a controlled humidity chamber over sulfuric-acid solutions (to maintain humidities). The samples initially at 8°C were brought to equilibrium with relative humidities ranging from 50% to 90%. Some samples were treated with propionic acids and some were prepared with binders (calcium lignosulphonate). Table 5.5 lists a summary of the results of Chaplin and Tetlow's experiments. Mold development has been found to be slower on pure protein than on pure starch, and much slower on pure fiber substrates. Compared with the starch grain, it appears that mold growth on forages is slower. The use of a binding agent had little effect on the mold growth. Propionic acid sprayed on samples improved the storage time. The molds were identified on most samples were from the *Aspergillus glaucus* group. *A. ruber*, *A. repens*, and *A. chevalier* also were identified on a few samples.

Chaplin and Tetlow [9] also studied the moisture uptake of wafers in bulk by exposing them to a change in humidity of the air. First a bulk of wafers 0.3 m was exposed to 90%

Table 5.5. Equilibrium moisture content and number of days before the first appearance of mycelium and spores for various treatments at 21°C

Treatments	Moisture content	8.2	10.0	13.3	—	18.1	27.9
	Relative humidity	49.6	58	70	75	79	89
No treatment	Spores			350	83	20	
	Mycelium			222	69	—	—
Binding agent	Spores			210	75	20	
	Mycelium			141	45	—	
Fungus inhibitor	Spores					141	
	Mycelium					102	

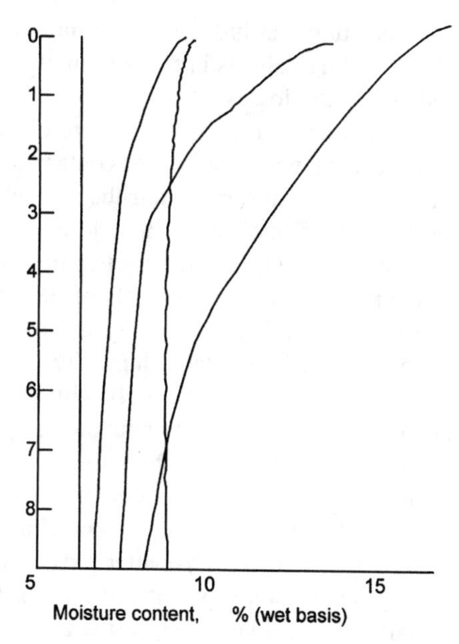

Figure 5.2. Moisture profile in a bulk of wafers (cubes).

humidity (11°C) for 6 days followed by a drop in humidity to 45% for the following 6 days. Their findings are plotted in Fig. 5.2, showing that the top layer gradually absorbs moisture as the humidity penetrates into the pile. Following a drop in relative humidity the top tends to return to a low moisture content. However, from the figure, it appears that the moisture of the pile increases gradually.

Dry hay is stored in the field or inside a storage building, barn, or silage silos. Under good weather conditions, baled hay and long loose hay can be stored in stacks outside for several months. In order to minimize the effect of rainfall on stacked long hay, care is taken to develop a thatched layer on top with a stacking sweep rake or manually. Stacks as long as 6 m and weighing about 6 tons can be made with stacking rakes of different forms. Machines are also available for loading, transporting, and unloading loose hay stacks.

After full or partial field curing and baling operations, bales can be collected, loaded, and transported to storage with a bale elevator and tractor–trailor bale harvester or an automatic bale wagon. Special machines also are available for picking, transporting, and stacking large round bales.

Rectangular bales are normally stacked in the field in a staggered form to facilitate air movement. Round bales are best stored horizontally instead of end-to-end to minimize the effect of rain. A bale-wrapping machine wraps the circumference of round bales with plastic materials for protection against rain. Stacked rectangular bales can also be protected with plastic. However, wrapped bales do not store better than unprotected bales. Round bales wrapped in plastic have greater total feed losses (7.5%) than bales stored in a single row unprotected from the weather (6.4%). The tight plastic wrap around the circumference prevents moisture that has penetrated into the bales from escaping. Consequently, considerable moisture collects in the bottom of wrapped bales, resulting in deterioration of hay and dry-matter losses.

Dry-matter and quality losses occur during the storage of hay in the field and in buildings.The losses are dependent on the moisture content of the hay entering the storage, and the environmental conditions surrounding the hay. Under normal conditions, loose and baled hay below 30% to 35% moisture content lose very little dry matter due to respiration during storage. Molds or fungi are only active in storage when the relative humidity of the air in the interstitial spaces exceeds about 70%, and the temperature is sufficiently high. Bacteria require a relative humidity of at least 95% to grow. The equilibrium moisture content of hay at 70% relative humidity is about 18%. In temperate regions in North America and Europe, hay, particularly with baled crops, can be stored in the field at a moisture content of 20% to 25% without molding for several months due to a combination of temperature, which for a greater part of the storage period is lower than 20°C, and storage conditions that allow free air circulation.

The higher the relative humidity, the moisture content of the hay, and the ambient temperature in the storage, the greater are the dry-matter and nutrient losses. For properly field-cured and barn-dried hay, the moisture content of the hay is low, and even relatively high humidities and temperatures are unlikely to cause the growth of molds and bacteria.

Chemical additives such as organic phosphate compounds and propionic acid cause a reduction of 5% to 25% in dry-matter and nutrient losses, and an increase in the drying rate and the suppression of respiration and of mold growth during field and barn storage of hay. The chemical additives are used primarily to suppress respiration and mold growth during storage but have the additional benefit of allowing the hay to be baled at a higher moisture content (up to 35%). A reduction in the dry matter loss of 600 kg/ha has been recorded by using propionic and acetic acids and ammonium buyrate.

As discussed previously, hay can be baled at a moisture content of 35% to 40% if it is to be dried by continuous ventilation with heated or unheated air. Intermittent ventilation is applied during subsequent storage. If the exhaust air temperature is not greater than the inlet air temperature when ventilation is applied during storage (1 to 2 d after drying), then the stack is sufficiently dry for safe storage. If the exhaust temperature is higher than inlet temperature, ventilation should be supplied for a reasonable length of time (1 to 3 d) to remove the hot spots in the hay.

5.6 Storage of Alfalfa Pellets

Cubes and pellets at moisture contents of 7% to 10% are susceptible to moisture absorption if exposed to humid conditions during transportation to and storage at export destinations. The process of moisture absorption by agricultural materials often results in the loss of quality. The rate at which the quality change occurs in stored products is dependent on the affinity of a material for moisture and on the ambient conditions.

Mold, yeast, and bacterial and other microbial activities increase rapidly at relative humidities above 70%. Nutrients in stored products are utilized by microorganisms for respiratory purposes. The heat of sorption can be an indicator of the product's affinity for moisture.

Alfalfa pellets and cubes initially at 7% to 10% moisture content can absorb up to 6% to 8% moisture when stored at 70% to 90% relative humidity and 10°C to 40°C. This is typical of the conditions existing in the Pacific regions that import alfalfa pellets and cubes. Moisture absorption also occurs when a temperature difference exists between the storage silo (and transit container) and the surroundings. A volume increase of 20% may result if pellets and cubes absorb about 6% to 8% of moisture. This has obvious implications for the economics of handling and storage of pellets and cubes since the potential expansion and moisture uptake of pellets and cubes affects the design of storage silos and transit containers. Also, the storability of pellets and cubes at 14% to 18% moisture content is limited.

Forage products have a higher affinity for moisture sorption than cereal grains, and thus more stringent measures than presently used for the commercial handling and storage of grains should be taken to prevent moisture sorption by forage pellets, cubes, and wafers.

Headly published a set of moisture content and relative humidity of dehydrated alfalfa pellets (6 mm diameter). Table 5.6 shows that at 21°C and rh of 76% the pellets are at 14% m.c. which may be considered a safe moisture content for storage. When the rh increases to 80% at the same temperature, the moisture content of the pellets increases to 16%. The equilibrium moisture content of pellets were higher at cooler temperatures.

Headly [9] also investigated the moisture uptake by pellets in a humid environment and the resulting increase in the volume of the pellets. Table 5.7 shows the percent expansion of dehydrated .025-in pellets initially at 7% to 9% moisture content. The storage temperature for these pellets was 11°C. The original graph presented by Headly indicated that stored at 58% and 75% relative humidity, pellets approached the maximum expansion after 2 weeks. The percent expansion at relative humidity of 93% after 2 weeks was within 90% of the asymptote value.

Headly investigated the effect of storing pellets at high humidity and the development of mold on the pellets. He observed that storing pellets with 7% to 9% moisture content in relative humidity greater than 75% and temperature of 11°C resulted in moldiness.

Table 5.6. Equilibrium moisture content and relative humidity of dehydrated alfalfa pellets

Temp, °C							
11	r.h., %	25.0	40.0	58.0	75.0	82.0	93.0
	m.c., %	6.9	7.5	14.1	17.3	20.4	25.7
21	r.h., %	23.0	35.0	55.0	76.0	80.0	
	m.c.,%	6.8	7.1	12.0	14.3	16.3	
32	r.h., %	22.0	32.0	51.0	76.0		
	m.c., %	5.9	8.8	12.0	14.2		

Source: [9]

Table 5.7. Volume expansion of the dehydrated alfalfa pellets stored in 11°C environment

	Volume Expansion (%)	
Relative Humidity (%)	1-Week Storage	3-Week Storage
93	16	30
75	5	10
58	5	6

Note. Initial moisture content of pellets 7%–9%
Source: [9]

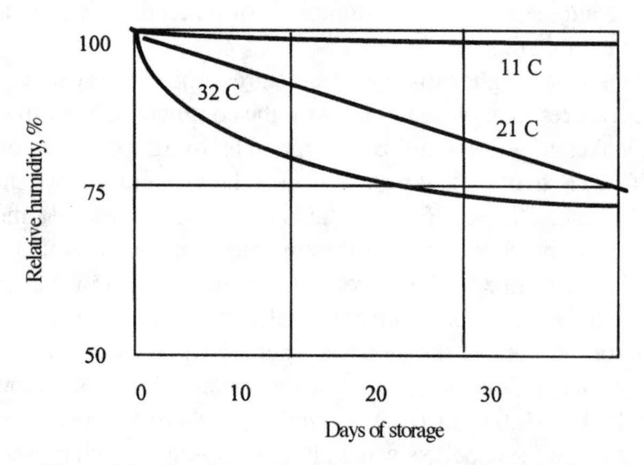

Figure 5.3. Conditions for safe storage of pellets initially at 7% to 9% m.c. The criteria is mold.

Figure 5.3 shows that regardless of the temperature, when pellets are stored at relative humidities less than 75%, the probability of mold growth is nil. In the same study, Headly [9] performed a limited number of durability tests on pellets stored in different environments. The data showed that storing at relative humidity between 55% to 75% produced most durable pellets. Pellets stored at higher or lower relative humidities were less durable and thus broke more easily.

5.7 Storage of Loose Hay

Unlike for dense hay, generally extensive research work has been conducted on the equilibrium moisture content of loose hay, particularly alfalfa. Early data provided by Zink [10] and Dexter et al. [11] have been used extensively. Table 5.8 is a summary of the combined data from these researchers showing that, on the average, the equilibrium moisture content of the second-cut hay is less than that of the first cut. The second-cut hay has more stems than the first cut and thus is less hygroscopic. The first-cut hay is more leafy than the second cut and contains more protein. Proteins not only absorb moisture readily but also degrade easily. An interesting feature of the equilibrium moisture-content data as shown in Table 5.8 is the sudden increase in moisture content

Table 5.8. Equilibrium moisture content for the alfalfa hay at 22°C

	Moisture Contant (%)						
	20% Relative Humidity	40% Relative Humidity	50% Relative Humidity	60% Relative Humidity	70% Relative Humidity	80% Relative Humidity	90% Relative Humidity
First cut (dried)	10.2	11.7	11.0	12.0	16.0	17.8	20.7[a]
Second cut (dried)	5.7	8.5	9.0	10.6	15.0	14.9	18.4[a]
Fresh cut (undried)	7.4	9.8	10.6	12.8	15.7	18.6[a]	36.8[b]
Fresh cut (dried)	6.4	9.4	—	13.3	—	19.0[a]	23.5[b]
Stems (dried)	—	—	9.4	—	12.5	15.8	21.7[a]
Leaves (dried)	—	—	10.4	—	13.8	17.5	24.8[a]

[a] Moldy.

[b] Very moldy.

when the relative humidity increases from 60% to 70%. A plot of relative humidity against moisture content shows that the inflection point in the isotherm curve is between 50% and 70%.

Dexter et al. [11] also tested the moisture-absorbing characteristics of freshly cut alfalfa. They observed that the undried fresh-cut alfalfa becomes moldy very quickly, and this results in a substantial dry-matter loss. Comparing the stems and leaves, the data in Table 5.6 show that leaves have much more affinity to water absorption than the stems. In all cases relative humidities higher than 70% result in abrupt increase in moisture content.

One may compare the equilibrium moisture content of loose alfalfa hay (Table 5.8) with that of dense alfalfa pellets (Table 5.6). It seems that the first-cut loose hay is more hygroscopic than the processed dense hay (pellets). However, pellets absorb more moisture than the second-cut loose hay when exposed to the same environment.

5.8 Cube Spoilage During Transport

The condition of regular-size cubes loaded in a container was monitored during transport from a plant in Alberta to Taiwan [12]. The container was 2.34 m wide and 12.2 m long and held about 40 tons of alfalfa cubes. Temperatures, relative humidity, and the formation of dew as a result of condensation were recorded continuously throughout the ocean transport. The monitored temperatures consisted of the ceiling temperature, the air inside the container, the temperature on the surface of the cube pile, and temperature 60 cm deep in the cube pile. The relative humidity of the air inside the container was recorded. The events representing condensation or no condensation were measured with a resistive sensor.

Figure 5.4 shows that the temperature inside a container was about 20°C at the time of loading for the May shipment. After a period of daily fluctuations, the temperature of the air inside the container remained relatively constant at about 20°C when the container was on the sea. The temperature in the container started increasing as the vessel approached

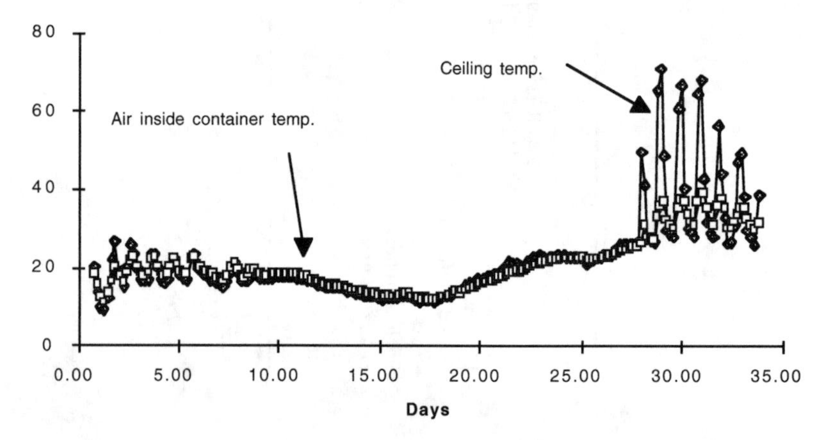

Figure 5.4. The record of air temperature inside the container during transport from Alberta to Taiwan in May 1993.

its destination. Temperature fluctuations were again observed when the container was unloaded at the port of destination. The container ceiling temperature reached as high as 60°C during the day and dropped as low as 20°C during the night at the port of destination.

The data for a July shipment is summarized in Table 5.9. Roof temperature reached a high of 32.2°C while in transit. The cube temperature at the 60-cm depth in the pile did not fluctuate rapidly but gradually increased from 19.4°C to 26.9°C at the port of destination. Maximum relative humidity was attained at minimum temperature. In Canada, when the temperature was high (30.0°C), the humidity in the room was low (63%). Conversely, when the temperature was low (18.1°C) the relative humidity was high (86%). These conditions are not conducive to mold growth. At the destination, however, the minimum temperature of 27°C was observed at a high relative humidity of 79%, which according to Fig. 5.3 could provide an ideal environment for mold growth.

During the transit the temperature of the cube remained low (about 17.8°C), and consequently the mold development was minimal. The temperature of cubes 60 mm deep remained relatively low, and thus the cubes at this depth remained mold-free for the greater part of the journey. The temperature of the cubes at 60 cm rose to about 27°C while in transit near the destination, and this condition is conducive to mold development.

Data on condensation showed that condensation on the ceiling of the container started 7 days after loading, and the water droplets remained on the ceiling during the entire trip. It is assumed that the humidity sensor functioned properly. The condensation data on the cube surface indicated that condensation occurred almost daily after loading the container. However, the condensation disappeared during the ocean trip. The condensation cycle reappeared when the load arrived at the port of destination.

Figure 5.5 is a plot of temperature versus corresponding relative humidity inside the container. The safe temperature and relative humidity can be superimposed on this chart

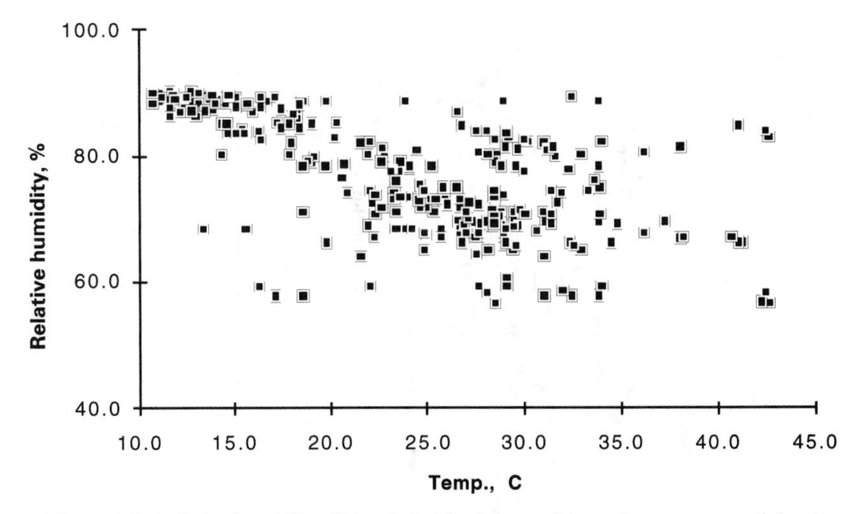

Figure 5.5. Relative humidity of the air inside the container and temperature of the air inside the container for the July shipment.

Table 5.9. Temperature and humidity history of a shipment of cubes from Alberta to Taiwan in July 1993

Location		Ceiling Temperature (°C)	Inside Air Temperature (°C)	Cube Surface Temperature (°C)	60 cm in the Cube Temperature (°C)	Relative Humidity (%)
In Canada (8 days)	Average	25.2	23.5	21.2	19.4	75
	Maximum	37.2	30.0	22.1	19.6	86
	Minimum	16.4	18.1	20.5	19.2	63
In transit (14 days)	Average	19.7	19.8	19.4	17.8	80
	Maximum	22.1	22.1	20.2	18.3	82
	Minimum	17.8	18.1	18.8	17.4	79
In destination (8 days)	Average	32.2	31.4	29.4	26.9	73
	Maximum	43.9	35.8	30.0	27.4	79
	Minimum	27.2	28.2	28.7	26.6	66

to show the potential for cube spoilage during transport. This graph indicates that the incidence of the unfavorable storage conditions was primarily during shipment.

References

1. Arinze, E. A., Sokhansanj, S., and Schoenau, G. J. 1993. Simulation of natural and solar-heated air hay drying system. *Computers and Electronics in Agriculture* 8:325–345.
2. Sokhansanj, S., Arinze, E. A., and Schoenau, G. J. 1993.Forage drying and storage. In *Handbook of Crop Drying and Storage* (ed. Bakker-Arkema and Maier). In press to be published by Marcel Decker. (Copies of the chapter are available at the Department of Agricultural and Bioresource Engineering University of Saskatchewan Saskatoon, SK S7N 5A9 Canada)
3. Church, D. C. 1991. *Livestock Feeds and Feeding*. Toronto: Prentice-Hall Canada.
4. Bruhn, J. C., and Oliver, J. C. 1978. Effect of storage on tocopherol and carotene concentration in alfalfa hay. *Journal of Dairy Science* 61:980–982.
5. Buckmaster, D. R., Rotz, C. A., and Martens, D. R. 1989. A model of alfalfa hay storage. *Transactions of the ASAE* 32:30–36.
6. Rider, A. O., Bachhelder, D., and McMurphy, W. 1979. Effects of long term storage on round bales. ASAE Paper no. 79-1538. St. Joseph, MI: ASAE.
7. Hesslop, L. C., and Bilanski, W. K. 1986. Economic benefit of weather protection for large round bales. *Canadian Agricultural Engineering* 28(2):131–135.
8. Schodt, R. 1989. Wrapped round bales store no better than unprotected bales. PAMI Notes. Prairie Agricultural Machinery Institute, Portage, Laprairie, Manitoba (1-800-567-PAMI).
9. Headly, V. E. 1969. Equilibrium moisture content of some pelleted feeds and its effect on pellet durability index.*Transactions of the ASAE* 12(1):9–12.
10. Zink, F. J. 1935. Equilibrium moistures of some hay. *Agricultural Engineering* 16:451–452.
11. Dexter, S. T., Sheldon, W. H., and Waldron, D. I. 1947. Equilibrium moisture content of alfalfa hay. *Agricultural Engineering*. 28(7):295–296.
12. Sokhansanj, S. M. H., Khoshteghaza, W. J., Crerar, E. Z., Jan, A., McNeil and A., Penner. 1996. Moisture migration in containerized alfalfa cubes during overseas shipment. Special report, Dept. of Agric. & Bioresource Eng. Univ. Saskatchewan Saskatoon, Canada, 57 p.
13. Chaplin, R. V., and R. M., Tetlow. 1971. Storage of dried grass wafers: moisture relationship, safe storage periods and changes in durability. *Journal of Stored Products Research* 7:171–180.

6 Waste Management and Recycling of Organic Matter

6.1 Waste Management

David Moffitt

Manure management is a critical concern in many parts of the world where the expanding livestock and poultry industry is concentrating. In past literature manure and related animal-production residues often have been referred to or classified as "waste." In this presentation, the term waste is replaced by *manure* and *residues* as much as possible in an effort to maintain the concept that manure is a resource to be used rather than a product to be wasted.

Much of the discussion that follows was adapted from the United States Department of Agriculture, Natural Resources Conservation Service's Agricultural Waste Management Field Handbook [1], to which the author was a major contributor.

6.1.1 Effects of Manure on the Water Resource

This section focuses on the effects that agricultural manure can have on water, air, and animal resources. Special emphasis is placed on the reactions of particular contaminants within the aquatic environment (how they change and how they affect aquatic life and human health).

Animal manure contains a number of contaminants that can adversely affect surface and ground water. In addition, certain constituents in manure can impact grazing animals, harm terrestrial plants, and impair air quality. However, where manure is applied to agricultural land at acceptable rates, crops can receive adequate nutrients without the addition of commercial fertilizer. In addition, soil erosion can be substantially reduced and the water-holding capacity of the soil can be improved if organic matter from animal manure is incorporated into the soil.

The principal constituents of animal manure that impact surface and ground water are organic matter, nutrients, and fecal bacteria. Manure may also increase the amount of suspended material in the water and affect the color either directly through the manure itself or indirectly through the production of algae. Indirect effects on surface water can also occur if sediment enters streams from feedlots or overgrazed pastures and from eroded streambanks at unprotected cattle crossings. The impact that these contaminants have on the aquatic environment is related to the amount and type of each pollutant entering the system and the characteristics of the receiving water.

All organic matter contains carbon in combination with one or more other elements. All substances of animal or vegetable origin contain carbon compounds and are, therefore, organic.

When plants and animals die, they begin to decay. The decay process is simply various naturally occurring microorganisms converting the organic matter—the plant and body tissue—to simpler compounds. Some of these simpler compounds may be other forms of organic matter or they may be nonorganic compounds, such as nitrate and orthophosphate, or gases, such as nitrogen gas (N_2), ammonia (NH_3), and hydrogen sulfide (H_2S).

In a natural environment the breakdown of organic matter is a function of complex, interrelated, and mixed biological populations. However, the organisms principally responsible for the decomposition process are bacteria. The size of the bacterial community depends on its food supply and other environmental factors including temperature and pH. Theoretically, the bacterial population doubles with each simultaneous division of the individual bacteria; thus, one divides to become two, two becomes four, four becomes eight, and so forth. The generation time, or the time required for each division, may vary from a few days to less than 30 minutes. One bacterium with a 30-minute generation time could yield 16,777,216 new bacteria in just 12 hours.

The principal nutrients of concern in the aquatic environment are nitrogen and phosphorus. An understanding of how these nutrients react in the environment is important to understanding the control processes discussed in subsequent sections.

Nitrogen occurs throughout the environment—in the soil, water, and surrounding air. In fact, 78% of the air we breathe is nitrogen. It is also a part of all living organisms. When plants and animals die or when manure products are excreted, nitrogen returns to the environment and is cycled back to the land, water, and air and eventually back to other plants and animals.

Figure 6.1 depicts the nitrogen cycle. It shows the flow from one form of nitrogen to another. The various forms of nitrogen can have different effects on our natural resources—some good and some bad. The conversion from one form of nitrogen to another is usually the result of bacterial processes. Some conversions require the presence of oxygen (aerobic systems), while others require no oxygen (anaerobic systems). Moisture content of the manure or soil, temperature, and pH speed or impede conversions.

In water-quality analyses, total nitrogen includes the organic, total ammonia ($NH_3 + NH_4$), nitrite (NO_2), and nitrate (NO_3) forms. Total Kjeldahl nitrogen includes the total organic and total ammonia nitrogen. The ammonia, nitrite, and nitrate forms of nitrogen may be expressed in terms of the concentration of N (NO_3-N or NH_4-N) or in terms of the concentration of the particular ion or molecule (NO_3 or NH_4). Thus, 45 mg/L of NO_3 is equivalent to 10 mg/L of NO_3-N.

Nitrogen in fresh manure is mostly in the organic form (60%–80% of total N). In an anaerobic lagoon, the organic fraction is typically 20% to 30% of total N. Organic nitrogen in the solid fraction (feces) of most animal manure is usually in the form of complex molecules associated with digested food, while that in the liquid fraction is in the form of urea.

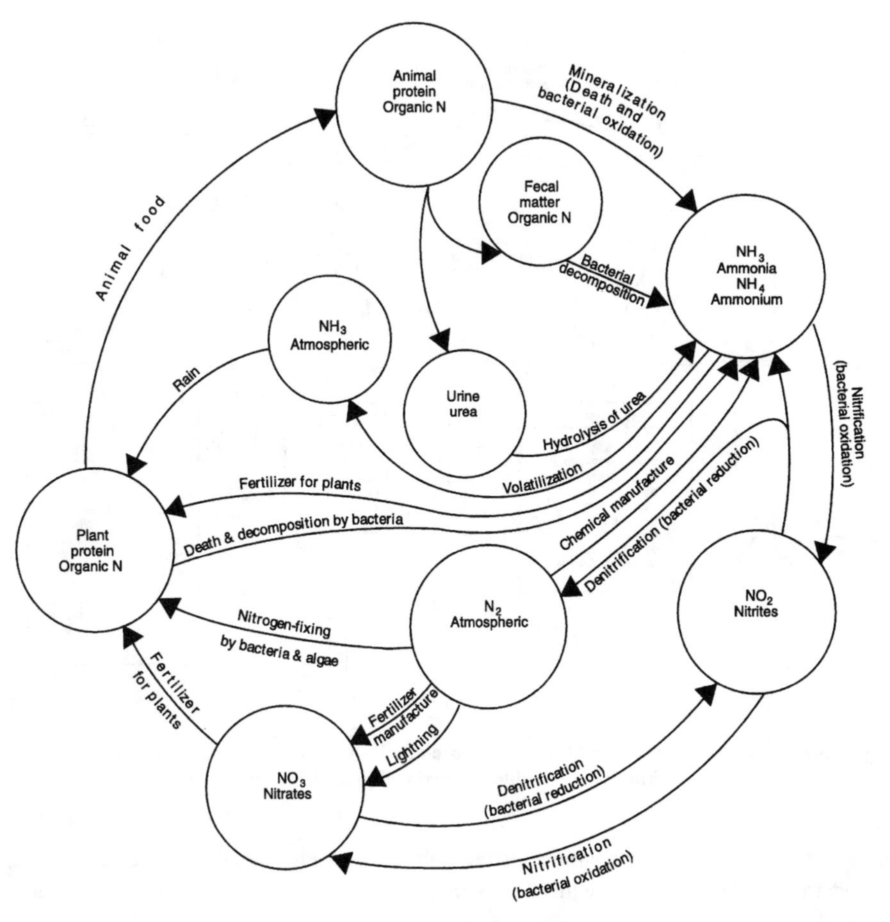

Figure 6.1. The nitrogen cycle.

Ammonium nitrogen is relatively immobile in the soil. The positively charged NH_4 tends to attach to the negatively charged clay particles and generally remains in place until converted to other forms.

Nitrite is normally a transitory phase in the nitrification and denitrification processes. Very little NO_2 is normally detected in the soil or in most natural waters.

The nitrate form of nitrogen is the end product of the nitrification process (the conversion of N from the ammonia form to nitrite and then to nitrate under aerobic conditions). The nitrate form of N is soluble in water and is readily used by plants.

The United States Environmental Protection Agency [2] established a criterion of 10 mg/L of NO_3-N for drinking water because of the health hazard that nitrites present for pregnant women and infants. Unborn babies and infants can contract methemoglobinemia, or "blue-baby syndrome", from ingesting water contaminated with nitrates. In extreme cases, this can be fatal. Blue-baby syndrome generally effects only infants who are less than 6 months old. The disease develops when nitrate is converted to

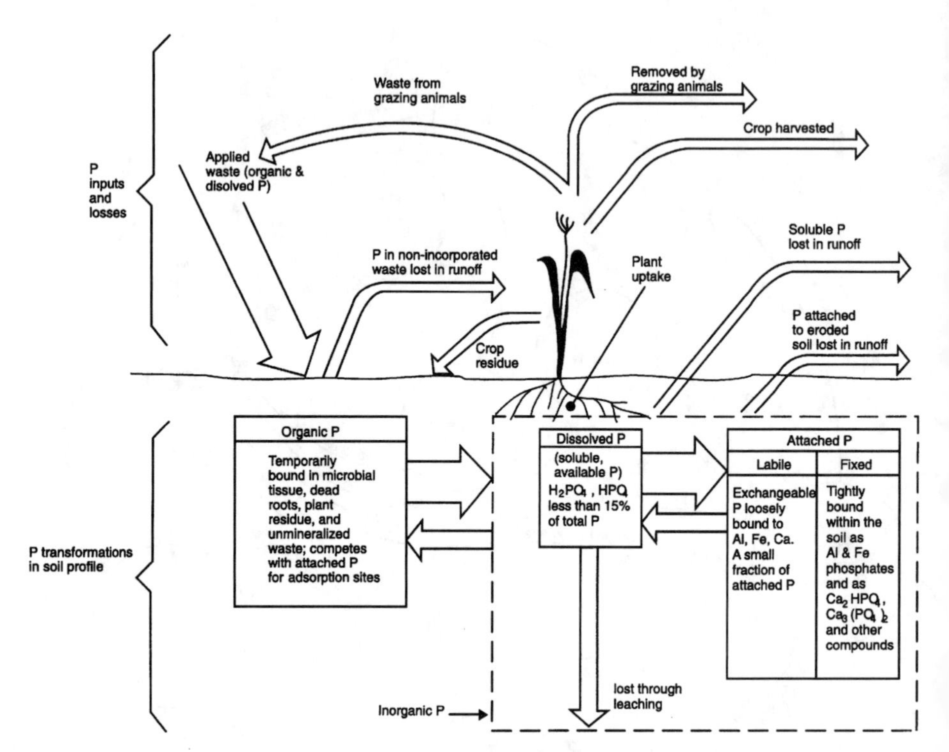

Figure 6.2. Phosphorus inputs and losses at a waste-application site and phosphorus transformation within the soil profile (abbreviated phosphorus cycle).

nitrite in the alkaline environment of the baby's stomach. The nitrite then enters the bloodstream and interacts with the hemoglobin, converting it to methemoglobin [3–5].

Phosphorus is one of the major nutrients needed for plant growth, whether the plant is terrestrial or aquatic. Because phosphorus is used extensively in agriculture, the potential for pollution from this source is high. Water samples are often analyzed for only total phosphorus; however, total phosphorus can include organic, soluble, or "bound" forms. An understanding of the relationship among these forms is important to understanding the extent to which phosphorus can move within the environment and the methods for its control. Figure 6.2 depicts the relationship between the phosphorus forms and illustrates ways that phosphorus can be lost from manure application sites.

Organic phosphorus is a part of all living organisms, including microbial tissue and plant residue, and it is the principal form of P in the metabolic byproducts (manures) of most animals. About 73% of the phosphorus in the fresh manure of various types of livestock is in the organic form.

Soluble phosphorus (also called *available* or *dissolved* P) is the form used by all plants. It is also the form that is subject to leaching. The soluble form generally accounts for less than 15% of the total phosphorus in most soils.

Attached or mineral phosphorus includes those compounds that are formed when the anionic (negatively charged) forms of dissolved P become attached to cations, such as iron, aluminum, and calcium. Attached phosphorus includes labile, or loosely bound, forms and those that are "fixed," or tightly adsorbed, on or within individual soil particles. It should be noted that the P that is loosely bound to the soil particles (labile P) remains in equilibrium with the soluble P. Thus, when the concentration of soluble P is reduced because of the removal by plants, some of the labile P is converted to the soluble form to maintain the equilibrium.

A number of factors determine the extent to which phosphorus moves to surface or ground water. Nearly all of these factors relate to the form and chemical nature of the phosphorus compounds. Some of the principal factors affecting movement of P to surface and ground waters are: degree of contact with the soil, soil pH, soil texture, and amount of manure applied. Other factors are the presence of erosion-control measures, degree of phosphorus retention in soils, and the presence or absence of aerobic conditions.

The Environmental Protection Agency [6] reported the following regarding phosphorus in natural water:

- High phosphorus concentrations are associated with accelerated eutrophication of water, if other growth-promoting factors are present.
- Aquatic plant problems develop in reservoirs and other standing water at phosphorus values lower than those critical in flowing streams.
- Reservoirs and lakes collect phosphates from influent streams and store part of them within consolidated sediment, thus serving as phosphate sinks.
- Phosphorus concentrations critical to noxious plant growth vary, and nuisance growths may result from a particular concentration of phosphate in one geographic area, but not in another.

The excreta from warm-blooded animals have countless microorganisms, including bacteria, viruses, parasites, and fungi. Some of the organisms are pathogenic (disease-causing), and many of the diseases carried by animals are transmittable to humans, and vice versa. Table 6.1 lists some of the diseases and parasites transmittable to humans from animal manure.

One indicator organism used widely to check for the presence of pathogens is a family of bacteria known as the coliforms. The whole group of coliforms is associated with both the feces of warm-blooded animals and with soils. However, the fecal coliform group represents a part of the total coliforms and is easily differentiated from the total coliforms during testing. A positive test for fecal coliform bacteria is a clear indication that pollution from warm-blooded animals exists. A high count indicates a greater probability that pathogenic organisms are present.

6.1.2 Effects of Manure on the Air Resource

Livestock-production facilities can be the source of gases, aerosols, vapors, and dust that, individually or in combination, can create such air quality problems as nuisance odors, health problems for animals in confined housing units, "greenhouse" effect [7], corrosion of materials, and the generation of deadly gases that can affect animals and humans.

Table 6.1. Diseases and organisms spread by animal manure

Disease	Responsible Organism
Bacterial	
Salmonella	*Salmonella* spp.
Eptospirosis	*Leptospiral pomona*
Anthrax	*Bacillus anthracis*
Tuberculosis	*Mycobacterium tuberculosis*
Mycobacterium avium	
Johnes disease, paratuberculosis	*Mycobacterium*
Brucellosis	*Brucella abortus*
Brucella melitensis, brucella suis	Ringworm
Listerosis	*Listeria monocytogenes*
Tetanus	*Clostridium tetani*
Tularemia	*Pasturella tularensis*
Erysipelas	*Erysipelothrix rhusiopathiae*
Colibacilosis	*E. coli* (some serotypes)
Coliform mastitis metritis	*E. coli* (some serotypes)
Rickettsial	
Q fever	*Coxiella burneti*
Viral	
New Castle	Virus
Hog Cholera	Virus
Foot and Mouth	Virus
Psittacosis	Virus
Fungal	
Coccidioidomycosis	*Coccidoides immitus*
Histoplasmosis	*Histoplasma capsulatum*
Various microsporum and trichophyton	
Protozoal	
Coccidiosis	*Eimeria* spp.
Balantidiasis	*Balatidium coli*
Toxoplasmosis	*Toxoplasma* spp.
Parasitic	
Ascariasis	*Ascaris lumbricoides*
Sarcocystiasis	*Sarcocystis* spp.

Different gases are produced as animal manure is degraded by microorganisms. Under aerobic conditions, carbon dioxide is the principal gas produced. Under anaerobic conditions, the primary gases are methane and carbon dioxide. About 60% to 70% of the gas generated in an anaerobic lagoon is methane, and about 30% is carbon dioxide. However, trace amounts of more than 40 other compounds have been identified in the air exposed to degrading animal manure. Some of these include mercaptans (this family of compounds includes the odor generated by skunks), aromatics, sulfides, and various esters, carbonyls, and amines.

Odors are associated with all livestock-production facilities. Animal manure is a common source of significant odors, but other sources, such as poor-quality or spoiled

Table 6.2. Properties and physiological effects of the most important gases produced from manure

Gas	Lighter than Air	Odor	Class	Comments
Ammonia	Yes	Sharp, pungent	Irritant	Irritation of eyes and throat at low concentrations; asphyxiating, could be fatal at high concentrations with 30- to 40-min exposure
Carbon dioxide	No	None	Asphyxiant	$<$ 20,000 mg/L = safe level; increased breathing, drowsiness, and headaches as concentration increases; could be fatal at 300,000 mg/L for 30 min
Hydrogen sulfide	No	Rotten eggs	Poison	Headaches, dizziness at 200 mg/L for 60 min; nausea, excitement, insomnia at 500 mg/L for 30 min; unconsciousness, death at 1000 mg/L
Methane	Yes	None	Asphyxiant, flammable	Headaches at 500,000 mg/L

feed and dead animals, can also be at fault. Freshly voided manure is seldom a cause of objectionable odor, but manure that accumulates or is stored under anaerobic conditions does develop unpleasant odors. Such manures can cause complaints at the production facility when the manure is removed from storage or when it is spread on the fields. Manure-covered animals and ventilation air exhausted from production facilities can also be significant sources of odor. The best insurance against undesirable odor emissions is manure management practices that quickly and thoroughly remove manures from production facilities and place them in treatment or storage facilities or apply them directly to the soil [8].

The gases of most interest and concern in manure management are methane (CH_4), carbon dioxide (CO_2), ammonia (NH_3), and hydrogen sulfide (H_2S). Table 6.2 provides a summary of the most significant characteristics of ammonia, carbon dioxide, hydrogen sulfide, and methane.

Methane

When manure is digested under anaerobic conditions, microbial fermentation produces methane. Methane is flammable, and in recent decades interest in using it as a source of energy on the farm has increased. Because methane is also explosive, extreme care is required when attempting to generate and capture this gas for on-farm use.

The emissions of methane from manure are driven by the quantity of manure produced, how it is handled, and the temperature. Emissions vary from system to system and throughout the year. In the United States, approximately 10% of the anthropogenic emissions of methane come from livestock manure, with swine and dairy accounting for 80% of the emissions [7].

Carbon Dioxide

Carbon dioxide can be an asphyxiant if it displaces normal air in a confined facility. Because CO_2 is heavier than air, it remains in a tank or other well-sealed structure, gradually displacing the lighter gases.

Ammonia

Ammonia is primarily an irritant and has been known to create health problems in animals in confinement buildings. Irritation of the eyes and respiratory tract are common problems from prolonged exposure to this gas. It is also associated with soil-acidification processes.

Hydrogen Sulfide

Hydrogen sulfide is deadly. Humans and farm animals have been killed by this gas after falling into or entering a manure tank or being in a building in which a manure tank was being agitated. Although only small amounts of hydrogen sulfide are produced in a manure tank compared with the other major gases, this gas is heavier than air and becomes more concentrated in the tank over time.

When tanks are agitated in preparation for pump-out, hydrogen sulfide can be released to the area overhead. If a tank is located beneath the animals in a building, forced-air ventilation in the building is imperative before operating the agitation equipment. An exhaust system should also be provided within the tank during agitation and pumping.

Hydrogen sulfide has the distinct odor of rotten eggs. At the first hint of this odor, the area around the tank should be immediately evacuated of all humans. H_2S deadens the olfactory nerves (the sense of smell); therefore, if the smell of rotten eggs appears to have disappeared, this does not indicate that the area is not still contaminated with this highly poisonous gas.

A person should never enter a manure-storage tank even to help rescue someone else who has succumbed to the hydrogen sulfide. Several lives have been lost attempting such rescues. If a tank must be entered, the air in the tank should first be evacuated using a forced-air ventilation system. Self-contained breathing apparatus, safety lines, and sufficient personnel to man the lines are needed in all cases. A mechanical hoisting device is preferable.

6.1.3 Effects of Manure on the Animal Resource

Grazing animals can be adversely affected if animal manure is applied to forage crops at an excessive rate [9]. Studies indicate that grass tetany, fescue toxicity, agalactia, and fat necrosis appear to be associated, in part, with high rates of fertilization from poultry litter on cool-season grasses (especially fescue). Highlights of these disease problems are provided in subsequent sections. Additional details on the clinical signs of these diseases and methods to reverse or prevent their occurrence should be discussed with a veterinarian.

6.1.4 Manure Characteristics

Manure and other livestock and poultry production residue described in this section are of an organic nature and agricultural origin. Some other residues of nonagricultural origin that may be managed within the agricultural sector are not included. Information and data presented can be used for planning and designing manure-management systems and system components and for selecting manure-handling equipment.

In most cases a single value is presented for a specific manure characteristic. This value is presented as a reasonable value for facility design and equipment selection for situations in which site-specific data are not available. Manure characteristics are subject to wide variation; both greater and lesser values than those presented can be expected. Therefore, much attention is given in this section to describing the reasons for data variation and to giving planners and designers a basis for seeking and establishing more appropriate values if justified by the situation [10].

Onsite manure sampling, testing, and data collection are valuable assets in manure-management system planning and design and should be used if possible [11]. Such sampling can result in greater certainty and confidence in the system design and in economic benefit to the owner. However, caution must be exercised to assure that representative data and samples are collected. Characteristics of "as excreted" manure are greatly influenced by the effects of weather, season, species, diet, degree of confinement, and stage of the production/reproduction cycle. Characteristics of stored and treated manures are strongly affected by such actions as sedimentation, floatation, and biological degradation in storage and treatment facilities.

Definitions of Manure Characterization Terms

Table 6.3 gives definitions and descriptions of manure-characterization terms. It includes abbreviations, definitions, units of measurement, methods of measurement, and other considerations for the physical and chemical properties of manure and residue.

Manures are often given descriptive names that reflect their moisture content, such as liquid, slurry, semi-solid and solid. Manures that have a moisture content of 95% or more exhibit qualities very much like water and are called *liquid manure*. Manures that have moisture content of about 75% or less exhibit the properties of a solid and can be stacked and hold a definite angle of repose. They are called *solid manure*. Manures that have between about 75% and 95% moisture content—25% and 5% solids—are *semi-liquid* (*slurry*) or *semi-solid*. Because manures are heterogeneous and inconsistent in their physical properties, the moisture content and ranges indicated must be considered generalizations subject to variation and interpretation.

The terms *manure*, *wastes*, and *livestock* and *poultry production residues* are sometimes used synonymously. In this section manure refers to combinations of feces and urine, and livestock and poultry production residues include manure plus other material, such as bedding, soil, spilled feed, and water that is spilled or used for sanitary and flushing purposes. Small amounts of spilled feed, water, dust, hair, and feathers are unavoidably added to manure and are undetectable in the production facility. These small additions must be considered to be a part of manure and a part of the "as excreted" characteristics presented. Litter is a specific form of poultry manure that results from "floor" production of birds after an initial layer of a bedding material, such as wood shavings, is placed on the floor at the beginning of and perhaps during the production cycle.

Table 6.3. Definitions and descriptions of manure characterization terms

Term	Abbreviation	Units of Measure	Definition	Method of Measurement	Remarks
Weight (mass)	Wt.	kg	Quantity or mass	Scale or balance	
Volume	Vol.	m^3, L	Space occupied in cubic units	Place in or compare to container of known volume; calculate from dimensions of containment facility	
Moisture content	MC	%	That part of a manure material removed by evaporation and oven drying at 217°F (103°C)	Evaporate free water on steam table and dry in oven at 103°C for 24 h or until constant weight	Moisture content (%) plus total solids (%) equals 100%
Total solids	TS	%, % w.b., % d.w.	Residue remaining after water is removed from manure material by evaporation dry matter	Evaporate free water on steam table and dry in oven at 103°C for 24 h or until constant weight	Total of volatile and fixed solids; total of suspended and dissolved solids
Volatile solids	VS, TVS, % d.w.	%, % w.b.	That part of total solids driven off as volatile (combustible) gases when heated to 1112°F (600°C); organic matter	Place total solids residue in furnace at 600°C for at least 1 h	Volatile solids determined from difference of total and fixed solids
Fixed solids	FS, TFS	%, % w.b., % d.w.	That part of total solids remaining after volatile gases are driven off at 1112°F (600°C); ash	Determine weight (mass) of residue after volatile solids have been removed as combustible gases when heated at 600°C for at least 1 h	Fixed solids equal total solids minus volatile solids

Dissolved solids	DS, TDS	% % w.b., % d.w.	That part of total solids passing through the filter in a filtration procedure	Pass a measured quantity of manure material through 0.45 micron filter using appropriate procedure; evaporate filtrate and dry residue to constant weight at 103°C.	Total dissolved solids may be further analyzed for volatile solids and fixed dissolved solids parts
Suspended solids	SS, TSS	%, % w.b., % d.w.	That part of total solids removed by a filtration procedure, dissolved solids	May be determined by difference between total solids and analyzed for volatile and fixed suspended solids parts	Total suspended solids may be further

Note. % w.b., percent measured on a wet basis; % d.b., percent measured on a dry basis.

Because of the high moisture content of as-excreted manure and treated manure, their specific weight is very similar to that of water—1 kg/L. Some manure and residues that have considerable solids content can have a specific weight of as much as 105% that of water. Some dry residues, such as litter, that have significant void space can have specific weight of much less than that of water. Assuming that wet and moist manures weigh 0.96 to 1.04 kg/L is a convenient and useful estimate for planning manure-management systems.

Units of Measure

Manure production from livestock is expressed in kilograms per day per 1000 kg of livestock live weight (kg/d/1000 kg). Volume of manure materials is expressed in cubic meters per day per 1000 kg of live weight (m^3/d/1000 kg).

The concentration of various components in manure is commonly expressed as milligrams per liter (mg/L). One milligram per liter is 1 milligram (weight) in 1 million parts (volume); for example, 1 L. Occasionally, the concentration is expressed in percent. A 1% concentration equals 10,000 mg/L. Very low concentrations are sometimes expressed as micrograms per liter (μg/L). A microgram is one millionth of a gram.

Various solid fractions of manure or residues, expressed in kilograms per day or as a concentration, generally are measured on a wet-weight basis, a percentage of the "as is" or wet weight of the material. In some cases, however, data are recorded on a dry-weight basis, a percentage of the dry weight of the material. The difference between these two values for a specific material is most likely very large. Nutrient and other chemical fractions of a manure material, expressed as a concentration, may be on a wet-weight or dry-weight basis, or expressed as kilograms per 1000 L of manure.

Amounts of the major nutrients, nitrogen, phosphorus, and potassium are always presented in terms of the nutrient itself. Only the nitrogen quantity in the ammonium compound NH_4 is considered when expressed as ammonium nitrogen (NH_4-N). Nutrients are discussed in more detail in the next section.

Characteristics

Whenever locally derived values for manure characteristics are available, this information should be given preference over the more general data used in this section. Carbon-to-nitrogen ratios were established using the percent ash content (dry-weight basis) to determine the carbon.

Daily as-excreted manure-production data are presented if possible in kilograms per day per 1000 kg livestock live weight (kg/d/1000 kg) for typical commercial animals and birds. Units of liters per day per 1000 kg live weight (L/d/1000 kg) allow manure production to be calculated on a volumetric basis. Moisture content and total solids are given as a percentage of the total wet weight of the manure. Total solids are also given in units of kg/d/1000 kg. Other solids data and the nutrient content of the manure are presented in units of kilograms per day per 1000 kg on a wet-weight basis.

As-excreted manure characteristics are the most reliable data available. Manure and manure properties resulting from other situations, such as flushed manure, feedlot manure, and poultry litter, are the result of certain "foreign" materials being added or some manure components being lost from the as-excreted manure. Much of the variation in livestock manure characterization data in this section and in other references results largely from the uncertain and unpredictable additions to and losses from the as-excreted manure.

Livestock manure produced in confinement and semi-confinement facilities are of primary concern and are given the greatest consideration in this section. Manure from unconfined animals and poultry, such as those on pasture or range, are of lesser significance because handling and distribution problems are not commonly encountered.

Foreign materials commonly added to manure in the production facility are bedding (litter), spilled feed and water, flush water, rainfall, and soil. These are often added in sufficient quantities to change the basic physical and chemical characteristics of the manure. Dust, hair, and feathers are also added to manure in limited amounts. Hair and feathers, especially, can cause clogging problems in manure-handling equipment and facilities, although the quantities may be small. Other adulterants are various wood, glass, and plastic items and dead animals and birds.

Spilled feed has a great influence on the organic content of manure and residues. Feed consumed by animals is 50% to 90% digested, but spilled feed is undigested. A kilogram of spilled feed results in as much manure equivalents as 2 to 10 kg of feed consumed. Small quantities, about 3%, of spilled feed are common and very difficult to see. Wastage of 5% is common and can be observed. Obvious feed wastage is indicative of 10% or more being spilled.

Manure characteristics for lactating and dry cows and for heifers are listed in Table 6.4. These data are appropriate for herds of moderate to high milk production. Quantities of dairy manure vary widely from small cows to large cows and between cows at low and high production levels. Dairy feeding systems and equipment often allow considerable

Table 6.4. Dairy manure characterization—as excreted[a]

		Cow		
Component	Units	Lactating	Dry	Heifer
Weight (Mass)	kg/d/1000 kg	80.00	82.00	85.00
Volume	L/d/1000 kg	81	81	81
Moisture	%	87.50	88.40	89.30
TS	% on a wet basis	12.50	11.60	10.70
	kg/d/1000 kg	10.00	9.50	9.14
VS	kg/d/1000 kg	8.50	8.10	7.77
FS	kg/d/1000 kg	1.50	1.40	1.37
COD	kg/d/1000 kg	8.90	8.50	8.30
BOD5	kg/d/1000 kg	1.60	1.20	1.30
N	kg/d/1000 kg	0.45	0.36	0.31
P	kg/d/1000 kg	0.07	0.05	0.04
K	kg/d/1000 kg	0.26	0.23	0.24
TDS		0.85		
C:N ratio		10	13	14

[a] Increase solids and nutrients by 4% for each 1% feed manure more than 5%.

feed spillage, which in most cases is added to the manure. Feed spillage of 10% can result in an additional 40% of total solids in a dairy manure. Dairy-cow stalls are often covered with bedding materials that improve animal comfort and cleanliness. Virtually all of the organic and inorganic bedding materials used for this purpose eventually will be pushed, kicked, and carried from the stalls and added to the manure. The characteristics of these bedding materials are imparted to the manure.

Milking centers—the milk house, milking parlor, and holding area—can produce about 50% of the manure and residues volume, but only about 15% of the total solids in a dairy enterprise. Because this very dilute wastewater has different characteristics than the manure from the cow yard, it is sometimes managed by a different procedure.

About 20 to 40 L of fresh water per day for each cow milked are used in a milking center in which flushing of manures is not practiced. However, if manure flush cleaning and automatic cow washing are used, water use can be 600 L/d per cow or more. Dairies employing flush cleaning systems use water in approximately the following percentages for various cleaning operations.

Parlor—cleanup and sanitation: 10%
Cow washing: 30%
Manure flushing: 50%
Miscellaneous: 10%

Table 6.5 lists characteristics of as-excreted beef manure. Beef manure of primary concern is that from the feedlots. The characteristics of these solid manures vary widely because of such factors as climate, diet, feedlot surface, animal density, and cleaning frequency. The soil in unsurfaced beef feedlots is readily incorporated with the manure because of the animal movement and cleaning operations. Spilled feed is an important

Table 6.5. Beef manure characterization—as excreted[a]

		Feeder, Yearling 340 to 500 kg			
Component	Units	High-Forage Diet	High-Energy Diet	205 to 340 kg	Cow
Weight (mass)	kg/d/1000 kg	59.10	51.20	58.20	63.00
Volume	L/d/1000 kg	59	51	58	62
Moisture	%	88.40	88.40	87.00	88.40
TS	% on a wet	11.60	11.60	13.00	11.60
	kg/d/1000 kg	6.78	5.91	7.54	7.30
VS	kg/d/1000 kg	6.04	5.44	6.41	6.20
FS	kg/d/1000 kg	0.74	0.47	1.13	1.10
COD	kg/d/1000 kg	6.11	5.61	6.00	6.00
BOD5	kg/d/1000 kg	1.36	1.36	1.30	1.20
N	kg/d/1000 kg	0.31	0.30	0.30	0.33
P	kg/d/1000 kg	0.11	0.094	0.10	0.12
K	kg/d/1000 kg	0.24	0.21	0.20	0.26
C:N ratio	kg/d/1000 kg	11	10	12	10

[a] Average daily production for weight range noted. Increase solids and nutrients by 4% for each 1% feed manure more than 5%.

Table 6.6. Beef manure characterization—feedlot manure

Component	Units	Unsurfaced Lot[a]	Surfaced Lot[b] High-Forage Diet	Surfaced Lot[b] High-Energy Diet
Weight (mass)	kg/d/1000 kg	17.50	11.70	5.30
Moisture	%	45.00	53.30	52.10
TS	% on a wet basis	55.00	46.70	47.90
	kg/d/1000 kg	9.60	5.50	2.50
VS	kg/d/1000 kg	4.80	3.85	1.75
FS	kg/d/1000 kg	4.80	1.65	0.75
N	kg/d/1000 kg	0.21		
P	kg/d/1000 kg	0.14		
K	kg/d/1000 kg	0.03		
C:N ratio		13		

[a] Dry climate (annual rainfall less than 60 cm); annual manure removal.
[b] Dry climate; semiannual manure removal.

Table 6.7. Swine manure characterization—as excreted[a]

Component	Units	Grower 18–100 kg	Replacement Gilt	Sow Gestation	Sow Lactation	Boar	Nursing/ Nursery Pig 0–18 kg
Weight	kg/d/1000 kg	63.40	32.80	27.20	60.00	20.50	106.00
Volume	L/d/1000 kg	62	33	27	60	21	106
Moisture	%	90.00	90.00	90.80	90. 00	90.70	90.00
TS	% a wet basis	10.00	10.00	9.20	10. 00	9.30	10.00
	kg/d/1000 kg	6.34	3.28	2.50	6.00	1.90	10.60
VS	kg/d/1000 kg	5.40	2.92	2.13	5.40	1.70	8.80
FS	kg/d/1000 kg	0.94	0.36	0.37	0.60	0.30	1.80
COD	kg/d/1000 kg	6.06	3.12	2.37	5.73	1.37	9.80
BOD5	kg/d/1000 kg	2.08	1.08	0.83	2.00	0.65	3.40
N	kg/d/1000 kg	0.42	0.24	0.19	0.47	0.15	0.60
P	kg/d/1000 kg	0.16	0.08	0.06	0.15	0.05	0.25
K	kg/d/1000 kg	0.22	0.13	0.12	0.30	0.10	0.35
TDS		1.29					
C:N ratio		7	7	6	6	6	8

[a] Average daily production for weight range noted. Increase solids and nutrients by 4% for each 1% feed waste more than 5%.

factor in the characterization of beef manures. Table 6.6 provides some representative values of manure as collected from a feedlot surface.

Swine manure and manure-management systems have been widely studied, and much has been reported on swine-manure properties. Table 6.7 lists characteristics of as-excreted swine manure from feeding and breeding stock. Spilled feed significantly changes manure characteristics. A 10% feed spillage increases manure total solids by 40%. Ration components can make a significant difference in manure characteristics.

Table 6.8. Poultry manure characterization—as excreted[a]

Component	Units	Layer	Pullet	Broiler	Turkey	Duck
Weight (mass)	kg/d/1000 kg	60.50	45.60	80.00	43.60	
Volume	L/d/1000 kg	58	46	79	43	
Moisture	%	75.00	75.00	75.00	75.00	
TS	% on a wet basis	25.00	25.00	25.00	25.00	
	kg/d/1000 kg	15.10	11.40	20.00	10.90	12.0
VS	kg/d/1000 kg	10.80	9.70	15.00	9.70	7.0
FS	kg/d/1000 kg	4.30	1.70	5.00	1.25	5.0
COD	kg/d/1000 kg	13.70	12.20	19.00	12.30	9.5
BOD5	kg/d/1000 kg	3.70	3.30	5.10	3.30	2.5
N	kg/d/1000 kg	0.83	0.62	1.10	0.74	0.7
P	kg/d/1000 kg	0.31	0.24	0.34	0.28	0.3
K	kg/d/1000 kg	0.34	0.26	0.46	0.28	0.5
TDS		2.89				
C:N ratio		7	9	8	7	6

[a] Increase solids and nutrients by 4% for each 1% feed waste more than 5%.

Corn, the principal grain in swine rations, has a high digestibility (90%). Values in Table 6.7 were developed for corn-based rations. If a grain of lower digestibility, such as barley (79%), is substituted for 50% of the corn in the ration, the total solids of the manure increase 41% and the volatile solids increase 43% above those of a ration based on corn. Spilled feed further increases the necessary size of storage units and lagoon facilities needed for manure from rations of lower digestibility.

Because of the high degree of industry integration, standardized rations, and complete confinement, layer and broiler manure characteristics vary less than those of other species. Turkey production is approaching the same status. Table 6.8 presents manure characteristics for as-excreted poultry manure.

Table 6.9 lists data for poultry flocks that use a litter (floor) system. Bedding materials, whether wood, crop, or other residue, are largely organic matter that has little nutrient component. Litter moisture in a well managed house generally is in the range of 25% to 35%. Higher moisture levels in the litter result in greater weight and reduced levels of nitrogen.

6.1.5 Manure-Management Systems

A manure-management system is a planned system in which all necessary components are installed and managed to control and use by-products of agricultural production in a manner that sustains or enhances the quality of air, water, soil, plant, and animal resources. Manure-management systems must be developed using the total systems approach. A total system accounts for all the manure associated with an agricultural enterprise throughout the year from production to utilization. In short, it is the management of all the manure, all the time, all the way.

The primary objective of most agricultural enterprises is the production of marketable goods. To be successful the farm manager must balance limited resources among

Table 6.9. Poultry waste characterization—litter

Component	Units	Layer High-Rise[a]	Broiler	Turkey Breeder[b]	Broiler	Duck[b]
Weight	kg/d/1000 kg	24.00	35.00	24.30		
Moisture	%	50.00	24.00	34.00	34.00	11.20
TS	% w.b	50.00	76.00	66.00	66.00	88.80
	kg/d/1000 kg	12.00	26.50	16.10		
VS	kg/d/1000 kg		21.40			58.60
FS	kg/d/1000 kg		5.10			30.20
N	kg/d/1000 kg	0.425	0.68	0.88	1.06	2.31
NH_4-N	kg/d/1000 kg			0.01		
P	kg/d/1000 kg	0.275	0.34	0.40	1.32	
K	kg/d/1000 kg	0.30	0.40	0.45	1.19	
C:N ratio			9			14

[a] No bedding or litter material added to waste.
[b] All values % on a wet basis.

many complicated and interdependent systems, such as cropping systems, livestock-management systems, irrigation and drainage systems, nutrient-management systems, and pest-control systems. For a manure-management system to be practical, it must interface with these other systems.

Manure Consistency

Manures of different consistencies require different management techniques and handling equipment. Manure may be in the form of a liquid, slurry, semi-solid, or solid. It can change consistency throughout the system or throughout the year. The total solids concentration of manure is the main characteristic that indicates how the material can be handled.

Factors that influence the total solids concentration of excreted manure include the climate, type of animal, amount of water consumed by the animal, and feed type. In most systems the consistency of the manure can be anticipated or determined. The total solids concentration of the manure can be increased by adding bedding to the manure, decreased by adding water, and stabilized by protecting the manure from additional water. Figure 6.3 illustrates how varying the total solids concentration for different animal manures affects consistency.

Manure-Management Functions

A manure-management system consists of six basic functions: production, collection, storage, treatment, transfer, and utilization. For a specific system these functions may be combined, repeated, eliminated, or arranged as necessary.

Production is the function of the amount and nature of manure generated by an agricultural enterprise. The manure requires management if quantities produced are sufficient enough to become a resource concern. A complete analysis of production includes the kind, consistency, volume, location, and timing of manure produced. The manure-management system may need to accommodate seasonal variations in the rate of production.

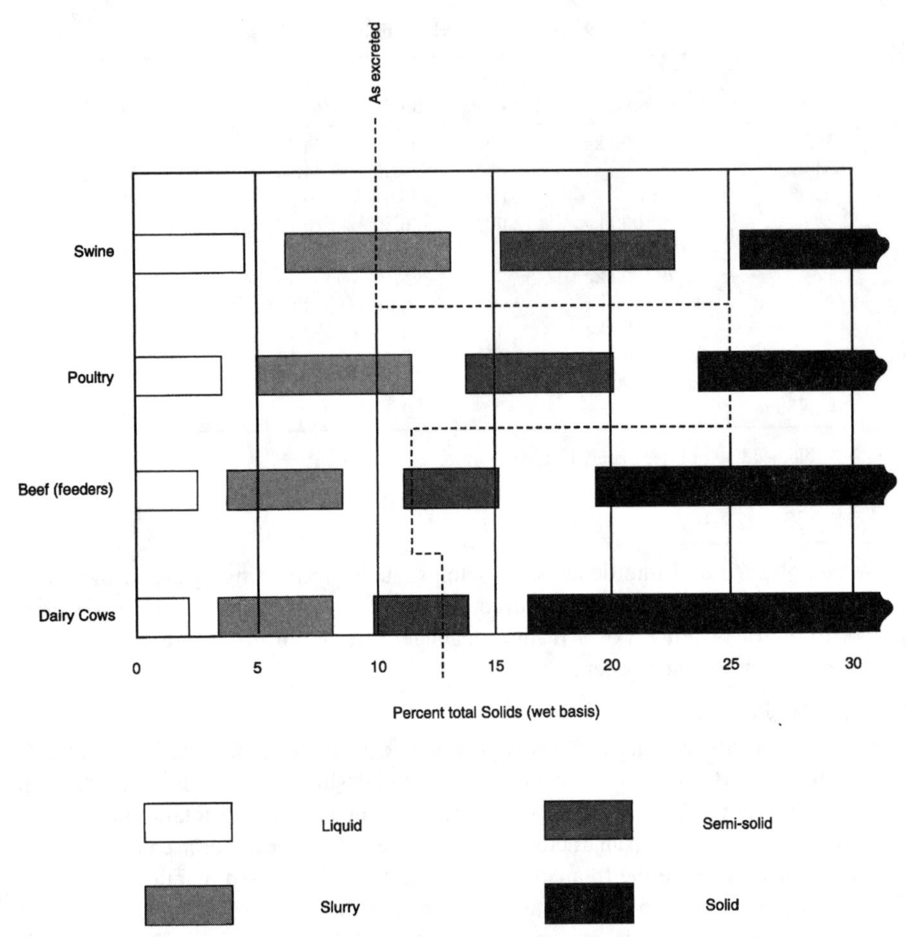

Figure 6.3. Relative handling characteristics of different kinds of manure and percent total solids.

Collection is the initial capture and gathering of the manure from the point of origin or deposition to a collection point. Design considerations include the volume and timing of manure and other livestock residues, the location and availability of power supplies, and the opportunity to use topographic features.

Storage is the temporary containment of the manure. The storage facility of a manure-management system is the tool that gives the manager control over the scheduling and timing of the system functions. For example, with adequate storage the manager has the flexibility to schedule the land application of the manure when the spreading operations do not interfere with other necessary tasks, when weather and field conditions are suitable, and when the nutrients in the manure can best be used by the crop. Design considerations should include providing adequate capacity to contain the daily production, such as manure, bedding, and wastewater, and normal runoff for the storage period, and some volume to capture runoff from rarer storm events. The storage period should be determined by the utilization schedule.

Treatment is any function designed to reduce the pollution potential of the manure, including physical, biological, and chemical treatment. It includes activities that are sometimes considered pretreatment, such as the separation of solids. Design considerations include the "strength" of the manure, its consistency, and system objectives such as the need to reduce odors or the desire to market the final product.

Transfer is the movement and transportation of the manure throughout the system. It includes the transfer of the manure from the collection point to the storage facility, to the treatment facility, and to the utilization site. The manure may require transfer as a solid, liquid, or slurry, depending on the total solids concentration. An appropriate design carefully weighs the economy of various transfer alternatives.

Utilization includes recycling reusable manure products and reintroducing nonreusable manure products into the environment. Manures may be used as a source of energy, bedding, animal feed, mulch, organic matter, or plant nutrients [12]. Properly treated, they can be marketable.

A common practice is to recycle the nutrients in the manure through land application. A complete analysis of utilization through land application includes selecting the fields; scheduling applications; designing the distribution system; selecting necessary equipment; and determining application rates and volumes, value of the recycled products, and installation and management costs associated with the utilization process.

Typical Agricultural Manure-Management Systems

The following discussion describes components of typical manure-management systems. Space does not permit the discussion of design procedures for specific components; however, those details are contained in references such as [1].

Dairy

Dairy operations vary, and each operation presents its own unique problems. Many older dairy operations were not designed with sufficient consideration given to manure management. As a result, the design of a manure-management system may require major modifications or alterations of existing facilities.

The dairy industry generally is concerned with the overall appearance of the dairy farms. Dairy operations require high standards of sanitation and must prevent problems associated with flies. Operations near urban areas must manage the manure in a manner that minimizes odors.

Dairy animals are typically managed on pastures in partial confinement, although the industry in many locations is moving toward total confinement of the milking string. While animals are on pasture, their manure should not be a resource concern if stocking rates are not excessive, grazing is evenly distributed, manure from other sources is not applied, and grazing is not allowed during rainy periods when the soils are saturated [13]. To prevent manure from accumulating in feeding, watering, and shade areas, the feeding facilities can be moved, the number of watering facilities can be increased, and the livestock can be rotated between pastures. To reduce deposition of manure in streambeds, access to the stream may be restricted to stable stream crossings and access points.

The manure in paved holding areas generally is easier to manage, and the areas are easier to keep clean. If the holding areas are unpaved, the traffic of the livestock tends to form a seal on the soil that prevents the downward movement of contaminated water. Care must be taken when removing manure from these lots so that damage to this seal is minimized.

Manure associated with dairy operations includes contaminated runoff, milk-house manure, bedding, and spilled feed. The collection methods for dairy manure vary depending on the management of the dairy operation. Dairy animals may be partly, totally, or seasonally confined. Manure accumulates in confinement areas and in areas in which the dairy animals are concentrated before and after milking.

Unroofed confinement areas must have a system for collecting and confining contaminated runoff. This can be accomplished by using curbs at the edge of the paved lots and reception pits where the runoff exits the lots. Paved lots generally produce more runoff than unpaved lots. On unpaved lots, the runoff may be controlled by diversions, sediment basins, and underground outlets. The volume of runoff can be reduced by limiting the size of the confinement area, and uncontaminated runoff can be diverted if a roof runoff management system and diversions are used.

The manure and associated bedding accumulated in roofed confinement areas can be collected and stored as a solid. The manure can also be collected as a solid in unroofed lots in humid climates in which the manure is removed daily and in unroofed lots in dry climates. Manure can be removed from paved areas by a flushing system. The volume of contaminated water produced by the system can be greatly reduced if provisions are made to recycle the flush water.

Manure, wash water from the milking house, and contaminated runoff should be stored as a liquid in a storage pond or structure. Manure may be stored as a slurry or liquid in a manure storage pond designed for that purpose or in a structural tank. It can be stored as a semi-solid in an unroofed structure that allows for the drainage of excess water and runoff or as a solid in a dry stacking facility. In humid areas the stacking facility should have a roof.

Liquid manure can be treated in an aerobic or anaerobic lagoon, or other suitable liquid-manure treatment facilities. Solids in the manure often are composted or otherwise processed. Often solids are separated from the manure slurries before storage or treatment.

The method used to transfer the manure depends largely on the consistency of the manure. Liquid and slurry manures can be transferred through open channels, pipes, or in a portable liquid tank. Pumps can be used to transfer liquid and sometimes slurry manure as needed. Solid and semi-solid manure can be transferred by mechanical conveyance equipment, in solid manure spreaders, and by pushing it down curbed concrete alleys. Semi-solid manure has been transferred in large pipes through the use of gravity, piston pumps, or air pressure.

Dairy manure is used as bedding for livestock, marketed as compost, and used as an energy source, but the most common form of utilization is through land application. Manure may be hauled and distributed over the land in a dry or liquid manure spreader. Liquid manure can be distributed through an irrigation system. Slurries may be distributed through an irrigation system equipped with nozzles that have large openings.

Beef

Beef brood cows and calves less than 1 year old are usually held on pastures or range. The calves are then finished in confined feeding facilities. While the animals are on pastures, their manure should not become a resource concern if the stocking rates are not excessive and the grazing is evenly distributed [13]. To prevent manure from accumulating in feeding, watering, and shade areas, the feeding facilities can be moved, the number of watering facilities can be increased, and the livestock can be rotated between pastures. To reduce deposition of manure in streambeds, access to the stream may be restricted to stable stream crossings and access points.

Wastes associated with confined beef operations includes manure, bedding, and contaminated runoff.

Beef cattle can be confined on unpaved, partly paved, or totally paved lots. If the cattle are concentrated near wells, adequate protection must be provided to prevent well contamination. Because much of the manure is deposited around watering and feeding facilities, paving these areas, which allows frequent scraping, may be desirable.

On unpaved lots, the traffic of the beef tends to form a seal on the soil that prevents the downward movement of contaminated water. Care must be taken when removing manure from these lots so that damage to this seal is minimized. The seal tends to break down after livestock are removed from the lot. To prevent possible contamination of ground water resources, all the manure should be removed from an abandoned lot.

Unroofed confinement areas must have a system for collecting and confining contaminated runoff. On unpaved lots the runoff can be controlled by using diversions, sediment basins, and underground outlets. Paved lots generally produce more runoff than unpaved lots, but curbs at the edge of the lots and reception pits where the runoff exits the lots help to control the runoff. Solid/liquid separators or settling basins can be used to recover some of the solids in the runoff. The volume of runoff can be reduced by limiting the size of the confinement area, and uncontaminated runoff can be excluded by use of diversions.

The manure in confinement areas that have a roof can be collected and stored as a solid. It may also be collected as a solid or semi-solid from open lots where the manure is removed daily and from open lots in a dry climate.

Manure can be stored as a bedded pack in the confinement area if bedding is added in sufficient quantities. Manure removed from the confinement area can be stored as a liquid or slurry in an earthen pond or a structural tank, as a semi-solid in an unroofed structure that allows drainage of excess water and runoff to a manure-storage pond, or as a solid in a dry stacking facility designed for storage. In areas of high precipitation, dry stacking facilities should be roofed. Contaminated runoff must be stored as a liquid in a manure-storage pond or structure. All earthen storage and treatment facilities are subject to seepage if not properly constructed and maintained [14].

Treatment of the manure in a lagoon is difficult for some livestock systems because of the volume of solids in the manure, but many of the solids can be removed before treatment. Liquid manure may be treated in an aerobic lagoon, an anaerobic lagoon, or other suitable liquid-manure treatment facilities. Solid manure can be composted. Constructed wetlands are another option for manure effluent treatment. Wetlands reduce the biochemical oxygen content of the effluent and other constituents such as nitrogen compounds [15].

The method used to transfer the manure depends largely on the consistency of the manure. Liquid manure and slurries can be transferred through open channels or pipes or in a portable liquid tank. Pumps can be used as needed. Solids and semi-solids may be transferred by using mechanical conveyance equipment, by pushing the manure down curbed concrete alleys, and by transporting the manure in solid manure spreaders. Piston pumps or air pressure can be used to transfer semi-solid manure through large pipes.

Beef-cattle manure can be used as bedding for livestock or as an energy source, or it can be marketed as compost, but the most common form of utilization is land application. The manure can be hauled and distributed over the land in appropriate spreading devices. Liquid manure can be distributed through an irrigation system, and slurries can be applied using irrigation equipment with nozzles that have large openings.

Swine

Open systems (pastures, woodlots, and wetlands), feedlot systems, confinement systems, or a combination of these are used for raising swine. Raising hogs in an open system may have a low initial investment but often results in animal-health and pollution-control problems. Even if sufficient land is available, hogs tend to congregate and concentrate their manure. Total-confinement systems eliminate the need to manage contaminated runoff and may allow for more automation in manure management.

Undesirable odors often are associated with swine operations. A swine manure-management system should incorporate odor-control measures if possible. A clean, neat appearance, an efficient management system, and positive public relations with those affected by the odors eliminates many complaints [8].

Waste associated with swine operations includes manure and possibly contaminated runoff. In some systems provisions must be made to manage flush water. Hogs tend to play with watering and feeding facilities, which can add to the manure load. The disposal of dead pigs may be a resource concern in some operations.

Swine manure can be collected by scraping or flushing. Scraped manure is collected as a solid or slurry; flushed manure must be handled as a liquid. The flush water should be recycled if possible so that the volume of contaminated water is kept to a minimum. The collection process can use automated equipment, or it can be as simple as raising swine on slatted floors over manure storage pits.

Swine manure can be stored as a solid, slurry, or liquid. If stored as a solid, it should be protected from precipitation. Above- or below-ground tanks or an earthen manure-storage pond can be used to store slurries or liquid manure.

Liquid manure from a swine operation is commonly treated in an anaerobic lagoon, but it can also be treated in an aerobic lagoon or oxidation ditch; however, these alternatives normally have a prohibitive cost. Solid manure and dead pigs can be composted.

The method used to transfer the manure depends largely on the consistency of the manure. Liquid manure and slurries may be transferred through open channels, pipes, or in a portable liquid tank. Pumps can transfer liquid manure as needed. Solids and semi-solids can be transferred by mechanical conveyance equipment. Piston pumps or air pressure can be used to transfer semi-solid manure through smooth pipes.

Swine manure is used as a feed supplement and an energy source through methane production. With proper ventilation and sufficient bedding, the solid manure can be composted in confinement facilities, and the heat generated from the composting process can be used to supplement heat in the buildings.

The most common use of the nutrients in swine manure is through land application. The manure can be hauled and distributed over the land by spreading devices. If odors are a problem, liquid manure can be injected below the soil surface. It can also be distributed through an irrigation system. Slurries can be distributed through an irrigation system equipped with nozzles that have large openings.

Poultry

The two basic poultry-confinement facilities are those used to raise turkeys and broilers used for meat and those used to house layers. Broilers and young turkeys are grown on floors on beds of litter shavings, sawdust, or peanut hulls. Layers are confined to cages. Fly control around layers is important to prevent spotting of the eggs. Disease control is important in both systems.

Waste associated with poultry operations include manure and dead poultry. Depending upon the system, waste can also include litter, wash–flush water, and spilled feed.

The manure from broiler and turkey operations is allowed to accumulate on the floor, where it is mixed with the litter. Near watering facilities the manure–litter pack forms a "cake" that generally is removed between flocks. The rest of the litter pack generally has low moisture content and is removed once a year. The litter pack can be removed more frequently to prevent disease transfer between flocks.

In layer houses, the manure that drops below the cage collects in deep stacks or is removed frequently using either a shallow pit located beneath the cages for flushing or scraping or belt scrapers positioned directly beneath the cages.

Litter from broiler and turkey operations is stored on the floor of the housing facility. When it is removed, it can be transported directly to the field for land application. If field conditions are not suitable or spreading is delayed for other reasons, the litter must be stored outside the housing facility. In some areas the litter may be compacted in a pile and stored in the open for a limited time; however, it generally is better to cover the manure with a plastic or other waterproof cover until the litter can be used. If the spreading is to be delayed for an extended period of time, the litter should be stored in a roofed facility.

If the manure from layer operations is kept reasonably dry, it can be stored in a roofed facility. If it is wet, it should be stored in a structural tank or an earthen storage pond.

Broiler and turkey litter can be composted. This stabilizes the litter into a relatively odorless mass that is easier to market and also helps to kill disease organisms so that the litter can be reused as bedding or supplemental feed to livestock. The litter can also be dried and burned directly as a fuel.

Liquid manure may be placed into an anaerobic digester to produce methane gas, or it can be treated in a lagoon. The high volatile solid content of the layer manure may require an anaerobic lagoon of considerable size. If odors are a problem, aeration is an alternative, albeit costly.

The method used to transfer the manure depends on the total solids content of the manure. Liquid manure can be transferred in pipes, gutters, or tank wagons, and dried litter can be scraped, loaded, and hauled as a solid. If the distances between the poultry houses and the fields for application are great, the litter may require transportation in a truck.

The manure from poultry facilities is often applied to the land. If the owners of a poultry house do not have enough land suitable for application, they can often arrange to apply the manure to a neighbor's land. Because of the high nutrient value of the litter, many landowners are willing to pay for the litter to be spread on their land. Whether on the owner's land or the neighbor's land, the manure must be spread for proper utilization. Poultry manure can also be used for the production of methane gas, burned directly as a fuel, reused as bedding, or used as a feed supplement to livestock [16].

Because of the large numbers of dead birds associated with large poultry operations, the disposal of dead birds is a resource concern. Poultry facilities must have adequate means for disposal of dead birds in a sanitary manner. To prevent spread of disease, the dead birds often are collected daily by hand. Disposal alternatives include incineration, rendering, burial, dropping into a buried disposal tank, or composting, in which the dead birds are mixed with litter and straw and composted, and the composted material is stored until it can be applied to the land.

References

1. United States Department of Agriculture, Natural Resources Conservation Service. 1992. Agricultural Waste Management Field Handbook. Washington, DC.
2. United States Environmental Protection Agency. 1986. Quality criteria for water. EPA 440/5-86-001. Washington, DC.
3. Buchholz, D., DeFelice, K., and Wollenhaupt, N. 1989. Agricultural sources of contaminants in ground water. University of Missouri, Columbia, MO.
4. Daniel, T. C., Edwards, D. R., Wolf, D. C., Steele, K. C., and Chapman, S. L. Undated. Nitrate in Arkansas groundwater. Cooperative Extension Service, University of Arkansas, Fayetteville, AR.
5. Dixon, M. L., Tyson, A., and Segars, B. 1992. Your drinking water: Nitrates. University of Georgia, Athens, GA.
6. United States Environmental Protection Agency. 1984. Technical support manual: Waterbody surveys and assessments for conducting use attainment. Washington, DC.

7. United States Environmental Protection Agency. 1993. Anthropogenic methane emissions in the United States; Estimates for 1990. Report to Congress, Office of Air and Radiation, Washington, DC.
8. Rieck, A., Carman, D., VanDevender, K., and Lansston, J. Undated. Managing a livestock operation to minimize odors. Cooperative Extension Service, University of Arkansas, Fayetteville, AR.
9. Davis, G. V. Jr., Hankins, B. J., Chapman, S. L., and Williams, R. 1989. Nitrate poisoning in cattle. Cooperative Extension Service, Fayetteville, AR.
10. Barth, C. L. 1985. Livestock manure characterization: A new approach. In *Agricultural Manure Utilization and Management, Proceedings of the Fifth International Symposium on Agricultural Manures*. St. Joseph, MI: ASAE.
11. Rieck, A., VanDevender, K., and Langston, J. Undated. Liquid animal waste sampling. Cooperative Extension Service, University of Arkansas, Fayetteville, AR.
12. Westerman, P. W., Safley, L. M. Jr., Barker, J. C., and Chescheir, G. M. III. 1985. Available nutrients in livestock manure. In *Agricultural Manure Utilization and Management, Proceedings of the Fifth International Symposium on Agricultural Manures*. St. Joseph, MI: ASAE.
13. Welch, T. G., Knight, R. W., Caudle, D., Garza, A., and Sweeten, J. M. 1991. Impact of grazing management on nonpoint source pollution. Texas Agricultural Extension Service, College Station, TX.
14. Brichford, S., Janssen, C., Yahner, J., and Eigel, J. 1995. Livestock waste storage: Indiana farmstead assessment for drinking water protection. Cooperative Extension Service, Purdue University, W. Lafayette, IN.
15. Riech, A., Langston, J., and VanDevender, K. Undated. Constructed wetlands: An approach for animal waste management. Cooperative Extension Service, University of Arkansas, Fayetteville, AR.
16. Stephenson, A. H., McCaskey, T. A., and Ruffin, B. G. 1989. Treatments to improve the feed nutrient value of deep stacked broiler litter. *Journal of Dairy Science* 67(Suppl. 1):441.

6.2 Recycling of Organic Matter

Jean Claude Souty

6.2.1 Land Application

Environmental Challenges

For a very long period, organic matter coming from livestock effluents, either mixed or not with other materials such as straw, was considered only as an interesting fertilizer for cultivated soils. However, in developed countries, intensification of most animal production, combined with new housing systems using no litter has contributed for a number of years—often in addition to other human activities—to the pollution of underground and subsurface waters. It is well known nowadays that too high a content of nitrogen in drinking water may be a danger to human health, especially where babies are concerned. An excess of phosphorus in subsurface water may also lead to eutrophication and thus to the destruction of many fish populations.

To those drawbacks of the overuse of organic matters must be added those more recently put into light in relation to new methods of animal housing and manure spreading: air pollution by ammonial and other volatile emissions, and conflicts between farmers and other citizens, due to odors.

In many countries, authorities have been led to enforce various types of regulations, in some cases very sophisticated, to eliminate or at least reduce these hazards and drawbacks. Following such regulations, the farmers have to pay more and more attention to the environmental impacts of their activities and to improving their practices for a more sustainable agriculture. The main points to be considered in this matter are storage capacities for animal waste, distances from water sources and dwellings, at appropriate fertilization allowing a sufficient supply of nutrients (but not more) applied only at relevant periods for crops.

Different Types of Animal Wastes

Regarding land spreading there are four main types of wastes:

- Solid manure, which is generally a mixture of dung with enough straw for being stored in dung-hill on stockyard
- Liquid manure, which is either a mixture of dung, urine and water or a slurry resulting from draining solid manure from the stockyard
- Sewage from the milking parlor, which includes washing water for milking machines and exersize areas inside the parlor; these types of effluents, which represent often a large percentage of wastes in dairy farms, can either be stored with liquid manure—but in this case storage capacities are more or less greatly increased—or treated separately
- Seepage from silos, which is only to be considered for silage with less than 25% of dry matter (grass silage mainly); such seepage is never in great quantities but is very pollutant and cannot be allowed to run out directly from the silos to the fields or the surface water; it has therefore to be spread with liquid manure, but it can be stored in the same tank or separately

Storage Capacities

One of the most important points is to have enough storage capacity for livestock wastes in order to avoid spreading when periods are not favorable for crop growth. These capacities have to be determined considering total wastes coming from animals, from equipment such as milking parlors, and from various other origins (silage, rainfall on uncovered exersize areas, or washing water). Of course, sources of uncontrolled quantities of water (drinking bowls, roofs without eaves) must be eliminated.

These quantities have to be established for the duration required by national regulation or after analysis of the agronomic and climatic context. In France, the minimum duration of storage has been determined by national authorities to be 4 months, but local authorities are allowed to increase this minimum, often to 6 months or more and sometimes to 9 months or even 1 year, as in northern countries of the European Union.

Basic data concerning quantities of animal wastes production—as excreted—are given in Tables 6.4 to 6.9. For calculation of storage capacities it is necessary to take into

Table 6.10. Examples of liquid manure storage capacities needed for 4 months

Animal	Housing	Capacity (m^3)
Dairy cattle	Loose housing	
(7000 L of milk per year)	Cubicles entirely covered	7.5
	Cubicles partially covered	
	Silage distributed	7.8[a]–9[b].
	Silage in self-feeding	8.4[a]–10.8[b]
	With strawyard	
	concrete exercise yard covered	5.4
	concrete exercice yard uncovered:	
	Silage distributed	5.6[a]–6.4[b]
	Silage in self-feeding	6.2[a]–8.8[b]
	Cowsheds without straw	7.2
Beef cattle	On slatted floor	3.2
	With strawyard and concrete	
	exercice yard covered	2.1
Pigs	Suckling sow (per place)	2.4
	Fattening pigs on slatted floor (per place)	
	Fed with lactoserum	1.20
	Fed with concentrates and drinking bowl	0.80
	Soup feeder	0.48
	Piglets	0.32

[a] Rainfall: 60 mm/month
[b] Rainfall: 120 mm/month

account a number of parameters concerning the type of housing, husbandry, rainfall, and so forth. Such data are available in reference [1], and Table 6.10 gives some examples of such capacities.

Fertilization Balance

Sufficient storage capacities are a necessary condition to avoid non–point source pollution (as well as direct pollution resulting from overflowing manure). But it is also necessary to match the quantities of manure—especially nitrogen and phosphorus—that are to be spread and the capacity of the soils and crops to utilize and transform these elements.

If this capacity is not sufficient, animal wastes must be either spread on the land of other farms or treated in order to reduce their nutrient contents. The first solution is not always feasible, and treatments of animal wastes are generally very expensive. If there are mainly pigs or poultry on the farm, a third solution may be considered, which consists of lowering the amount of nitrogen produced by the animals themselves through altering the protein content of the feed during animal growth.

In any case, it is necessary to know as precisely as possible the quantities of nutrients that are to be spread and those likely to be absorbed by cultivated soils. The nutrient that needs special attention regarding pollution hazards is nitrogen; hence it is strongly recommended to carry out at least one nitrogen balance. However, more and more often attention must also be paid to phosphorus.

Table 6.11. Nitrogen production per year by livestock

Type of Animal	kg of Nitrogen Produced
Cattle	
Dairy cow	73
Suckling cow	51.1
Heifer (<1 year)	21.9
Heifer (1–2 years)	43.8
Heifer (>2 years)	58.4
Beef cattle (<1 year)	21.9
Beef cattle (1–2 years)	43.8
Beef cattle (>2 years)	51.1
Veal calf	2.1
Pigs	
Sow or boar	17.5
Piglet (after weaning)	0.7
Poultry	
Laying hen	0.5
Point of lay pullet	0.125
Broiler	0.04
Turkey (female)	1
Future reproducing turkey	0.25
Duck	0.069
Sheep	
Adult ewes (>1 year)	10
Lambs (6 months)	3
Ewe lamb	5
Rabbits	
Female in cage	4.5
Rabbit produced	0.084

Source: [2, 3].

Data concerning animal-wastes nitrogen contents are given in Table 6.11. These data allow an estimation of the quantities of nitrogen present in farm wastes. Some examples of plant requirements are given in Table 6.12.

Calculations on such bases allow one to determine if nitrogen is structurally in excess or not for the total surface of the farm likely to receive livestock effluents. For a first and more simple approach to this issue, in France we use the relevant European regulation that requires one not to exceed 210 kg/ha of nitrogen coming from animal wastes.

Conditions for Spreading

Independent of crop types and their yields, two sorts of parameters have to be considered regarding possibilities of spreading manure without risking pollution water directly and making neighbors uneasy. These parameters may reduce, sometimes strongly, the total spreadable area: distances from dwelling houses and water surfaces or water catchments (Table 6.13 gives examples of what is required in France, but distances are often some what different in other countries), and slopes, which must not be too strong.

Table 6.12. Nutrient composition of crops

Crop		N	P_2O_5	K_2O
Kg/9 harvested				
Soft corn		1.9	0.9	0.7
Oat		1.9	0.8	0.7
Barley	GRAIN	1.5	0.8	0.7
Rye		1.4	1.0	0.6
Maize		1.5	0.7	0.5
Colza		3.5	1.4	1.0
Sunflower		1.9	1.5	2.3
Soft corn		2.5	1.1	1.7
Oat		2.5	1.1	1.9
Barley	GRAIN	2.1	1.0	1.9
Rye	+	2.0	1.3	1.8
Maize	STRAW	2.2	0.9	2.3
Colza		7.0	2.5	10.0
Sunflower		3.7	2.5	10.0
Potato		0.35	0.17	0.65
Kg/T of DM		N	P_2O_5	K_2O
Maize silage		12.5	5.5	12.5
Grass silage		20.0	6.0	25

Table 6.13. Minimum distances for spreading liquid manure (French regulation)

	Water Captage for Human Consumption	Beach	Fish Breeding
100*	50	200	500

* 50 m if odour reduction by treatment and 10 m if immediate burying by injection

In any case liquid manure may have to be spread on soils covered with snow or when it rains strongly. More generally manure spreading has to be done when organic nitrogen may be well mineralized and when crops are able to absorb quickly this nitrogen. So good decisions have to be taken in considering climate, crops, and soils and following agronomist advices. On the other hand, spreading manure during periods in which tourists are numerous in these areas must be avoided.

Equipment for Spreading

There are three main sorts of equipment for spreading animal wastes: spreaders, vacuum tankers, and irrigation systems. They can be used according the dry-matter content of the product to be spread.

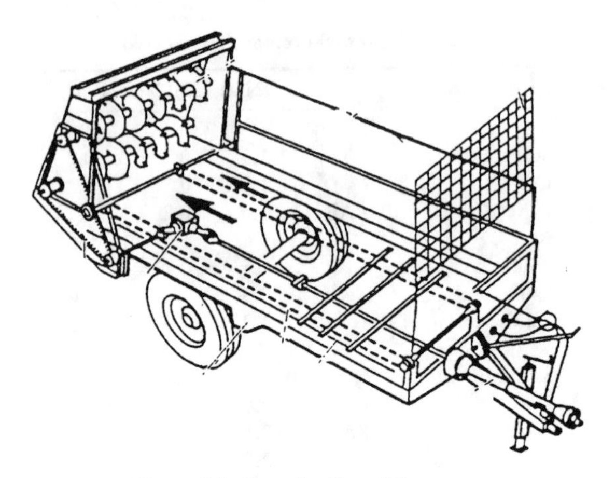

Figure 6.4. Spreader for solid manure.

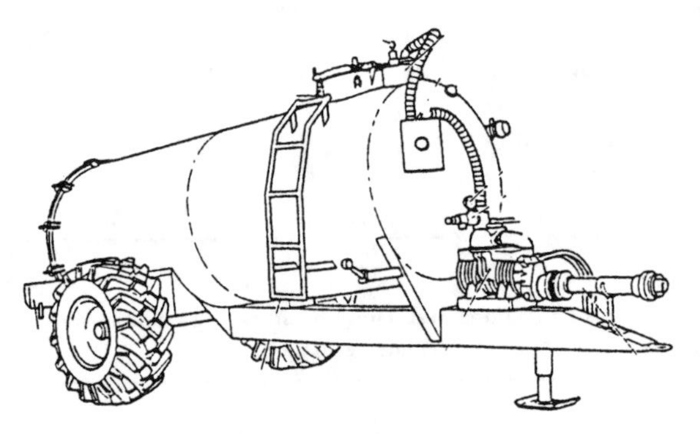

Figure 6.5. Vacuum tanker.

Spreaders

Only solid manure can be spread by this type of equipment (Fig. 6.4), because the percentage of dry matter must be at least 13%. Between 8% and 13% we find manure with small quantities of straw, which requires same type of equipment but equipped with a watertight tank.

Vacuum Tankers

This equipment (Fig. 6.5) is used for spreading slurry or liquid manure with no more than 8% dry matter. Such tankers may be equipped with systems allowing one to bury liquid manure in the soil; the advantages of burying are both avoiding losses of NH_3 in the air and increasing spreadable land, because there are no odors while spreading and so it may be done nearer to dwelling areas.

Irrigation Systems

If it is very diluted (less than 3%), liquid manure may be spread with equipment similar to those used for irrigation, in order to avoid evaporation.

Table 6.14. Estimated manure production rates and biogas generation potential from animal wastes

	Dairy Cattle	Beef Cattle	Swine	Poultry
Manure production* (kg/d)	39	26	23	27
Total solids (kg/d)	4.8	3.4	3.3	7.9
Volatile solids (kg/d)	3.9	2.7	2.7	5.8
Digester efficiency	35	50	55	65
COD/VS ratio	1.05	1.12	1.19	1.28
Biogas production L/453 kg animal/d[a]	860	870	1020	2860

* Estimated output in kg per 453 kg live weight.
[a] Based on theoretical gas production rate of 831 L/kg (VS) destroyed.

6.2.2 Energy Production

Biogas Production

Principles

Biogas is produced by various bacteria contained in animal wastes in anaerobic conditions and temperatures preferably above 35°C. Its composition ranges from 55% to 70% CH_4, the other gas being mainly CO_2. Table 6.14 gives indications about biogas production according to manure production rates for different types of animals [4].

Economical and Technical Conditions

Methane cannot be stored on the farm under high pressure, the cost of such storage being very high, so it has to be used very soon—practically no more than 2 days after having been produced. Biogas can be used in these conditions directly for heat-using purposes, preferably ones constant through the year—cooking or lighting, for instance. In developing countries (India, China), several hundred thousands of rural reactors, for instance are functioning for these purposes. But increasing use of electricity and industrial gases will probably contribute to making them disappear more quickly.

In developed countries the use of biogas is limited to some big units (mainly piggeries) in which there are needs for heating all year round, or to situations in which heat from biogas may be economically transformed to electricity. Figure 6.6 shows an example of a plant constructed in Denmark [4].

Heat from Poultry Manure

Some big plants for incineration of solid manure originating from broilers units are now functioning in Europe. Heat produced by this method is converted into electricity. The advantages of this are both to allow no spreading of effluents and to use renewable energy. For instance, a plant treating 200,000 tons of manure per year, corresponds with 1.3 million square meters of poultry, may produce 100 GWh/y.

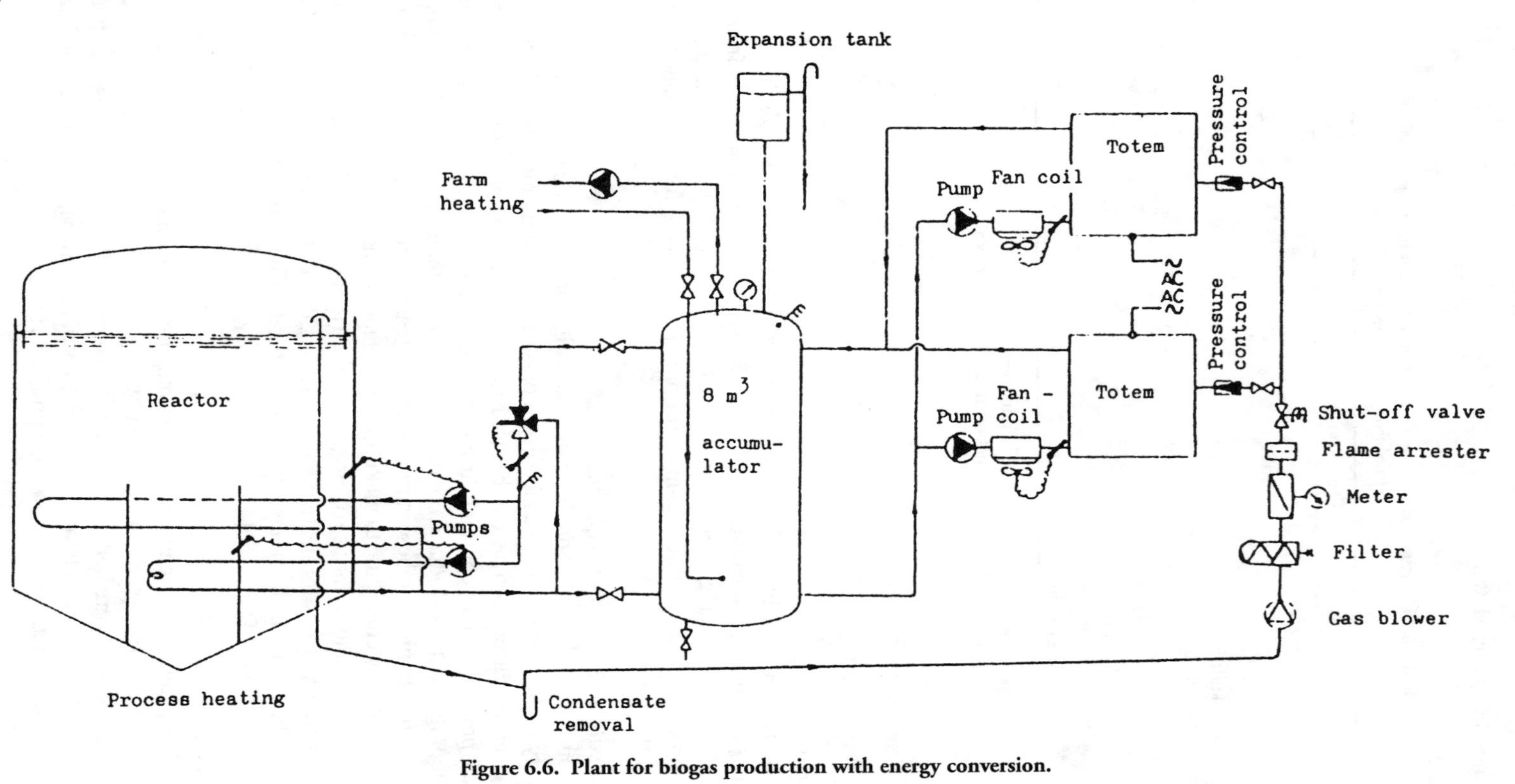

Figure 6.6. Plant for biogas production with energy conversion.

References

1. Bâtiments d'élevage bovin, porcin et avicole. Réglementation et préconisations relatives à l'environnement, 1996. 140 p. PARIS: Ministère de l'agriculture de la pêche et de l'alimentation.
2. CORPEN, 1996.-Estimation des rejets d'azote et de phosphore des élevages de porcs. Impact des modifications de conduite alimentaire et des performances techniques, 24 p. Paris : Ministère de l'agriculture de la pêche et de l'alimentatio, Ministère de l'environnement.
3. CORPEN, 1988 - Bilan de l'azote à l'exploitation, 35 p. Ministère de l'agriculture et de la pêche, Ministère de l'environnement.
4. STAFFORD, D. A., WHEATLEY, B. I., HUGHES, D. E., 1980.-Anaerobic digestion, 528 p.- Londres: applied science publishers ltd.

Bibliography

AFGR, 1993. Bâtiments d'élevage et environnement. Enjeux et perspectives. Actes du colloque, Nantes, 140 p.

Biskupek, B. Dohler, H. 1993. Environmentally acceptable manure utilization. KTBL seminar, Waltersdorf- Thüringen, 1992/11/26-27. KTBL Arbeitspapier, n° 182, 153 p.

Chevalier, D. Wiart, J., 1992. Les matériels d'épandage (valorisation agricole des boues d'épuration et autres engrais de ferme: fumiers,lisiers, fientes) 50 p. Paris: Agence de l'environnement et de la maîtrise de l'énergie.

Coillard, J., 1997. Procédés de traitement des lisiers de porc étudiés en France. Principales techniques adaptées à la gestion des lisiers en zone d'excédent, Ingénieries (10):17–33.

Coillard, J., Pradal, G., Buatier, C., 1995. Le traitement à la ferme des lisiers de porc. Etude d'un procédé extensif de type "lagunage naturel". CEMAGREF, TECHNI-PORC (1):17 p.

Corpen, 1988. Bilan de l'azote à l'exploitation, 35 p. Ministère de l'agriculture et de la pêche, Ministère de l'environnement.

Costs of agricultural production of biogas. 1992. KTBL symposium, Darmstadt, 1992/11/25-26. KTBL Arbeitspapier, N° 185, 138 p.

Dewi, I., Axford, R. F. E., Marai, I. F. M., Omed, H. M., 1993. Pollution in livestock production systems]. International symposium, University College of Wales, Bangor, 1992/09, 463 p. Wallingford, UK: CAB International.

Galanos, E. Gray, K. R. Biddlestone, A. J. Thayanithy, K. 1995. The aerobic treatment of silage effluent: effluent characterization and fermentation. J. Agric. Engng. Research 62(4):271–279.

Jones, D. I. H. Jones, R. 1995. The effect of crop characteristics and ensiling methodology on grass silage effluent production. J. Agric. Engng. Research 60, (2):73–81.

Jones, D. I. H. Jones, R. 1996. The effect of in-silo effluent absorbents on effluent production and silage quality]. J. Agric. Engng. Research 64(3):173–186.

Maîtrise et prévention des pollutions dues aux élevages, 1994-Colloque SIMA, Villepinte, 16 février 1994, 145 p. Antony: Cemagref.

Martinez, J. 1997. Solepur: A soil treatment process for pig slurry with subsequent denitrification of drainage water. J. Agric. Engng. Research 66(1):51–62.

Michel-Combe, D., Loudec, J. L., Merillot, J. M., 1995. Le lisier de porc. Les méthodes et les finalités du traitement à la ferme, 48 p. Paris : Ministère de l'Agriculture et de la Pêche.

Nitrogen flow in pig production and environnemental consequences. Symposium proceedings, Wageningen, 1993/06/8-11. Wageningen: PUDOC.

Stafford, D. A., Wheatley, B. I., Hughes, D. E., 1980. Anaerobic digestion, 528 p. Londres: Applied Science Publishers Ltd.

Taiganides, E. P., 1992. Pig waste management and recycling: the Singapore experience, 368 p. Ottawa, Canada: International Development Research Centre.

Texier Cl., 1997. Elevage porcin et respect de l'environnement, 110 p. Paris: Institut Technique du porc.

Tillie, M., Billon, P., Houdoy, D., Fevrier, D., 1995. Elevage bovin et environnement: Prévenir les risques de nuisance et de pollution, 103 p. Paris: Ministère de l'Environnement, Institut de l'Élevage.

Tillie, M., Capdeville, J., 1992. Etude sur les déjections bovines: détermination de stockage et composition des déjections bovines à la ferme. 84 p. PARIS: Institut de l'Élevage.

Tveitnes, S. 1993. Animal wastes, 119 p. As, Norvège: Statens Fagteneste for Landbruket.

University of Illinois, Department of Agricultural Engineering Cooperative Extension Service. 1991. Livestock waste management conference. Livestock waste: maximum use with minimum problems. Urbana-Champaign, Illinois, USA, 1991/03/19 .73 p.

Ziegler, D., Heduit, M., 1991. Engrais de ferme, valeur fertilisante, gestion, environnement, 35 p. Paris: ITP, ITCF, ITEB.

7 Draught Animals

David H. O'Neill

7.1 Utilization

7.1.1 Principles of Animal Traction

Two major activities in crop production are land preparation and weed control. For most subsistence and smallholder farmers the energy, or power, needed to prepare the soil for sowing or planting can be a major limitation on how much they can grow. Also, the time required to control the weeds, which compete with the crop for nutrients (and moisture in the drier areas), can be a major constraint on the crop yield.

For both land preparation and weeding tasks, the use of draught animal power (DAP) can improve the timeliness of farmers' operations by supplementing the farmers' own muscle power (and that of their families) by providing the effort needed to pull a plough or cultivator and to complete the task more quickly than would be possible by human labor alone. There is, therefore, a double benefit in that not only is drudgery reduced but improved timeliness of ploughing and weeding leads to better crop yields. It is generally considered that the landholding of a typical smallholder farmer (0.5 to 3.0 ha) would be insufficient for a tractor to be economically feasible or sustainable.

The are no ostensible physical differences between draught animals and others of the same species, although certain breeds are more renowned for their draught capabilities. Almost any bovine (including buffalo), equid, or camelid can become a draught animal, provided that it is reasonably healthy and responds to training. In general, the most important criterion is body mass, or live weight. This gives an indication of the amount of muscle on the animal and, thus, its potential to exert a force or, more specifically, a pull that the farmer can utilize through soil-working implements. A basic guide is that an animal can pull about 10% to 15% of its weight for a working period of around 4 h. This is a useful starting point for drawing up the basic specification for a draught animal–based cultivation system, although there is some variation among species, breeds and working conditions. For example, bovines would be nearer to the 10% end of the range, while equids and camelids would be at the higher end.

When putting animals to work together, good harnessing systems are essential to avoid unnecessary wastage of animals' efforts and energy, and to reduce injuries, which further impair work output. There are basically four types of harness (Fig. 7.1): the head, or neck, yoke (lashed to the animals' horns); the withers, or shoulder, yoke (better suited

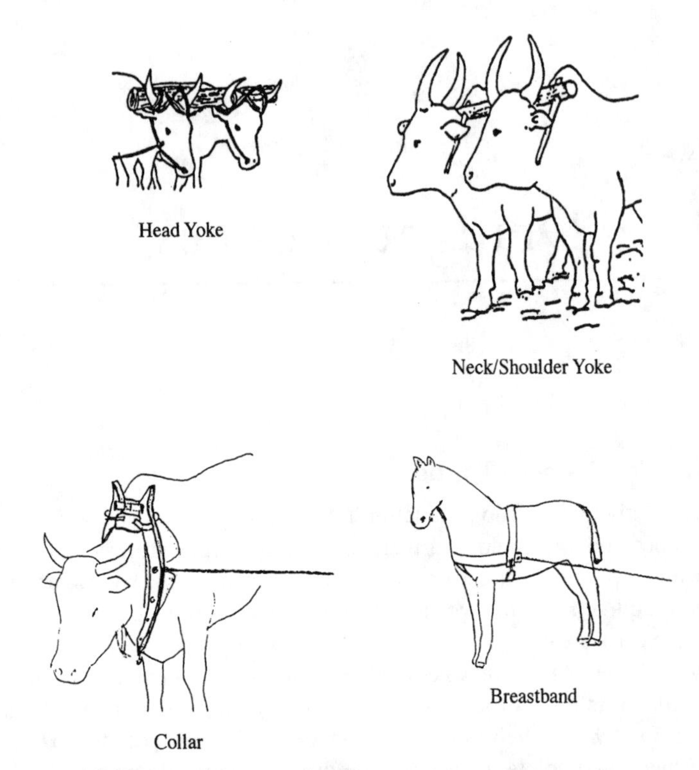

Figure 7.1. Examples of common types of harness.

to *Bos indicus* as it rests against the hump as well as the shoulders); the collar harness; and the breast-band harness. Generally, collar-type harnesses are the most suitable for all species but they tend to be the most expensive. Yokes are simple and relatively cheap but are not recommended for equids. Harnessing animals together increases the draught capability and work output, but losses are incurred. The addition of each extra animal brings about a loss of around 5% to 10% of the combined individual capabilities. Thus, harnessing together more than 10 to 12 animals would be counterproductive unless the harnessing system is first-class. Several guides on the design and use of harnessing and hitching systems have been published, to which the reader is referred for further information (see References).

In mechanical terms, the work done by a draught animal is the product of the pulling force (usually the horizontal component is used) and the distance moved while pulling (Fig. 7.2).

The force resisting the forward movement of the plough (R) is related to the force, or tension (T), in the draught chain or beam, as shown by the following equation:

$$R = T \cdot \cos \alpha \tag{7.1}$$

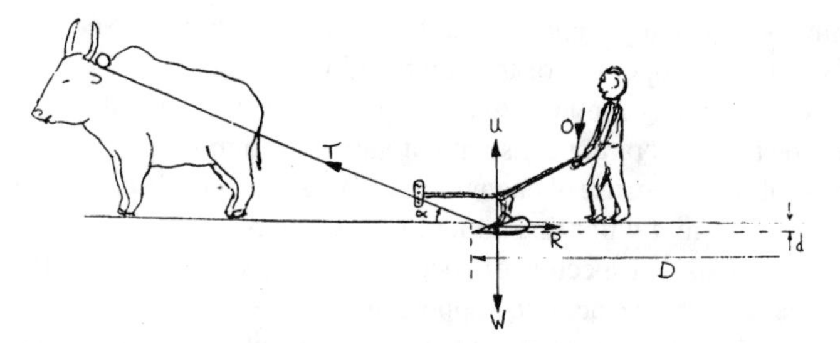

Figure 7.2. Diagrammatic representation of an ox pulling a plough.

assuming that the force applied through the handles by the operator (O) is negligible. (If an implement is properly set, this is the case.) The vertical forces may be resolved in a similar way, again assuming O is negligible:

$$T \cdot \sin \alpha = W - U \tag{7.2}$$

where W is the weight of the implement and U is the resultant uplifting force exerted by the soil.

The mechanical energy expended (E), or work done, is given by

$$E = T \cdot \cos \alpha \cdot D \tag{7.3}$$

where D is the distance moved.

The energy expended can be assessed in two ways: either as the product of the average draught force and the total distance moved, or as the sum of the instantaneous products of force and distance. Generally speaking, the former is easier to measure with simple instruments (e.g., spring balance or dynamometer and tape measure); the latter is more accurate but requires more complex instrumentation.[1]

If the power consumption (P) is required, this (by definition) is the rate of energy expenditure and can be calculated as the quotient of energy and time:

$$P = E/t = T \cdot \cos \alpha \cdot D/t \tag{7.4a}$$

or

$$P = T \cdot \cos \alpha \cdot v \tag{7.4b}$$

where t is the time spent working and v is the speed. For greater accuracy, Eq. (7.4b) should be used, with the mean of the instantaneous products of force and speed computed.

If the force through the handles cannot be considered negligible, Eqs. (7.1) and (7.2) must be corrected by including terms for O_H and O_V respectively. In practice, this occurs most commonly when the operator presses down on the handles (O_H has the same algebraic sign as W) to increase the working depth (d) in compensation for an incorrectly set implement. It should be noted that this effort makes control of the

[1] Such as the Silsoe Animal Traction Logger or the CTVM Ergometer—see section 7.1.4.

implement involved (usually a plough) hard work for the operator. Such a situation can usually be avoided by proper use of the depth and width regulators.

The depth of working is another important parameter, as it has a direct effect on both R and U. In practice, d is rarely constant but varies continuously (by as much as 25%) during a period of work because of changes in soil quality and field conditions. Generally speaking, an increase in d will bring about increases in both R and U.

For draught animals, irrespective of species, to be used successfully, the local farming system must have the capacity to support livestock and to sustain them in a healthy condition. In addition, the farmers must have some insight or predisposition toward the care, husbandry, and training of animals. Furthermore, the fabrication, or at least maintenance, of animal-drawn implements must be within the capabilities of local blacksmiths. Despite the obvious attractions of DAP for smallholder farmers, not all farming systems are suited to the application of DAP, and caution should be exercised if recommending DAP as part of a mechanization strategy. The most obvious exceptions are tsetse-infected areas (even with trypano-tolerant cattle) and low-yielding arid or semiarid production systems.

7.1.2 Tasks Performed

The greatest benefit of the use of DAP for smallholder farmers is land preparation, particularly primary cultivation at the start of the cropping cycle. The energetically demanding soil-manipulation tasks usually can be made less onerous to the farmer if he or she uses animal traction. Primary land preparation is usually effected with a plough, most widely in the form of an ard. In some parts of southern Africa and Latin America, where the legacy of European colonizers remains, the mouldboard plough is more popular, for its apparently superior weed control. A useful catalogue of animal-drawn ploughs (and other implements) is available in *Tools for Agriculture* [1]. A smaller but more fully annotated listing, giving performance specifications, of animal-drawn cultivating implements is available [2].

As discussed previously, weeding is a major constraint on crop production and yield optimization. The use of animal-drawn weeders (also called *cultivators* and *hoes*) can expedite the task of weeding (which, ideally, may be performed up to four times as a crop reaches maturity) and thereby enhance farmers' outputs. Weeding is not usually an energy- or draught-intensive task and can be carried out by a single ox or a pair of relatively light animals (e.g., donkeys), but it is entirely dependent on the crop being planted in rows of suitable spacing. Broadcast crops are not suitable for animal-drawn weeding equipment and, as a consequence, animal-drawn seeders are an attractive proposition to farmers. However, such devices are not very popular with farmers, as they tend to be mechanically unreliable. They are also relatively expensive in comparison with ploughs and cultivators because of the metering and spacing mechanisms required. Adding to their complexity, by incorporating fertilizer dispensers for example, which improves time economy has rarely proved successful.

The most successful approach to the use of DAP is to regard it as part of the continuum of available power sources, from human labor to tractor power, and to employ each power source where and when it is most appropriate. There are smallholder farmers who utilize a combination of human, animal, and engine power but use engine power only for the operations demanding the highest draught. Thus, a farmer might hire or borrow a tractor to open up a fallow plot and then use a team of oxen for the subsequent operations to create the seed bed. Weed control operations would then be carried out both by animals pulling a cultivator and by hand hoe. The cultivator would be preferred for interrow weeding, provided that the crop were not damaged, and the hand hoe would be used for interplant, or in-row, weeding, if the farmer chose to do so. As the draught requirement for a plough is likely to exceed that for a cultivator (on the same plot), the farmer may be able to use a smaller team for weeding than for ploughing. For most smallholder farmers harvesting is generally carried out by manual labor and is an activity to which the whole family contributes.

The equipment typically used in draught-animal operations is listed in Table 7.1, together with relevant comments. Ploughs are the most universally owned and used implements, and it is not uncommon to find farmers who own a plough (or ploughs) yet do not own draught animals. Farmers who do not own cultivators are likely to use their ploughs for weeding as well as land preparation. If the farmer owns a mouldboard

Table 7.1. Equipment suitable for draught-animal power applications

Implement	Comments
Traditional/ard ploughs	Cut furrows and disturb soil without inverting it Usually made at local level, generally of wood but with a metal tip
Mouldboard ploughs	Cut the soil and turn the furrow, burying weeds and trash Made of steel, usually in factories, but some blacksmiths capable of fabrication; commonest share and mould board sizes for DAP are 6, 8, and 10 in (150, 200, and 250 mm) Should have adjustable hitch point and may have wheel or skid to improve stability
Puddlers	Rotating blades for wet-land preparation to disturb the surface (by inversion) and break clods
Seeders	Comprise furrow opener (tine), seed dispenser (if seed metered, called a drill), followed by furrow closer; may be combined with fertilizer dispenser; Complex and expensive for smallholder farmers
Cultivators	For interculture in row crops; tines vary in number (usually 3 to 5 for DAP) and shape (e.g., flat, curved, spring rigid, sweep, duckfoot) and may be fitted with width-adjusting mechanism Some rotary cultivators suitable for DAP available
Ridgers	For making ridges, bunds, or raised beds by the action of mouldboards or discs
Harrows	Comprise a number (usually 20 to 40 for DAP) of steel teeth/spikes mounted on a metal frame (triangular, square, diamond, etc.)
Tool bars	Metal frame, usually with land wheel, to which different soil-working tools (e.g., tines, shares mouldboards, ground-nut lifters) can be attached
Wheeled tool carriers	Platform with axle to which a wide range of tools can be attached; can often be converted into a cart Rather expensive for smallholder farmers

plough (rather than an ard), she or he may weed using just the share with the mouldboard removed. However, some recent research by Riches *et al.* [3] in Zimbabwe has suggested that, for maize, weeding is more effective if the mouldboard is not removed, as the soil being pushed over the in-row weeds smothers them.

Animals can also be used to power stationary equipment, such as for water-raising, or to expedite postharvest operations such as threshing, milling, or crushing (sugar cane). However, because of the complexity and cost (use of gears, robustness) and inconvenience, farmers favor the use of electric motors or small internal combustion engines and only resort to animal draught if there is no alternative. There is not a strong economic argument for using animals to power stationary equipment even if a farmer already owns them and they are not in use elsewhere.

Animals are widely used for transporting goods and people and a wide range of carts (two-wheel) and wagons (four-wheel) designed to be pulled by animals is in use around the world. All species of working animals are used to pull carts, but the main pack animals are equids and camelids, with bovines (including buffalo) rarely used for this purpose. Sources of more information on transport are given in the References.

7.1.3 Species

Any large domestic animal may be used for draught, and farmers will choose the species that are available, fit in with their system of farming and are affordable. For the crop and livestock farmers of sub-Saharan Africa who, typically, own cattle, oxen would be the first choice. But, in the areas of southern Africa that have been most seriously affected by droughts in the past decade where many cattle have perished, donkeys are increasingly being used for crop production because, for smallholders, they are becoming more available and affordable than cattle. In the drier Sahelian region, camels are the first choice but for how much longer smallholder farmers will be able to afford them is questionable. In southeast Asia, particularly in the hot humid regions, buffalo are popular for draught work. In Latin America and eastern Europe, horses, mules and oxen are the preferred draught animals. In India, which probably has the largest population of draught animals in the world, oxen (usually called *bullocks*) are the most popular, but in the more arid areas, such as Gujarat, camels are widely used.

7.1.4 Work Performance

As discussed previously the force an animal can exert is dependent mainly on its live weight, and the work done, or energy expended, is calculated from the force applied and the distance moved. Summaries of the approximate work capabilities of different species are given in Tables 7.2 and 7.3. These are taken from different reference sources and indicate the wide range of values that can be expected.

The total amount of work that can be done depends on the animal's energy reserves and the onset of fatigue, which, in turn, depend on the animal's condition and nutritional status. The energy cost of work can be found by measuring the animal's metabolic rate while working. The most direct determinant of metabolic rate is the animal's oxygen

Table 7.2. Sustainable power of individual animals in good condition

Animal	Typical weight (kg)	Typical draught force (N)	Typical working speed (m/s)	Power output (W)	Working hours/day	Energy output/day (MJ)
Ox	450	500	0.9	450	6	10
Buffalo	550	650	0.8	520	5	9.5
Horse	400	500	1.0	500	10	18
Donkey	150	200	1.0	200	4	3
Mule	300	400	1.0	400	6	8.5
Camel	500	650	1.0	650	6	14

Note. For animals of different weight the power output and energy output per day may be adjusted proportionately.
Source: [4].

Table 7.3. Draught capability and power outputs of various animals

Animal	Average Weight (kg)	Approximate Draught Capability (N)	Average Speed (m/s)	Power Developed (W)
Bullock	500–900	600–800	0.56–0.83	560
Cow	400–600	500–600	0.70	340
Water buffalo	400–900	500–800	0.80–0.90	560
Light horse	400–700	600–800	1.0	750
Mule	350–500	500–600	0.9–1.0	520
Donkey	200–300	300–400	0.70	260
Camel	450–500	400–500	1.1	500
Man	60–90	300	0.28	75

Source: [5].

consumption, but this is not easily measured, especially in field conditions. A less direct but more accessible determinant is heart rate, which varies linearly with metabolic rate. If an animal's resting and maximum heart rates are known, the heart rate while working is a good indicator of the stress caused by both the work and the thermal conditions. Furthermore, the heart-rate recovery during short breaks from work and after the end of work can indicate the fatigue induced by the work.

Equipment to measure work output in field conditions has been developed by Silsoe Research Institute—the Animal Traction Logger; by the Centre for Tropical Veterinary Medicine (University of Edinburgh), the Ergometer; and by Centre de Coopération Internationale en Recherche Agronomique pour le Développement Systèmes Agro-alimentaires et Ruraux, adapted from the Silsoe design. For measuring oxygen consumption, there is a modified version of the Silsoe device, and the Centre for Tropical Veterinary Medicine has developed a device based on the Oxylog.[2] The Silsoe Animal Traction Logger also has the ability to monitor working heart rate.

[2]Manufactured by P. K. Morgan Ltd., Kent, U.K., to monitor human oxygen uptake during exercise.

7.2 Husbandry and Work Performance

7.2.1 Body Condition

The physical/physiological state of an animal, including its fitness for work, can be judged reasonably effectively by visual examination. Experienced farmers who know their animals have, of necessity, developed such a skill, and now systems of body-condition scoring have been developed. These are mainly to help extension workers and researchers judge the condition of animals and have the added advantage of providing a consistency of appraisal. The body-condition score reflects, primarily, how healthy an animal appears and integrates its level of nourishment and the presence of any obvious disease or injury. One example is that proposed by Nicholson and Butterworth for oxen [6], which runs from 1 to 9, representing emaciated to obese. For work, body-condition scores of between 4 and 6 would seem to be optimum. Oxen with a body condition of more than 6 may be too overweight to give optimum performance and may be more susceptible to heat stress than leaner animals.

Working animals must be fed more than their nonworking counterparts because of their increased metabolic activity. A summary of the feed requirements of draught oxen is given in Table 7.4, which shows the work components separately.

The data in Table 7.4 have been in use for over 20 years, but more recent information has suggested that these are overestimates. Table 7.5 shows data from Pearson [8] on the estimated energy expenditure, as a proportion of the maintenance requirement, for various draught tasks.

It is clear that these requirements are less than those given in Table 7.4, in which even for the medium workload the recommended feeding ration is double that for maintenance.

A number of factors must be considered with regard to the feeding of working animals. Does the working animal have enough time to feed? Is enough feed available and is it of sufficient quality in terms of both energy content and digestibility? Does work affect appetite and the passage of food through the body? If so, in what way and to what extent? For more details, see Pearson [8]. For horses, oxen, and buffalo, most of the questions

Table 7.4. Feed requirements for work oxen (adapted from FAO, 1972)

		Light Work		Medium Work		Heavy Work	
Weight of ox (kg)	Maintenance Ration (MJ)	Work Ration (MJ)	Total Ration (MJ)	Work Ration (MJ)	Total Ration (MJ)	Work Ration (MJ)	Total Ration (MJ)
250	29.9	15.6	45.5	29.9	59.8	45.5	75.4
300	33.8	16.9	50.7	33.8	67.6	50.7	84.5
350	37.7	19.5	57.2	37.7	75.4	57.2	94.9
400	41.6	20.8	62.4	41.6	83.2	62.4	104.0
450	45.5	23.4	68.9	45.5	91.0	68.9	114.4
500	49.4	24.7	74.1	49.4	98.8	74.1	123.5

Note. 1 kg of maize or other cereal grain yields about 13 MJ of metabolizable energy.
Adapted from [7].

Table 7.5. Estimates of daily energy requirements of draught ruminants

Task	Animals		Hours Spent Working	Location	Estimated Energy Expenditure
	Type	Liveweight (kg)			
Ploughing wetland	Buffalo	—	4	Indonesia	1.24–1.37
Ploughing dryland	Cattle	—	3	Indonesia	1.71–1.76
Ploughing dryland	Cattle	250	5	Nepal	1.25–1.46
Ploughing dryland	Cattle	150	3–4	Bangladesh	1.40–1.50
Tine cultivation, dryland	Cattle	260	4–5	Gambia	1.60
Harrowing dryland	Cattle	620	5	Costa Rica	1.60–1.28
Carting loads	Cattle	620	5–6	Costa Rica	1.60–1.28
Carting loads	Buffalo	400	5–6	Nepal	1.76–1.80

Source: [8.]

relating to feed and feeding practices can be answered, but for donkeys and camelids less research has been done and recommendations are less well defined. For camels, in particular, the issues are further complicated by the camel's ability to dehydrate and thereby modify its metabolic processes.

7.2.2 Work Schedules

The physiological effects of work on animals are, in principle, very similar to the effects of exercise on humans. For more information, the reader should consult relevant sections of books on ergonomics or work physiology (see References). Detailed laboratory investigations of the physiological processes influencing the work and power output of oxen and horses also have been undertaken by Brody and colleagues whose publications may be consulted for thorough analysis (see References). Similarly, the principles for heat exchange also apply, but some of the thermoregulatory processes may differ (e.g., sweating, panting). Nevertheless, climatic conditions in which the potential for losing body heat (working muscles generate three units of heat energy for each unit of mechanical energy) is inhibited impose an additional burden on the cardiorespiratory system. Farmers, therefore, try to avoid working their animals at the hottest times of the day and aim to work for about 3 hours starting about sunrise and then do another hour or two just before sunset. This is a recommended work schedule for more temperate conditions too because the animals, by working two shorter periods, would have more opportunity to recover from the stresses of the work. Disadvantages from this schedule could arise if the animals have to walk long distances between the homestead and the fields or are left for periods without water and feed.

Draught animals do not usually have a constant pattern of work during the year except in farming systems growing three crops a year, and in these they are beginning to be replaced by motorized power sources such as power tillers. In a typical annual cycle, the heaviest demands on the animals are made when the land has to be prepared for the major cropping season. (Land preparation for the secondary season tends to be easier, as the land will have been more recently cropped.) This preparation usually involves ploughing, and it coincides with the situation of low or exhausted feed stocks when the animals are usually found to be in their poorest body condition. However, the extent to which body condition may deteriorate without compromising work output or efficiency is not yet fully understood. It would seem reasonable for farmers to save some feed during the year to give to their draught animals just before or during ploughing. However, this represents a rather idealistic and rarely achieved situation for most smallholder farmers, and the benefits of doing so have not been quantified. There is some evidence to suggest that the farmers who have been able to store feed would not give it to their draught animals anyway, as it would be used to greater economic advantage if fed to cows in-calf or with calves.

7.2.3 Multipurpose Animals

There is an almost universal complaint from smallholder farmers who rely on DAP that there is insufficient draught power to meet everyone's needs. Traditionally, farmers who have used animals for draught work have kept some males specifically for this purpose, but increasing pressures on land and feed, together with the underlying costs of maintaining and maybe purchasing them, have prompted a change. This is particularly the case for farmers who own cattle, who have tried to economize by using cows for work. It is still rare to find a farmer spanning cows exclusively, but it is not uncommon to find cows being spanned with oxen to make a team of two or four, where the farmer cannot afford to buy or maintain a complete team of males. This raises the question of how much work a female animal can do before her milk production or, more seriously, her fertility is affected. If a cow is worked to the point of infertility, the outputs of milk and calves (or foals) are lost and the farmer has lost an asset of greater potential value than a working animal. Recently, research has been undertaken to evaluate the effects of using cows for draught, but the results have not been conclusive. It is clear, however, that when milking cows are used for work supplementary feeding is essential.

The use of draught animals for transport is unlikely to be a problem, as the greatest demand for transport occurs at and after harvest when there are few calls on the animals for land preparation or weeding tasks. In the farming systems in which there is a secondary cropping season, the cultivation operations are less arduous and there should be an adequate supply of feed.

In sub-Saharan Africa donkeys are becoming more widely used for cropping operations. There are several possible reasons but the commonest is a shortage of suitable animals: either the farmers cannot afford the purchase price or they cannot afford to keep them.

7.3 Draught Animals in Farming Systems

The importance of draught animals in current food production in industrially developing countries, and their role in increasing food production in the future, is clear from the FAO statistics given in Table 7.6.

DAP provides 23% of the mechanical energy used for crop production compared with only 6% from tractors. Future improvements in crop production are likely to based on

Table 7.6. Proportional contributions to the total energy use for crop production in developing countries

Region	Contributions (%)		
	Human	Animal	Tractor
North Africa	69	17	14
Sub-Saharan Africa	89	10	1
Asia (excluding China)	68	28	4
Latin America	59	19	22
Overall	71	23	6

Note. Animal and tractor contributions estimated by converting to man-day equivalents per hectare.
Source: [9].

the transition from human labor to DAP, rather than from DAP to tractor power, but such a change can occur only if farmers find it profitable to adopt or extend DAP.

The successful use, or introduction, of draught animals depends on a number of factors. These include the intensity of agricultural production, the local population density, access to markets, soil type, and the length of the growing season [10]. As all these factors increase or improve, the likelihood of DAP being economically viable and sustainable also increases. The attitudes towards the adoption of DAP may be more favorable in cultures in which livestock have a social or socioeconomic function, such as for dowries.

References

1. *Tools for Agriculture*. 1992. London: Intermediate Technology Publications.
2. Mbanje, E., and O'Neill D. 1997. *Database for Animal Drawn Tillage Implements*. In Ellis-Jones, J. Pearson, A. O'Neill, D., and Ndlovn, L. (eds.) "Improving the Productivity of Draught Animals in Sub-Saharan Africa". Proceedings of a Technical Workshop, 25–27 February 1997, Harare. International Development Group Report IDG/97/7, Silsoe Research Institute, Silsoe UK pp. 77–119.
3. Riches C. R., Twomlow, S. J., and Dhliwayo, H. 1997. Low-input weed management and conservation tillage in semi-arid Zimbabwe. *Experimental Agricultural* 33: 173–187.
4. Inns, F. 1992. Field power. In *Tools for Agriculture*, pp. 9–18. London: Intermediate Technology Publications.
5. Campbell, J. K. 1990. *Dibble Sticks, Donkeys and Diesels*. Philippines: International Rice Research Institute, Manila.
6. Nicholson, M. J., and Butterworth, M. 1986. A guide to condition scoring of Zebu cattle. ILCA (International Livestock Centre for Africa). Addis Ababa, Ethiopia. 36 p.
7. CEEMAT. 1972. The employment of draught animals in agriculture. Rome: FAO.
8. Pearson, A. 1996. Feeding strategies for cattle and buffalo used for work. *World Animal Review* 87: 45–55.

9. FAO. 1987. *The Next 25 Years*. Rome: Author.
10. Ehui, S., and Polson, R. 1993. A review of the economic and ecological constraints on animal draft cultivation in Sub-Saharan Africa. *Soil and Tillage Research* 27: 195–210.

Bibliography

Bakkoury, M., and Prentis, R. A. (eds.). 1994. *Working equines*. Proceedings of Second International Colloquium, April 20–22, Rabat, Morocco. Rabat, Morocco: Actes Editions, Institut Agronomique et Veterinaire Hassan II.

Barwell, I., and Ayre, M. 1982. *The Harnessing of Draught Animals*. Intermediate Technology Publications, London.

Bordet, D. 1988. *Draft Animal Power Technology in French-Speaking West Africa*. CEEMAT, France.

Brody, S. 1964. *Bioenergetics and Growth*. New York: Hafner Publishing Company.

Conroy, D., and Barney, D. 1986. *The Oxen Handbook*. LaPorte, CO: Butler Publishing & Tools.

Dennis, R. 1996. Guidelines for the Design, Production and Testing of Animal-Drawn Carts. London: Intermediate Technology Publications.

Falvey, L. Undated. *An Introduction to Working Animals*. MPW, Melbourne, Australia.

Fielding, D., and Pearson, R. A. (eds.). 1990. Donkeys, mules and horses in tropical agricultural development. Proceedings of a Colloquium held September 3–6, Edinburgh, UK. Centre for Tropical Veterinary Medicine, University of Edinburgh, UK.

Hoffmann, D., Nari, J., and Petheram, R. J. (eds.). 1989. Draught animals in rural development: Proceedings of an international research symposium. July 3–7, 1989, Cipanas, Indonesia. ACIAR Proceedings No. 27, ACIAR, Canberra.

Jones, P. A., 1991. *Training Course Manual on the use of Donkeys in Agriculture in Zimbabwe*. Borrowdale, Harare, Zimbabwe: AGRITEX Institute of Agricultural Engineering.

Lawrence, P. R., and Pearson, R. A. 1985. Factors affecting the measurement of draught force, work output and power of oxen. *Journal of Agricultural Science* 105: 703–714.

Munzinger, P. 1982. *Animal Traction in Africa*. Eschborn: GTZ.

O'Neill, D. H., and Hendriksen, G. (eds.). 1993. Human and draught animal power in crop production. Proceedings of an International Workshop, January 18–22, Rome: FAO.

O'Neill, D. H., and Kemp, D. C. 1989. A comparison of work outputs of draught oxen. *Journal of Agricultural Engineering Research* 43: 33–44.

O'Neill, D. H., Wanders, A. A., and Mbanje, E. 1997. *Technology for Donkey Utilisation*. Paper to ATNESA (Animal Traction Network for Eastern and Southern Africa) Workshop "Improving Donkey Utilisation and Management", 5–9 May 1997. Debre Zeit Ethiopia. (Proceedings in press, ATNESA c/o AGRITEX Institute of Agricultural Engineering, Harare, Zimbabwe.)

Sims, B. G., and O'Neill, D. H. 1996. Draft animal power on fragile hillsides: Technology for environmental protection. Paper to International congress on Animal Traction, Havanna, Cuba, February 19–24.

Sims, B. G., and Starkey, P. 1996. Experiences with the introduction of draft animals and new implements. Paper to International congress on Animal Traction, Havanna, Cuba, February 19–24.

Srivastava, N. S. L., and Ojha, T. P. 1987. Utilisation and economics of draught animal power. Proceedings of the national seminar on status of animal energy utilisation held at CIAE, Bhopal, January 24–25. Bhopal: CIAE.

Starkey, P., 1989. *Harnessing and Implements for Animal traction*. Eschborn: GTZ.

Starkey, P., Mwenya, E., and Stares, J. (eds.). 1994. Improving animal traction technology. Proceedings of first workshop of the Animal Traction Network for Eastern and Southern Africa (ATNESA), January 18–23. Lusaka, Zambia. Wageningen, The Netherlands: Centre for Agriculture and Rural Cooperation.

Svendsen, E. D. 1986. *The Professional Handbook of the Donkey*. England: The Donkey Sanctuary.

Upadhyay, R. C. 1990. *Draught Animal Efficiency Limiting Factors*. Mathur, India: Sunil Publications.

Upadhyay, R. C., and Madan, M. L. 1985. Studies on blood acid-base status and muscle metabolism in working bullocks. *Animal Production* 40: 11–16.

Vall, E. 1996. Les animaux de trait au Nord-Cameroun: zebu, ane et cheval. Performances a l'effort et adaptations physiologiques. Document du travail du CIRAD-EMVT No.1, Montpellier, France.

Zerbini, E., Gemeda, T. O'Neill, D. H., Howell, P. J., and Schroter, R. C. 1992. Relationships between cardio-respiratory parameters and draught work output in F_1 crossbred dairy cows under field conditions. *Animal Production* 55: 1–10.

8 Aquacultural Systems

F. W. Wheaton and S. Singh

8.1 Introduction

According to Food and Agriculture Organization statistics, the 1994 world aquacultural production of finfish, shellfish, and aquatic plants was 25.5 million tons, valued at US $39.83 billion [1]. These figures represented overall increases of 11.8% and 10.3% over the 1993 production in weight and value, respectively. Asian countries contributed almost 90% of the total world aquacultural production in 1994. The Food and Agriculture Organization further estimated that the combined output of ocean fisheries and aquaculture will grow 10% to 30% by the year 2010 and that aquaculture will continue to increase its share in the total production due in part to declining or stabilized ocean harvests in many parts of the world.

Nash [2] in a recent comprehensive review of world aquaculture reports that carp and tilapia production represents about 46% of the total world aquacultural production on a weight basis, salmon and trout 4%, shrimp and prawn 2.2%, oysters and mussels 37%, and catfish 2.2%. Lawson [3], based on a survey of recent government publications, reports that the freshwater production accounts for 80% to 90% of the world's finfish production, and 95% of this is in ponds. Lawson [3] further reports that cage culture represents only 3% to 4% of freshwater finfish production, but about 40% of marine and brackish water finfish production.

The aquaculture industry is the fastest growing segment of the U.S. agriculture. U.S. aquaculture production has been dominated by finfish production, particularly catfish, which represents about half of total U.S. finfish production [3]. However, the expanding aquaculture industry is now paying close attention to other fish species and many of them are showing commercial potential. Tilapia and hybrid striped bass are two such species. Lawson [3] and Stickney [4] discuss current culture practices of several commercial aquatic species and the changing pattern of U.S. aquaculture production.

8.2 System Types

An aquacultural production system can be described simply as production of marketable aquatic organisms under controlled or semicontrolled conditions. Aquacultural systems are classified in a variety of ways depending upon the viewer's perspective and

Table 8.1. Aquacultural production from different systems

System Type	Production (% of total aquacultural production)
Ponds	41
Pens and cages	3
Raceways and silos	1
Estuaries and bays	26
Unspecified	29
Total	100 (7.9×10^9 kg over 6×10^6 hectares surface area)

Sources: Compiled from Brune [5] and Nash [6].

interest. Common classifications are based on construction type (pond, net pen, raceway, or tank-based), species reared (fish, crustacean, plant, or bivalve), intensity of production (extensive or intensive), number of species (monoculture or polyculture), culture water salinity (fresh, salt, or brackish water), culture water temperature (cold or warm), fish feeding (controlled feeding or naturally fed), economics (commercial or subsistence), and level of environmental control (open, semiclosed, or closed). Irrespective of the classification, an aquaculture system's prime requirement is adequate supply of suitable-quality water. Stickney [4] provides detailed classifications based on intensity of production and culture water temperature.

The constructional classification, followed in this chapter, directly relates to the engineering characteristics of aquacultural production systems and is the most widely used method, particularly by engineers, for distinguishing between two different systems. Table 8.1 shows the worldwide aquacultural production from different systems based on the data by Nash [2]. Brune [5] reviewed this data and revealed that the production from unspecified sources (about 29% of the total) is probably dominated by pond aquaculture. The following is a brief description of various types of aquacultural systems. Some very useful references used in preparing this chapter are mentioned at appropriate places. The reader is advised to refer to these references for more detailed information.

8.2.1 Ponds

Ponds, man-made bodies of water usually smaller than lakes, fall into the category of extensive or semi-intensive culture systems. A wide variety of aquatic organisms can be grown in ponds. In fact, it is estimated that 41% of global aquacultural production comes from ponds [2]. Pond sizes vary from a fraction of a hectare (for spawning or fingerling production) to several hectares (for grow-out). The pond depth varies from less than a meter to close to 3 m [5]. The pond depth selection depends on the penetration of sunlight. The depth should be such that the sunlight cannot reach the pond bottom; otherwise, rooted plants will grow in the bottom. The water supply for ponds, in the order of preference, can be from springs, wells, streams, run-off, or municipal sources. For a given production capacity, ponds require less water than raceways but more than tank-based recirculating systems. Seepage and evaporation are the principal water losses. Because ponds are outdoor systems, the most important decision for pond construction is the site selection. Ideally, the site should have an adequate supply of good quality water and clay-rich soil to reduce losses from the pond. Where needed, a variety of pond sealers are available to reduce seepage. They do, however, increase pond construction costs.

Two primary types of ponds are embankment and excavated ponds. Embankment ponds are formed by building up a dam, dike, or similar above-ground structure to hold water. Excavated ponds are constructed by removing soil from an area to form a depression that can be filled with water. Embankment ponds are the type most commonly used for fish culture, as they can be constructed in a wide range of topographic conditions. In most situations, they also have other significant advantages over excavated ponds including simpler construction, lower pumping cost, and ease of harvesting. Primary disadvantages of excavated ponds are their need to be constructed in an area of relatively flat topography and the cost of removing the necessarily large amounts of soil. Moreover, any water removed from an excavated pond must be pumped. Some ponds require both embankments and excavation to achieve the desired water-holding capacity. Considerations for pond construction include intensity of production, feeding and harvesting method, water supply and drainage, soil type, topography, conformation, dimensions and slopes, maintenance, and aeration.

The feeding method in aquacultural ponds may be either natural (cultured organisms derive their food from algae or other naturally growing aquatic organisms in the pond) or manual with commercial diets. A combination of the two methods can also be used. In densely stocked ponds, some type of mechanical aeration may be needed to maintain proper oxygen concentration.

Production per unit area varies by a factor of 10 for different practices and species. Low-intensity and minimum-management ponds yield 500 to 1000 kg/ha, but yields above 10,000 kg/ha are achievable using commercial practices [5].

8.2.2 Raceways

Raceways are single-pass, relatively shallow structures, typically with a rectangular cross-section, containing aquatic organisms. Significant water flow through a raceway is required to remove accumulated wastes, provide aeration, and maintain suitable environmental conditions for the species reared. Raceway culture is widely used in hatcheries because flowing water minimizes metabolite concentration and enhances oxygen transfer. Intensity of production can be much higher in raceways than in ponds because the water is changed continuously. Salmon and trout culture have been successfully practiced in raceways. Raceways can utilize gravity flow of water to minimize pumping costs, provided the water supply is a freshwater spring and the raceways are located on the downstream side.

There are two basic arrangements of raceways—series and parallel. The series arrangement requires less water than the parallel arrangement because the outlet water from one raceway is used by the next one in the series. Design considerations for raceways are a fish culture suitable for raceways, water supply for the raceway, shape, size, intensity, series or parallel flow, maintenance, and aeration. Design calculations include water-exchange rate, stocking density, water velocity through the raceway, oxygen concentration, and ammonia concentration. Soderberg [6] and Youngs and Timmons [7] illustrate these calculations.

8.2.3 Net Pens and Cages

Net pens and cages, terms often used interchangeably, are floating structures in open water and are usually located within a pond, lake, reservoir, river, estuary, or open ocean environments. Netting or plastic mesh material is used around a frame to form an enclosure for aquatic organisms intended for production. Beveridge [8] provides detailed information on various aspects of net-pen and cage aquaculture. Two primary types of net enclosure are used for fish culture. The first, the net pen, is employed in relatively shallow waters (usually less than 10 m). This system consists of rigid poles, pilings, or other fixed structures driven into the bottom [9]. The rigid framework supports the net that is stretched around it. The bottom of the enclosure is formed by the natural bottom of the water body. Net pens range in size from 1 m^2 to several hectares in surface area [8]. In deeper water the closeness of spacing and size of the framework elements needed to withstand forces imposed on them become excessive. Construction and materials costs also become prohibitive. Thus in deeper water anchored floating net enclosures or cages are used. Cages are generally smaller than net pens and range in size from 1 m^2 to over 1000 m^2 in surface area [8]. Primary design considerations of net retaining structures are shape, size, stocking intensity, water flow, management method, materials of construction, and loads.

The advantages of net-pen systems include low input and minimum-management. The main disadvantage is that the net pens are exposed to climatic disturbances that may cause severe structural damages and loss of fish. Also, net-pen systems may have significant detrimental effects on the natural environment. This often is reflected in the high cost of obtaining a permit for establishing net-pen operations in certain areas [10].

The fish are first produced in hatcheries and then transferred to cages as young fingerlings. One of the methods used by Beveridge [8] to classify cage culture is based on feed inputs, such as extensive (natural foods such as plankton or detritus), semi-intensive (feed containing less than 10% protein), and intensive (feed containing more than 20% protein). Harvesting of fish from net retaining structures is done either by lifting a cage above the water surface and emptying the cage or by crowding the fish into a small area.

8.2.4 Tanks and Recirculating Aquacultural Systems

Tank aquaculture has been in existence for a long time in the form of home aquariums, fingerling production units, and large public-display facilities. Although tank-based aquaculture represents a very small fraction of world aquacultural production at present, the future of such systems is promising. This is because tank-based systems offer several advantages over other aquacultural systems. Higher intensity of production, better environmental control, lower water usage, potential for low impact on the natural environment, control of market time and size, and site-independence are the main benefits. Because of significant advances in materials technology, a wide variety of tanks, ranging from wooden tanks with liners to prefabricated fiberglass structures, are available commercially in various shapes, sizes, and price ranges. Also, the availability of a wide variety of pipes and fittings makes installation of tank-based systems fast and easy. Tanks are smaller than ponds, for easy transportability and cleaning. Large tanks are usually assembled at the location.

Tank-based systems represent a significant shift in the way aquaculture is practiced by providing a higher level of control over production and management operations. Applications of sophisticated water treatment and electronic technologies in aquaculture are accelerating this transformation. Recirculating aquaculture systems are at the forefront of this technological transformation. National and international production figures for these systems are not available, as recirculating systems are still in the developmental stages and very few have been operating on commercial scale.

A recirculating aquaculture system is defined as an assemblage of parts and devices used for husbandry of aquatic organisms in which water is continuously cleaned and recycled. These systems are more complex in design and operation than any other system described here. They employ a variety of technologies to conserve water and provide high-intensity production. The schematic of a recirculating system is shown in Fig. 8.1. Timmons and Losordo [11] provide a comprehensive collection of articles by several experts on various engineering and management aspects of recirculating aquaculture systems.

The advantages offered by recirculating systems include control of fish environment for reduced stress, faster growth, better feed conversion ratio, reduced incidence of diseases, reduced water usage, lack of site limitations, year-round production, control of market time and size, and consistency of quality and quantity of product. Main disadvantages are that these systems are capital intensive; they need skilled management; and a limited number of treatment compounds are available for treating fish without harming the biofilters.

A typical recirculating aquaculture system is a configuration of several chemical, biological, and mechanical processes. Essential components include a fish-culture tank, a

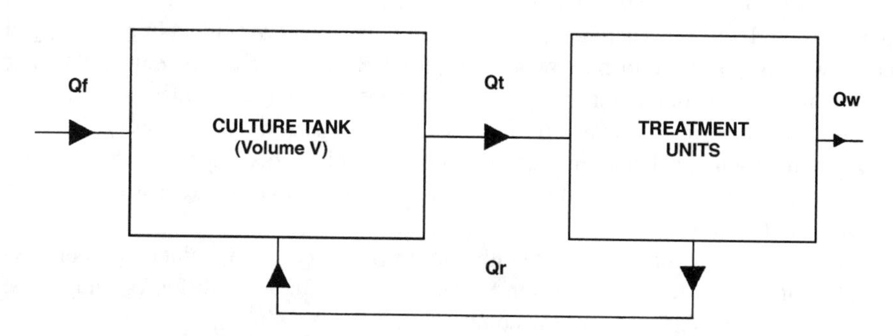

Figure 8.1. Schematic of a recirculating aquacultural system.
Note: Residence Time = Volume/Tank outflow = V/Qt; Replacement Time = Volume/Tank inflow = V/Qf; Recycle Ratio = Recycle flow/Tank outflow = Qr/Qt; Turnover Time = Tank outflow/Volume = Qt/V = 1/Residence Time; Wastewater Discharge = Qw.

biofilter for ammonia removal, a particulate removal device, and an aeration/oxygenation system. Optional components or accessories may include an ozone-application device for disinfection, color, or odor control; a degassing device for carbon dioxide removal; and monitoring and control devices. For indoor systems heating, ventilation, and air-conditioning equipment may also be required.

Desirable characteristics of an aquaculture tank include a smooth interior, good circulation, ease of cleaning and sterilization, durability and strength, a nontoxic surface, and inexpensiveness. Commonly available shapes are circular, oval, rectangular, and silo-shaped. Tanks can be made of a wide range of materials including concrete, plastics, wood with liner, steel, and fiberglass.

Particulates found in aquaculture systems are mostly feces, debris, feed, and bacterial mass. Particulate removal devices utilize one or more of the following properties of the particulates to remove the particles from the culture water: density, size, electrical charge, and chemical and magnetic properties. Most common methods are gravity separation by a sedimentation chamber, multitube clarifier, and hydroclone. Physical separation of particulates from the culture water is also done using strainer devices such as sand, screens (either fixed or rotary), or other fixed-bed media.

Dissolved metabolites in the culture water consist mainly of organic compounds and ammonia. Ammonia in its un-ionized form is extremely toxic to most aquacultural species. Detoxification and metabolite-removal devices usually depend on either chemical or biological methods.

Chemical filters work on the principle of adsorption (concentration of a substance on a surface or interface). Examples of chemical filters are granular activated carbon and ion-exchange filters. Filter breakthrough and filter regeneration are important concepts in the chemical filter design and operation.

Biofilters operate on the principle of biological conversion of toxic waste into nontoxic or much less toxic compounds by bacteria (heterotrophs and autotrophs) growing on a medium. Ideal biofilter media should have large surface area and high void fraction and be stable, durable, lightweight, and inexpensive. Commonly used biofilter media are sand, gravel, lime stone, crushed oyster shells, and plastics (PVC, Styrofoam, Polyethylene, Fiberglass, and Teflon). There are several different biofilter types including submerged bed (down-flow, up-flow, or horizontal flow), fluidized bed, bead filter, emerged bed or trickling filter, and combination (reciprocating or rotating biological contactors). Biofilter operation requires time on start-up to stabilize. After reaching steady-state, dissolved oxygen, temperature, pH, nutrients, and inhibiting chemicals become the most important operational parameters. The biofilter efficiency also depends on media surface area, mass loading, and hydraulic loading.

Aeration/oxygenation devices use either air or pure oxygen to maintain proper oxygen concentration in the culture water. Surface aerators, fine-pore diffusers, and u-tube oxygenation devices are some of the common ones.

Although expensive, ozone treatment is sometimes used for disinfection, fine-particle removal, dissolved-organics removal, or color removal. Dose, location, and time are important parameters for ozone application in recirculating aquaculture systems.

References

1. *Aquaculture Magazine Buyer's Guide '97*. 1997. pp. 6–26.
2. Nash, C. E. 1988. A global overview of aquaculture production. *Journal of the World Aquaculture Society*, 19(2).
3. Lawson, T. B. 1995. *Fundamentals of Aquacultural Engineering*. Chapman and Hall, New York.
4. Stickney, R. R. 1994. *Principles of Aquaculture*. John Wiley and Sons, New York.
5. Brune, D. E. 1991. Pond aquaculture. In: Engineering Aspects of Intensive Aquaculture: *Proceedings from the Aquaculture Symposium*. Northeast Regional Agricultural Engineering Service, Cornell University, Ithaca, NY, pp. 118–130.
6. Soderberg, R. W. 1995. *Flowing Water Fish Culture*. Boca Raton, FL: CRC Press, Inc. 147 pp.
7. Youngs, W. D. and M. B. Timmons. 1991. A historical perspective of raceway design. In: *Engineering Aspects of Intensive Aquaculture*. Northeast Regional Agricultural Engineering Service, Ithaca, NY, pp. 160–169.
8. Beveridge, M. 1996. *Cage Culture*. Fishing News Books, Oxford, England.
9. Wheaton, F. W. 1977. *Aquacultural Engineering*. John Wiley and Sons, New York.
10. Riley, J. 1991. Ocean net-pen culture of salmonids: engineering considerations. In: Engineering Aspects of Intensive Aquaculture: *Proceedings from the Aquaculture Symposium*. Northeast Regional Agricultural Engineering Service, Cornell University, Ithaca, NY, pp. 131–150.
11. Timmons, M. B. and T. M. Losordo. 1994. *Aquaculture Water Reuse Systems: Engineering Design and Management*. Elsevier Science, Amsterdam.

9 Environmental Requirements

S. Singh and F. W. Wheaton

9.1 Primary Constraints in Aquacultural Systems

9.1.1 Properties of Water

Water, the environment of all aquatic life, has some peculiar physical and chemical properties owing to its unique molecular structure. For example, the liquid nature of water is an anomaly among most inorganic compounds. Both melting and boiling points of water are unusual for liquids of similar molecular weight. Unlike most other liquids, the volume of a water mass increases by 11% to 12% upon freezing. Wheaton [1], Stickney [2], and Lawson [3] are good references for detailed information on the physical and chemical properties of water from an aquaculture point of view. Some important properties are summarized in Table 9.1 and discussed here.

Water temperature is the most influential parameter in aquatic systems because of its significant effect on both the physiological processes of the aquatic life (most are cold-blooded) and the environment surrounding the aquatic life. For example, the metabolic rate of catfish doubles for each 10°C rise in temperature over the range 0°C to 35°C [4]. This effect is called Q_{10} *factor* and applies to almost all cold-blooded aquatic organisms including bacteria, zooplankton, insects, and fish. Figure 9.1 demonstrates a general effect of temperature on an aquatic organism. Most aquatic organisms have a narrow temperature range for optimum growth. In open or natural systems, solar radiation is the prime source of thermal energy. The density of water varies with temperature and is maximum at 3.94°C. This property leads to the thermal stratification of the water column in aquaculture ponds, which can cause resistance to circulation and oxygen transfer. Specific heat of water as compared with other liquids is high (4.186 kJ/kg·K), and therefore it acts as a good heat-storage medium, protecting aquatic organisms against large temperature fluctuations in outdoor aquacultural systems. However, this property also requires large amounts of heat for raising water temperature in controlled-environment, warm-water aquacultural systems. Water has a high viscosity relative to other liquids, which decreases with an increase in temperature. Viscosity affects sedimentation or sinking rate of plankton in ponds, sedimentation rate of sediments, and energy expenditure by fish for swimming.

Table 9.1. Important properties of water

Property	Value	Significance
Chemical structure	Polar molecule with an angle of 104.5° between the two hydrogen atoms	Many unusual properties of water are a result of its unique molecular structure
Hardness	Soft: 0–0.055 ppt Slightly hard: 0.056–0.1 ppt Moderately hard: 0.101–0.2 ppt Very hard: 0.201–0.5 ppt	Classification used for freshwater
Salinity	Fresh water: 0–0.5 ppt Salt water: 30–37 ppt Brackish water: 2–30 ppt	Has significance in osmoregulation of salts in fish body
Latent heat of vaporization	2258 kJ/kg at 100°C and one atmospheric pressure	Affects evaporation losses
Latent heat of fusion	3336 kJ/kg at 0°C	Affects freezing
Specific heat	4186 kJ/kg·K	Protects aquatic life against sudden temperature changes
Density	At 0°C: 0.99987 g/cm^3 (ice) At 4°C: 1.00000 g/cm^3 (maximum) At 10°C: 0.99973 g/cm^3 At 25°C: 0.99707 g/cm^3	Seawater density is maximum at −1.4°C and 24.7 ppt salinity
Viscosity	Decreases with an increase in temperature	Affects sedimentation of particles and swimming speed of fish
Thermal conductivity	Decreases with an increase in salinity	Can be used as an estimate of salinity
Surface tension	Decreases with an increase in temperature	Used in designing some water treatment systems
Hydrogen ion concentration	Neutral: pH = 7 Acidic: pH < 7 Basic: pH > 7	Need to be maintained within a fairly narrow range in most aquacultural systems
Vapor pressure	Depends on temperature and salinity	
Osmotic pressure	Depends on temperature and salinity	Has significance in osmoregulation of salts in fish body
Transparency	Variable	A measure of turbidity
Color	Variable	May indicate presence of impurities in water
Refractive index	Depends on temperature and salinity	Influences color

Source: Compiled from Wheaton [1].

9.1.2 Oxygen as a Constraint

Like most organisms, aquatic animals need oxygen for their survival. Because they live in water they need oxygen in the dissolved form. Unlike terrestrial organisms, aquatic animals expend considerable energy for respiration, as water is much denser than air. In ponds and other outdoor systems, photosynthetic activity in the water during the day is the primary source of oxygen for aquatic life. However, at night, the photosynthetic plants compete with other aquatic organisms for oxygen. This, coupled with biodegradation of organic matter, often makes oxygen a limiting factor, if no artificial aeration is used in a pond, particularly a densely stocked pond.

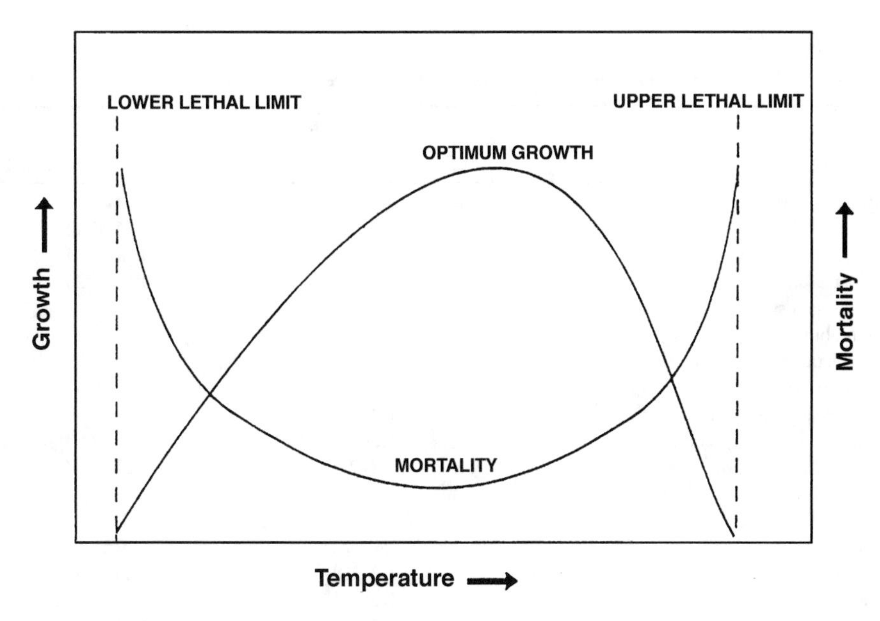

Figure 9.1. Hypothetical effect of water temperature on fish growth.

Oxygen transfer to water represents significant economic cost in most aquacultural systems. It can be achieved by several different methods such as mechanical or gravity aerators and pure oxygen injection. Table 9.2 lists oxygen-transfer rates of various oxygenation devices used in modern aquacultural systems. The oxygen-transfer rate into culture water directly depends on the saturation concentration of oxygen in the culture water. The saturation concentration is influenced by temperature, elevation, barometric pressure, salinity, hydrostatic pressure, and oxygen concentration in the feed gas. It decreases with an increase in water temperature, salinity, or elevation, and increases with an increase in barometric pressure, hydrostatic pressure, or oxygen concentration in the feed gas. Table 9.3 provides the saturation concentration of dissolved oxygen in water in contact with air at different values of temperature and elevation. The solubility of principal components of atmospheric air in freshwater is shown in Table 9.4 as a function of atmospheric pressure and gas composition. Some important observations can be made from this table. The solubility of oxygen doubles (from 10.08 mg/L to 20.32 mg/L) when the pressure is increased from 1 to 2 atm. Also, using pure oxygen as a feed gas, the oxygen solubility in the water can be increased almost five times that obtainable using air as a feed gas. Because nitrogen solubility is higher than that of oxygen, the use of air as a feed gas at higher pressures will result in water becoming quickly saturated with nitrogen, which is known to cause gas-bubble disease in fish.

9.1.3 Fish Waste

Water is considered a universal solvent (i.e., provided enough time, many substances including some undesirable waste materials will dissolve in it). This causes direct physical exposure of fish to their own waste. The dissolved portion of uneaten feed and

Table 9.2. Oxygen transfer rates of various devices used in aquaculture

Aeration Device	Transfer Rate (kg O_2/kWh) At 20°C and 0.0 mg DO/L	At 20°C and 6 mg DO/L
Diffused-air systems		
Fine-bubble	1.2–2.0	0.25–0.42
Medium-bubble	1.0–1.6	0.21–0.34
Coarse-bubble	0.6–1.2	0.13–0.25
Pure-oxygen systems		
Fine-bubble		1.2–1.8
Mechanical surface aeration		1.0–1.2
Turbine-sparger		1.2–1.5
Low-speed surface with or without draft tube	1.2–2.4	0.25–0.50
High-speed floating aerator	1.2–2.4	0.25–0.50
Floating rotor aerator	1.2–2.4	0.25–0.50
U-tube aerator		
Zero head available	4.5	0.95
30-cm head available	45.6	9.58
Gravity aerator	1.2–1.8	0.25–0.38
Venturi aerator	1.2–2.4	0.25–0.50
Static tube systems	1.2–1.6	0.25–0.34

Source: Compiled from Colt and Tchobanoglous [5].
Note: DO, dissolved oxygen.

Table 9.3. Solubility of oxygen in fresh water to water-saturated air at various temperatures and elevations above mean sea level

	Solubility by Elevation in Feet						
Temperature (°C)	0	500	1000	1500	2000	3000	4000
0	14.60	14.34	14.09	13.84	13.59	13.12	12.65
1	14.20	13.95	13.70	13.46	13.22	12.95	12.30
2	13.81	13.57	13.33	13.09	12.86	12.41	11.97
3	13.45	13.21	12.97	12.74	12.52	12.07	11.65
4	13.09	12.86	12.63	12.41	12.19	11.76	11.34
5	12.76	12.53	12.31	12.09	11.87	11.46	11.05
10	11.28	11.08	10.88	10.68	10.49	10.12	9.76
15	10.07	9.89	9.71	9.54	9.37	9.04	8.71
20	9.08	8.91	8.75	8.59	8.44	8.14	7.84
25	8.24	8.09	7.95	7.80	7.66	7.38	7.12
30	7.54	7.40	7.26	7.13	7.00	6.74	6.50
35	6.93	6.81	6.68	6.55	6.43	6.19	5.96
40	6.41	6.29	6.17	6.05	5.94	5.71	5.49

Source: Compiled from Weiss [6].

Table 9.4. Solubility of major components of air as a function of pressure and feed gas composition at 15°C

Pressure (atm)	Feed Gas Composition	Gas Solubility (mg/L)		
		Oxygen	Nitrogen	Carbon Dioxide
1	Air	10.08	16.36	0.69
1	Pure oxygen	48.14		
2	Air	20.32	33.00	1.38
2	Pure oxygen	97.02		

Source: Compiled from Colt and Orwicz [7].

fish excretory wastes are the main concern. Suspended waste particles of greater than about 100 μm in size can be easily removed by physical separation; however, dissolved compounds and fine waste particles require more sophisticated techniques (e.g., chemical and biological methods) for their removal from the water.

9.2 Environmental Needs of Aquatic Organisms

Maintaining an optimum growth environment in an aquacultural system is basic to its long-term commercial success. Meade [8] provides a general guideline for water quality parameters for aquaculture (Table 9.5). The physical, chemical, and biological factors of the aquatic environment have significant effect on survival, growth, and reproductive behavior of the aquatic organisms [9]. Physical factors, such as temperature, change more slowly than do chemical factors in outdoor systems. Chemical factors include dissolved gases, pH, nitrogenous metabolites, inorganic ions, hardness, alkalinity, and toxic compounds. The chemistry of an aquatic environment can change significantly within a short period of time, causing stress or death to cultured organisms. Intrinsic and extrinsic biological factors of the cultured organisms and the organisms growing in the same medium are of importance when designing aquacultural systems. However, very little control can be exercised on biological factors in isolation with physical or chemical factors [9].

In general, warmwater species of fish, such as tilapia, tolerate a wider range of environmental conditions than coldwater species such as trout [9]. Stickney [2] defines warmwater species as those with temperature optima at or above 25°C and coldwater as those with temperature optima below 20°C. Species that have their temperature optima within these two temperatures are considered midrange species. Table 9.6 provides optimum temperature ranges for common aquatic organisms of interest for aquacultural production.

Next to temperature, the most important water-quality parameter for aquaculture is dissolved-oxygen concentration. A minimum dissolved-oxygen concentration of 5 mg/L is recommended for warmwater species and 6 mg/L for coldwater species ([3], [1]). At lower concentrations than these, the fish will be stressed, resulting in slower growth. There are several factors that influence the oxygen-consumption rate of fish. Most

Table 9.5. General water quality criteria for aquaculture

Parameter	Concentration (mg/L except for pH)
Alkalinity	10–400
Aluminum	<0.01
Ammonia	<0.02
Total Ammonia	<1.0
Arsenic	<0.05
Barium	5
Cadmium	
Alkalinity <100 mg/L	0.0005
Alkalinity >100 mg/L	0.005
Calcium	4–160
Carbon dioxide	0–10
Chlorine	<0.003
Copper	
Alkalinity <100 mg/L	0.006
Alkalinity >100 mg/L	0.03
Dissolved oxygen	5 to saturation
Hardness, Total	10–400
Hydrogen cyanide	<0.005
Hydrogen sulfide	<0.003
Iron	<0.01
Lead	<0.02
Magnesium	<15
Manganese	<0.01
Mercury	<0.02
Nitrogen (N_2)	<110% total gas pressure <103% as nitrogen gas
Nitrite	0.1 in soft water
Nitrate	0–3.0
Nickel	<0.01
PCB's	<0.002
pH	6.5–8.0
Potassium	<5
Salinity	<5%
Selenium	<0.01
Silver	<0.003
Sodium	75
Sulfate	<50
Sulfur	<1
Total dissolved solids	<400
Total suspended solids	<80
Uranium	<0.1
Vanadium	<0.1
Zinc	<0.005
Zirconium	<0.01

Source: Compiled from Meade [8].

Table 9.6. Optimum temperature ranges for common aquatic species

Species	Temperature (°C)	Reference
Brook trout	7–13	Piper et al. [10]
Brown trout	9–16	Petit [11] and Piper et al. [10]
Rainbow trout	10–16	Petit [11] and Piper et al. [10]
Atlantic salmon	15	Petit [11]
Chinook Salmon	10–14	Piper et al. [10]
Coho salmon	9–14	Piper et al. [10]
Sockeye salmon	15	Petit [11]
Sole	15	Petit [11]
Turbot	19	Petit [11]
Plaice	15	Petit [11]
European eel	22–26	Petit [11]
Japanese eel	24–28	Petit [11]
Common carp	25–30	Petit [11]
Mullet	28	Petit [11]
Tilapia	28–30	Petit [11]
Channel catfish	21–29	Tucker and Robinson [12] and Piper et al. [10]
Striped bass	13–23	Piper et al. [10]
Red swamp crawfish	18–22	Romaire [13]
Freshwater prawn	30	Romaire [13]
Brine shrimp	20–30	Romaire [13]
Brown shrimp	22–30	Romaire [13]
Pink shrimp	>18	Romaire [13]
American Lobster	20–24	Romaire [13] and Hedgecock et al. [14]
American oyster	>8	Romaire [13]

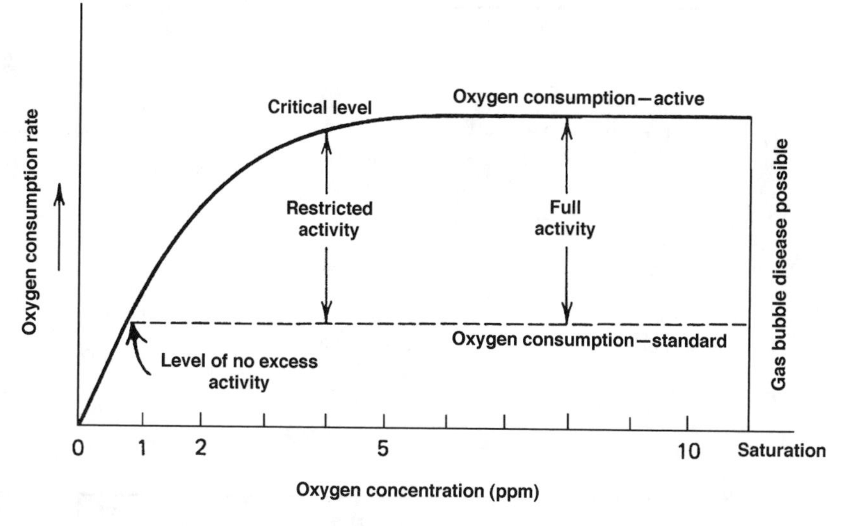

Figure 9.2. Dependence of fish activity and oxygen consumption rate on available oxygen concentration in water.

Table 9.7. Oxygen consumption rate and dissolved oxygen concentrations corresponding to activity levels

	Oxygen Consumption Rate		Dissolved-Oxygen Level and Activity		
Species	kg O_2/1000 kg fish·day	Size (g)	ppm	Temperature (°C)	Activity
Rainbow trout	7.2 at 15°C and unknown feeding level	100	2.4–3.7	16	Limiting
	7.2 at 15°C and production feeding level	100	1.05–2.06	18.5	Limiting
			2.5	13–20	Limiting
			5–5.5	20	Limiting
			9	20	Full
Sockeye salmon	5.6 at 15°C and unfed	28.6			
	6.6 at 15°C and 3% feeding level	28.6			
Channel catfish	13.4 at 30°C and unfed	100	1.0		Limiting
	19.5 at 30°C and satiation feeding	100	3.0–5.0		Full
Largemouth bass	2.7 at 20°C and unfed	35	2.3–4.8	0–4	Limiting
	2.7 at 20°C and unfed	150			
	15.3–19.9 at 20°C and standard feeding	1.6			
Blue tilapia			1.0	25	Critical
Malaysian prawn	36.0 at 24°C and unfed	0.5			
	43.0 at 24°C and satiation feeding	0.5			

Sources: Compiled from Beamish [15], McLarney [16], and Colt and Tchobanoglous [5].

Table 9.8. Nitrogen toxicity to selected finfish species

Species	96-hour LC_{50} (mg/L NH_3)	48- or 96-hour LC_{50} (mg/L NO_2-N)	96-hour LC_{50} (mg/L NO_3-N)
Pink salmon	0.08–0.1		
Mountain whitefish	0.14–0.47		
Brown trout	0.50–0.70		
Rainbow trout	0.16–1.1	0.19–0.39	1360
Largemouth bass	0.9–1.4	140	
Smallmouth bass	0.69–1.8	160	
Common carp	2.2	2.6	
Red shiner	2.8–3.2		
Fathead minnow	0.75–3.4	2.3–3.0	
Channel catfish	0.5–3.8	7.1–13	1400
Bluegill	0.55–3.0	80	420–2000
Cutthroat trout		0.48–0.56	
Chinook salmon		0.88	1310
Labyrinth fish		28–32	
Goldfish		52	
Green sunfish		160	
Guadalupe bass		190	1260
Guppy			180–200

Source: Compiled from Russo and Thurston [17].

Table 9.9. Un-ionized ammonia fraction (% NH_3) in total ammonia nitrogen at different temperatures and pH

Temperature (°C)	Ammonia Fraction (% NH_3) by pH Level								
	6.0	6.5	7.0	7.5	8.0	8.5	9.0	9.5	10.0
0	0.00827	0.0261	0.0826	0.261	0.820	2.55	7.64	20.7	45.3
1	0.00899	0.0284	0.0898	0.284	0.891	2.77	8.25	22.1	47.3
2	0.00977	0.0309	0.0977	0.308	0.968	3.00	8.90	23.6	49.4
3	0.0106	0.0336	0.106	0.335	1.05	3.25	9.60	25.1	51.5
4	0.0115	0.0364	0.115	0.363	1.14	3.52	10.3	26.7	53.5
5	0.0125	0.0395	0.125	0.394	1.23	3.80	11.1	28.3	55.6
6	0.0136	0.0429	0.135	0.427	1.34	4.11	11.9	30.0	75.6
7	0.0147	0.0464	0.147	0.462	1.45	4.44	12.8	31.7	59.5
8	0.0159	0.0503	0.159	0.501	1.57	4.79	13.7	33.5	61.4
9	0.0172	0.0544	0.172	0.542	1.69	5.16	14.7	35.3	63.3
10	0.0186	0.0589	0.186	0.586	1.83	5.56	15.7	37.1	65.1
11	0.0201	0.0637	0.201	0.633	1.97	5.99	16.8	38.9	66.8
12	0.0218	0.0688	0.217	0.684	2.13	6.44	17.9	40.8	68.5
13	0.0235	0.0743	0.235	0.738	2.30	6.92	19.0	42.6	70.2
14	0.0254	0.0802	0.253	0.796	2.48	7.43	20.2	44.5	71.7
15	0.0274	0.0865	0.273	0.859	2.67	7.97	21.5	46.4	73.3
16	0.0295	0.0933	0.294	0.925	2.87	8.54	22.8	48.3	74.7
17	0.0318	0.101	0.317	0.996	3.08	9.14	24.1	50.2	76.1
18	0.0343	0.108	0.342	1.07	3.31	9.78	25.5	52.0	77.4
19	0.0369	0.117	0.368	1.15	3.56	10.5	27.0	53.9	78.7
20	0.0397	0.125	0.396	1.24	3.82	11.2	28.4	55.7	79.9
21	0.0427	0.135	0.425	1.33	4.10	11.9	29.9	57.5	81.0
22	0.0459	0.145	0.457	1.43	4.39	12.7	31.5	59.2	82.1
23	0.0493	0.156	0.491	1.54	4.70	13.5	33.0	60.9	83.2
24	0.0530	0.167	0.527	1.65	5.03	14.4	34.6	62.6	84.1
25	0.0569	0.180	0.566	1.77	5.38	15.3	36.3	64.3	85.1
26	0.0610	0.193	0.607	1.89	5.75	16.2	37.9	65.9	85.9
27	0.0654	0.207	0.651	2.03	6.15	17.2	39.6	67.4	86.8
28	0.0701	0.221	0.697	2.17	6.56	18.2	41.2	68.9	87.5
29	0.0752	0.237	0.747	2.32	7.00	19.2	42.9	70.4	88.3
30	0.0805	0.254	0.799	2.48	7.46	20.3	44.6	71.8	89.0

Source: % NH_3 values were calculated using the equations proposed by Emerson et al. [18].

important of these include genetics of the fish, the water temperature, stress on the fish, age and species of the fish, the available oxygen concentration, and level of fish activity.

In aquaculture, maintaining dissolved oxygen in water at a level that is suitable for fish is an important activity for involved personnel. Fish oxygen-consumption rate increases with available oxygen concentration up to a critical concentration, as shown in Fig. 9.2. Above this critical concentration, fish engage in normal physiological activity. Dissolved oxygen level below critical concentration may result in reduced fish growth or even mortality. Table 9.7 shows oxygen-consumption rates of some common fish species and dissolved-oxygen concentrations corresponding to the fish activity levels.

Table 9.10. Toxicity of selected metals to aquatic species

Metal	96-hour LC_{50} (μg/L)	Safe Level (μg/L)
Cadmium	80–420	10
Chromium	2000–20,000	100
Copper	300–1000	25
Lead	1000–40,000	100
Mercury	10–40	0.10
Zinc	1000–10,000	100

Source: Compiled from Boyd [19].

Table 9.11. Toxicity of selected pesticides to aquatic species

Pesticide	96-hour LC_{50} (μg/L)	Safe Level (μg/L)
Aldrin/Dieldrin	0.20–16	0.003
BHC	0.17–240	4.0
Chlordane	5–3000	0.01
DDT	0.24–2	0.001
Endrin	0.13–12	0.004
Heptachlor	0.10–230	0.001
Toxaphene	1–6	0.005

Source: Compiled from Boyd [19].

Nitrogen is a principal component of amino acids, which form the building blocks of proteins available in food and feed of most living organisms. In aquaculture, protein metabolism by fish results in excretion of nitrogenous compounds, most of which dissolve in the culture water. Accumulation of some of these compounds, especially ammonia–nitrogen and nitrite–nitrogen, is a serious concern in aquaculture because they are highly toxic to fish. Table 9.8 shows the nitrogen-toxicity data for various fish species. In aquacultural waters ammonia–nitrogen is found in two states: ionized (NH_4^+) and unionized (NH_3). These two states form an equilibrium that depends on the pH and temperature of the culture water. Table 9.9 shows the effect of pH and temperature on this equilibrium. An increase in pH or temperature causes an increase in the concentration of the toxic form of ammonia (un-ionized). Sum total of these two states of ammonia is referred to as *total ammonia nitrogen concentration*, a routinely monitored parameter in most intensive-production systems. Oxidation of ammonia to nitrate–nitrogen, a relatively nontoxic substance, is a two-step process and is accomplished by chemoautotrophic bacteria (*Nitrosomonas* and *Nitrobacter*). This process, *nitrification*, is an acidifying process (causing destruction of alkalinity) and requires significant amount of oxygen. Nitrite–nitrogen, an intermediate product of this process, is very toxic to fish.

Some heavy metals and pesticides are also toxic to aquatic organisms. Depending on the water source, metals may be present in the water used for aquaculture. Pesticides and herbicides are often used around fish ponds to control pests and vegetation. Boyd [18] provides data on toxicity of selected metals and pesticides to aquatic life (Tables 9.10 and 9.11, respectively).

References

1. Wheaton, F. W. 1977. *Aquacultural Engineering*. John Wiley and Sons, New York.
2. Stickney, R. R. 1994. *Principles of Aquaculture*. John Wiley and Sons, New York.
3. Lawson, T. B. 1995. *Fundamentals of Aquacultural Engineering*. Chapman and Hall, New York.
4. Huner, J. V. and H. K. Dupree. 1984. *Third Report to the Fish Farmers*. Fish and Wildlife Service, U.S. Department of Interior, Washington, DC.
5. Colt, J. E. and G. Tchobanoglous. 1981. Design of aeration systems for aquaculture. In: L. J. Allen and E. C. Kinney (eds.), *Proceedings of the Bio-Engineering Symposium*. American Fisheries Society, Bethesda, MD, pp. 138–148.
6. Weiss, R. F. 1970. The solubility of nitrogen, oxygen, and argon in water and seawater. *Deep Sea Research* 17:721–735.
7. Colt, J. E. and C. Orwicz. 1991. Aeration in intensive culture. In: D. E. Brune and J. R. Tomasso (eds.), *Aquaculture and Water Quality*. World Aquaculture Society, Baton Rouge, LA.
8. Meade, J. W. 1989. *Aquaculture Management*. Van Nostrand Reinhold, New York.
9. Parker, N. C. and K. B. Davis. 1981. Requirements of warmwater fish. In: L. J. Allen and E. C. Kinney (eds.), *Proceedings of the Bio-Engineering Symposium for Fish Culture*. American Fisheries Society, Bethesda, MD.
10. Piper, R. G., I. B. McElwain, L. E. Orme, J. O. McCraran, L. G. Fowler, and J. R. Leonard. 1982. Fish Hatchery Management. U.S. Fish and Wildlife Service, Washington, DC.
11. Petit, J. 1990. Water supply, treatment and recycling in aquaculture. In: G. Barnabe (ed.), Aquaculture, vol. 1. Ellis Horwood, New York.
12. Tucker, C. S. and E. H. Robinson. 1990. *Channel Catfish Farming Handbook*. NY: Van Nostrand Reinhold, New York.
13. Romaire, R. P. 1985. Water quality. In: J. V. Huner and E. E. Brown (eds.), *Crustacean and Mollusk Aquaculture in the United States*. AVI Publishing, Westport, CT.
14. Hedgecock, D., K. Nelson, and R. A. Shleser. 1976. Growth differences among families of the lobster Homerus americanus. *Proceedings of World Mariculture Society* 7:347–361.
15. Beamish, F. W. H. 1970. Oxygen consumption of largemouth bass, *Micropterus salmoides*, in relation to swimming speed and temperature. Canadian Journal of Zoology 48:1121–1128.
16. McLarney, W. 1984. *The Freshwater Aquaculture Book*. Hartley and Marks, pp. 450–451.
17. Russo, R. C. and R. V. Thurston. 1991. Toxicity of ammonia, nitrite, and nitrate to fish. In: D. E. Brune and J. R. Tomasso (eds.), *Aquaculture and Water Quality*. World Aquaculture Society, Baton Rouge, LA.
18. Emerson, K., R. C. Russo, R. E. Lund, and R. B. Thurston. 1975. Aqueous ammonia equilibrium calculations: Effects of pH and temperature. *Journal of the Fisheries Research Board of Canada* 32:2379–2383.
19. Boyd, C. E. 1990. *Water Quality in Ponds for Aquaculture*. Auburn University/Alabama Agricultural Experiment Station, Auburn, AL.

10 Materials for Aquacultural Facilities

F. W. Wheaton and S. Singh

10.1 Considerations in Material Selection Process

Aquacultural facilities are profoundly different than traditional poultry or livestock production operations because of their unique rearing environment, water and/or salt water. Therefore, material selection for aquacultural facilities is an important aspect of the facilities construction. Due consideration of water quantity and quality is a prerequisite for the material selection process. The following constraints either limit the applicability of many materials for aquacultural purposes or require that special precautions and methods be used.

10.1.1 Weight of Water

More often than not aquacultural facilities handle large quantities of water. Because water is a relatively high-density fluid, it exerts considerable force on the structure enclosing it. This demands a careful consideration of materials strength. The components that hold static or flowing water require special attention. In open systems such as ponds and net pens, static and dynamic forces induced by the weight of the water and the prevailing winds and currents are very important criteria for site selection and pond construction. The soil type and the topography of the selected site will also determine pond-construction cost and influence the decision on pond size. In closed or semiclosed systems, tank or raceway walls and water conducts and associated fittings need to be properly sized for the pressure and water flow rate.

10.1.2 Corrosion

The potential of chemical reaction between the culture water and the materials used in the construction is a very serious concern in all aquacultural facilities. Important factors that may influence the chemical reaction include the environment (light, temperature, humidity, air, etc.), the presence of catalysts, oxygen concentration, and physical movement. The chemical reaction can potentially cause accelerated deterioration of the material and contamination of the culture water. This deteriorating effect on a material is called *corrosion*. Fresh-water systems can be expected to be considerably less vulnerable to corrosion than salt-water systems.

Fresh-Water Corrosion

Generally, fresh water from wells, lakes, rivers, streams, and municipal supplies is noncorrosive. In fresh-water systems, probably the most common form of corrosion is rusting of steel structures and components. Rusting is basically oxidation of wet metal (steel) surfaces into oxides. Steel structures and components are rust-prone and, therefore, special precautions should be taken with them. Rust-proof paints or coatings can be applied to ordinary steel surfaces, but this may result in an increased cost of components. Also, the rust-proof paints or coatings need to be compatible with the aquatic environment and strong enough to avoid chipping. If coated steel is undesirable other metals, such as aluminum or stainless steel, or plastics, such as polyvinyl chloride (PVC) or acrylic, should be considered. Additional corrosion resistance tends to increase materials cost.

Salt-Water Corrosion

Corrosion is a more common problem in salt-water systems. Salt water is very corrosive, and because of this special materials must be used for pipes, tanks, pumps, and other equipment coming into contact with the salt water. Most plastics, epoxy resins, and rubber compounds withstand saltwater very well. However, any materials used in aquaculture facilities growing or processing organisms for human consumption must satisfy government regulations with respect to toxicity, safety, and strength. Some of the epoxies, rubbers, and plastic materials contain compounds toxic to the cultured organism or to human consumers. These toxic compounds may be present either in the resin compound or in the hardener compound. Stainless steel may be used in most food-processing applications, but in production systems (particularly recirculating systems) one should make sure that chromium is not dissolved from the stainless steel and concentrated in the growing organisms. Shellfish are particularly sensitive to this, because they concentrate heavy metals in their bodies [1]. Most other metals are corroded very rapidly in salt water. Steel/iron structures and components that remain in contact with salt water for long periods of time may disappear altogether if proper coatings are not applied on them or corrosion rates are not allowed for by using heavier material thickness.

10.1.3 Biofouling

Biofouling, growth of unwanted attached aquatic organisms, is one of the most serious operational and maintenance problems encountered in most salt-water and some fresh-water culture systems. There are some 200 different species of marine-fouling organisms [2]. The size, diversity, and development pattern of the fouling community varies with the materials used and the site environmental conditions [3]. During the setting stage in their life cycle, many fouling organisms growing in water must attach to solid objects. The object may be a pipe, a tank, or a pump; in fact, it makes no difference to the organisms, but to the aquaculturist such nonselectivity can present a real management problem. Fouling rate is most rapid in areas in which water currents are slow and generally becomes less of a problem as salinity decreases. Thus, freshwater farms are less affected by fouling than marine farms.

Biofouling has multiple effects on the culture system. Biofouling deposits on net or float structures can increase weight and drag, reduce buoyancy, and severely limit water circulation through the culture space [4]. In a cage culture, the frequency and duration of the minimum water flow directly impacts the maximum stocking density that can be maintained. Therefore, biofouling can significantly reduce the production potential and create a stress-inducing environment for the cultured organisms. Also, if cage bags are fouled by algae and other organisms, the specific effective size of the mesh is decreased, increasing drag forces and adding to the weight of the cage. In addition, the decrease in mesh size reduces water exchange through the bag, reducing the rate of oxygen replenishment and waste removal. The increased drag forces cause the bag to deform, crowding the fish. Thus, fouling can have significant effects on unit productivity. Major net system failures, with fouling as a strong contributing factor, have already occurred in several large commercial marine aquaculture operations [4]. The rate of fouling is high in cages situated near thermal effluents. In land-based structures, drain pipes for effluent and other pipes and fittings can become clogged due to fouling. Fouling reduces the life of materials used in construction and requires frequent maintenance to reduce pressure losses in the system, thereby increasing both capital and operating costs of the system.

Biofouling Control Techniques

Wheaton [1] describes several methods that are available for dealing with fouling. These methods can be classified as chemical, mechanical, thermal, or constructional. Each method has its advantages and disadvantages.

In chemical sterilization, a chemical that is toxic to the fouling organisms may be added to the culture water. The chemical kills the attached organisms and prevents pipe clogging. Chlorine and ozone treatments of culture water are two examples of chemical sterilization. The toxic chemical can be applied in two different ways: the continuous addition of a low concentration of the toxic chemical to the water that prevents the organisms from ever becoming attached to the structure or equipment, or an intermittent process in which a high-concentration, short-duration chemical treatment is followed by a relatively long period during which no chemical is added. The cycle is repeated at a necessary time interval, and this procedure becomes a part of the regular management.

Chemicals used for fouling prevention can be, and generally are, toxic to the cultured aquatic organisms or aquacultural crop as well as to the fouling organisms. A chemical toxic to the crop could not be used with the low-concentration, continuous chemical treatment. However, because the concentration used in the continuous chemical method is low, the aquacultural crop may be able to grow in the water provided that the crop has a higher tolerance than the fouling organism. Proper care must be taken to ensure that the crop is not subjected to a sublethal concentration of a toxic chemical and the resulting stresses. The crop growth rate will almost always be affected under stressed conditions.

The intermittent process uses a high concentration of the treatment chemical, and the potential for toxicity to the crop is increased. However, with planning at the time of construction of a facility, bypass arrangements for the chemically treated water can be made to incorporate intermittent treatment as an integral part of the facility. In this arrangement, treated water will not come in contact with the crop but will pass through the bypass piping and plumbing system. After treatment and flushing to remove the chemical, water is again allowed to pass over the crop. However, it is difficult to remove fouling from the crop itself and from the structure in which the crop is held. Fouling can possibly be removed by treating the growing areas during the time between crops, provided that the fouling rate is slow enough. This procedure is usually not possible with partial harvests.

Another problem with the chemical treatment is that the chemicals may be nontoxic to the crop, but the crop may absorb some of the antifouling chemical. The crop may become unsuitable for human consumption, if the chemical is toxic to humans or is not cleared by health authorities for use on the crop [1].

A simple mechanical method of removing fouling may involve the use of hand labor and scraping or scrubbing devices. However, the labor cost and lack of physical access to certain areas such as inside of long pipes and other small components severely limit the practical application of this approach. An automated mechanical scraping system used for cleaning cooling tubes in power plants may be applied in aquacultural facilities. This method uses spherical balls having a diameter less than that of the smallest tube or pipe. The balls are introduced into the water passing through the pipes. As the solid balls are carried through the pipes, they scrape along the inside surface of the pipes, removing attached organisms. The water flow flushes the organisms away. The balls are collected after passing through the pipes and can be reused. This system could be employed in the piping of many aquaculture systems. However, it would be difficult to use in those components of the system in which the water velocity is low, such as culture tanks. The balls can potentially reduce the food supply entering the system by damaging or killing food organisms suspended in the water. Also, damage to the crop by mechanical abrasion is a possibility.

Thermal fouling-control methods are somewhat similar to chemical treatments. Temperature change, usually employed as an intermittent process, can be used to reduce or eliminate fouling. The temperature of the water flowing into the system is increased until it exceeds the lethal temperature of the fouling organisms, and the dead fouling agents are flushed away by water [1]. This method has limited commercial application because nearly always at least some fouling organisms have a lethal temperature above the lethal temperature for the crop being grown. Raising the temperature sufficiently to kill all fouling organisms will also kill the crop. Raising the water temperature is expensive in terms of heat energy. As with the chemical method, the thermal method can be used between the crops if the fouling rate is slow.

Fouling can also be prevented by disinfecting the water using ultraviolet light. However, water must flow past the ultraviolet light in a thin layer, because particles close to the light will shield particles further from the light. Fouling organisms getting through the treatment chamber alive will contaminate the system. Some problems also have been experienced because of variations in lamp intensity with age. In spite of the problems, ultraviolet techniques appear to be among the best available for aquacultural enterprises that do not depend on living organisms in the incoming water as food for the crop [1]. Treatment and maintenance costs for the ultraviolet units need to be considered.

Most fouling organisms cannot attach to a surface if the water velocity over the surface exceeds some minimum value. If incorporated during facilities construction, this characteristic can be utilized to prevent fouling in many parts of an aquacultural system. The water velocities in pipes, pumps, and other components are maintained above the highest velocity at which fouling organisms can set. This antifouling system is simple, but it is difficult to employ in large components like growing tanks and at changes in direction in pipes. Maintaining high velocities in tanks calls for large volumes of water because of the large cross-sectional area of the tanks. The higher water volume required to maintain the necessary velocity increases pumping costs. In piping systems the velocity can be easily maintained or increased by decreasing the pipe size. However, higher pipe velocities cause a greater pressure loss per unit length of the pipe. Pumping costs increase as the pressure drop increases. Because pressure must be increased at the water source to increase the velocity of flow, operating (pumping) costs of the system go up as the water velocity in the piping system increases. Also, the velocity on the inside of an elbow curve may be low, causing settling of fouling organisms, even if the average velocity in the piping system is high enough to prevent it. High water velocities in growing tanks or ponds can also be detrimental to the crop. For example, oysters reduce pumping or stop altogether if water velocity is too high [1]. Finfish expend more energy in swimming in high-velocity water, and this reduces food-conversion ratios because more energy is lost to the increased metabolic needs of the fish. The higher water velocities do help to remove waste materials and to maintain water quality in the growing area. Some raceway systems rely on high water-flow rates for waste removal, and therefore this method can sometimes be used with them.

Duplicate plumbing systems are often employed to prevent fouling in aquatic growing systems. The water flow is periodically switched from one system to the other. The unused system is drained and allowed to remain dry long enough to kill all fouling organisms in the system, or it is closed off and allowed to become anaerobic. By periodically switching from one plumbing or piping system to the other, both systems can be kept free of fouling organisms. Although the installation expense of this system is prohibitive in some situations, it often provides not only a pollution-free fouling control system but also a backup pumping or plumbing system in case of breakdown or routine maintenance requirements of the prime system. Often this aspect of the dual system makes installation costs justifiable even if it saves one crop during the life of the system [1]. Usually the volume or space requirements of the growing tanks or ponds make it impractical to extend the dual system to these areas. However, growing areas are much more accessible and, therefore, easier to maintain free of fouling than other parts of the system. For example, ponds having soil as the side and bottom material are virtually fouling free if correctly constructed. This technique is one of the most practical and effective available.

Some copper alloys, particularly copper/nickel, have significant resistance to fouling [4]. Galvanized steel construction with anticorrosion safeguards is also fouling-resistant [2]. These alloys can sometimes be used for structures that remain in contact with water in marine systems (cages and net pens).

10.1.4 Ozone as a Constraint in Material Selection

Ozone treatment of culture water can be used to improve water quality or disinfection. Because ozone is a strong oxidizing agent, any materials used in an ozone treatment system must be inert to ozone. Also, if an aquacultural facility uses ozone for water treatment, it is important that proper materials are selected for components that will hold ozonated water. Operating costs of an ozone delivery system increase as the delivery system degrades. Greater amounts of ozone must be produced in systems with substandard materials, because some of the ozone is lost in the oxidation of materials in the delivery system itself. Also, as the delivery system fails, it must be replaced, adding to the overall cost of the operation. Ozone is toxic to humans and any system leakage is a serious safety hazard. Therefore, it is important to construct the ozone system from materials that will not corrode.

10.2 System Components and Material Selection

The type of aquacultural facility, whether pond, raceway, net pen, or tank-based plays an important role in the material selection process. An aquacultural facility has several structural components. Each component serves a different purpose. Therefore, the material-selection process for an individual component should consider that particular component's role in the system. Wheaton [1], Lawson [3], Beveridge [2], and Huguenin et al. [4] provide useful insights into the material-selection process for the components of aquacultural facilities.

10.2.1 Tanks

Tanks can be constructed from a variety of materials, including wood, concrete, plastic, fiberglass, metal, and glass [3]. Aquaculture tanks should have the following characteristics: smooth interior surface to prevent abrasion, nontoxicicity, durability and portability, ease of cleaning and sterilizing, noncorrosiveness, structural strength, and easy affordability. The choice of material should include these characteristics.

10.2.2 Raceways

Although earthen raceways are sometimes used, the majority are constructed from poured concrete or concrete or cement blocks. Earthen raceways are sometimes lined with waterproof liners to reduce water loss through leakage. Many small, experimental raceways are fabricated from wood, metal, fiberglass, plastic, or other materials.

10.2.3 Waterproof Lining

Waterproof liners can be used with tanks fabricated from virtually any material. They are commonly used in place of expensive coatings or sealers. Liners are also used for sealing ponds. They eliminate the hazard of toxicity by heavy metals, paints, treated wood, or other substances. Liners can be custom-made to fit any tank or pond size and shape. They must be carefully handled to avoid tearing and have a useful life of about 5–10 years, depending on material and use. The life of a liner is significantly reduced by highly acidic or alkaline water. Many modern tank or pond liners are resistant to ultraviolet rays and therefore have a relatively long life when used outdoors. Cost, service conditions, and availability are factors that must be considered in material selection for pond sealers.

Waterproof materials such as polyethylene, vinyl, and butyl rubber are gaining acceptance as pond linings. These linings reduce seepage loss to essentially zero, provided that they are not broken or punctured. They also give a more dependable seal than well-compacted bentonite, but for many applications cost is a significant factor. Black polyethylene films are less expensive than vinyl and butyl rubber for a given thickness [1].

10.2.4 Screen Mesh

Screen mesh for particulate filtration is available in a wide variety of materials including galvanized steel, carbon steel, brass, stainless steel, nylon, and other fabrics. Mesh material selection is based on characteristics of the influent and desired mesh size. For example, galvanized steel would be a poor selection for use in salt water because it corrodes easily. Because some mesh sizes are available in a limited number of materials, mesh size desired may influence screen material selection.

10.2.5 Nets

Materials chosen for nets should have high resistance to sunlight and be able to withstand weathering well and provide the necessary strength. A net having specific gravity in excess of 1 will sink, usually a desirable trait in fixed nets. Cost of netting is also a major consideration. Polyvinylidene chloride or vinyl chloride netting is recommended by Shimozaki [5]. Polyvinyl alcohol may be used, but it has poor weathering resistance and low specific gravity. Polyethylene also sees limited use because it is low in cost and strong, but this material has low specific gravity. Nylon's unsatisfactory qualities such as low specific gravity and poor resistance to sunlight are usually outweighed by its low cost [2].

10.2.6 Ozone Unit

Hochheimer and Wheaton [6] discuss suitability of various materials for use with ozone in aquacultural applications. Table 10.1 presents the list of these materials commonly used in aquacultural facilities and their relative resistances to ozone. Polyvinyl chloride (PVC) can be used for temporary ozone contact, but it is not recommended for long-term application. Reinforced concrete should also be used with precautions, as ozone may corrode galvanized steel reinforcing bars in the structure. For best results, 304 or 316 stainless steels for flanged or threaded applications and 304L or 316L for welded (tungsten inert gas) piping joints are best for all wet and dry gas-piping components and flexible couplings [8]. Valves should be made of stainless steel (both body and face) with Viton, Teflon, or Hypalon membranes and gaskets [8]. Concrete joints made from Sikaflex-1A also are recommended for use with ozone systems [8].

10.3 Advantages and Disadvantages of Various Materials

In general, an ideal material for aquacultural facilities will have the following characteristics: high strength; low cost; easy availability; resistance to corrosion, weather, and fouling; light weight; ease of transport, construction, and repair; nontoxicity; reusability; and smooth texture. No single material posses all these characteristics. Specific materials presented here have certain advantages and disadvantages compared with each other and therefore find their use in conditions requiring their positive characteristics.

Table 10.1. Suitability of materials for use with ozone systems

Material	Resistance
Metals	
Chromium, nickel, silver	Minor effect
Brass	Minor effect
Aluminum	No effect
Pig iron	No effect
Galvanized steel	Minor effect
304 Stainless steel	Minor effect
316 Stainless steel	No effect
Copper	No effect
Elastomers	
Buna N (Nitrile)	Not recommended
EPDM	No effect
Hypalon	No effect
Kel-F	No effect
Natural rubber	No recommended
Neoprene	Moderate effect
Silicone	No effect
Viton	No effect
Plastics	
ABS	Minor effect
Acetel (Delrin)	Moderate effect
CPVC	No effect
Hytrel	Moderate effect
LDPE	Moderate effect
Nylon	Not recommended
Polycarbonate	No effect
Polypropelene	Minor effect
PTFE (Teflon)	No effect
PVC	Minor effect
PVDF (Kynar)	No effect
Others	
Concrete	No effect
Glass	No effect
Ceramics	No effect

Source: Compiled from Damez [7].

10.3.1 Masonry

Concrete is widely used for constructing large tanks or pools [1, 3]. It is easy to work with and can be formed into any shape. Properly reinforced with steel bars, concrete will last indefinitely for use with either fresh or salt water. Because of its weight and expense, however, concrete is used for the construction of permanent facilities. The interior surfaces of concrete tanks should be smoothed to avoid abrasion of fish skin and scales. Also, rough or porous surfaces are harder or impossible to clean and disinfect. Interior surfaces can be coated with a sealer, many of which are commercially available. Uncoated concrete surfaces are suitable for most aquaculture purposes, but only after a sufficient curing period [3]. Harmful substances can leach from newly poured concrete. Thus, concrete tanks should be flushed with clean water for several days prior to use [3]. Concrete and cement blocks are widely used for raceway construction, as they offer cost and strength advantages over other materials for handling relatively large quantities of water.

10.3.2 Metals

Metals have several uses in aquacultural applications. One can find both large structures such as tanks and smaller components such as pipes and fittings made of commonly used metals. Durability and smoothness are two main advantages of metallic components. However, metallic components, especially tanks, are heavy and therefore difficult to transport. Also, metal parts may cost more than plastic parts and be corrosion-prone. Metals may become toxic in waters that are poorly buffered [1]. Metal bars, rods, and sheets can be used as supports to plastic or concrete structures. Galvanized metal tanks should not be used for aquaculture purpose unless they are coated or lined. Zinc leaches from the galvanized coating and can cause heavy-metal poisoning. Salmonids are generally quite susceptible to zinc toxicity. The 96-hour LC_{50} for zinc ranges from 0.09 mg/L to 41 mg/L for various species [9]. It is difficult to get most paint coatings to adhere to galvanized metal; therefore, it is best to avoid use of this material altogether, or a liner should be used.

Steel

Steel can be used for tank and other large structural constructions in freshwater systems. Steel often is the least expensive means of providing structural support to large components such as tanks. Rust and heavy weight are potential problems with steel. With proper coatings, steel can provide good service in both fresh-water and salt-water systems. Coatings may increase cost and, particularly for saltwater systems, steel rarely is the best choice because of corrosion. Sectional steel tanks can be easily bolted together to construct a large tank, offering the advantages of easy construction, transportation, and dismantling [3].

Aluminum

Aluminum serves well in fresh water, and certain aluminum alloys provide good service in salt-water systems. In closed-cycle or recirculating systems, all metals including aluminum should be used with care because even limited erosion over long periods can cause an ion buildup in the culture water. Aluminum and steel are two metals that are commonly used in the fabrication of small tanks popular in hatcheries for rearing small fry and fingerling. If used with caution, some aluminum alloys can be used for brackish or salt-water culture. Water with a pH well into the acidic range will cause aluminum to become soluble and leach into the water [3].

Stainless Steel

Stainless steel provides good service in fresh water and fairly good service in salt water. However, even the corrosion-resistant 300 series stainless steels will eventually corrode in saltwater systems. Stainless steel is expensive, but for screens and similar applications it is a good alternative if plastic materials are not available or do not provide the necessary strength. Small stainless steel tanks are generally safe to use with either fresh or salt water, particularly 316 stainless steel. However, stainless steel tanks are too expensive for large-scale use.

Copper Alloys

Many copper alloys, particularly copper–nickel (90%–10%), have considerable biofouling resistance and very low corrosion rates and can, under many circumstances, be used in areas with only modest water flow without any detrimental effects on culture organisms [4]. Copper-based antifouling paints are commonly applied to synthetic-fiber fish cages [2, 4]. Metallic copper-alloy meshes can be used in rigid structures to give long life and low-maintenance service and have previously been successfully used on a smaller scale for marine fish containment [4]. However, copper can be absorbed by shellfish, a characteristic that may be harmful to marketing efforts.

10.3.3 Plastics/Rubber Compounds

The use of synthetic, hydrocarbon-based materials is increasing rapidly in aquacultural facilities, as these materials offer several advantages over traditional materials. However, they can be expensive and vulnerable to industrial pollution. In aquacultural applications, plastics are most vulnerable to the ultraviolet component of solar radiation. The term *plastic* represents a variety of polymers including acrylics, polyethylene, polypropylene, vinyl, PVC, fiberglass, and similar materials. Each has its own set of good and bad features. Most plastics can be used in both fresh-water and salt-water systems. Lawson [3] and Wheaton [1] explain several uses of plastics in aquacultural systems. Plastics are durable, smooth, light, inert, and easily formed into desired shapes. A major advantage of plastic tanks over other tanks is that they are lightweight and thus easily portable. Repairs to plastic tanks are also easier. At present, almost all aquacultural structures and parts are available in some form of synthetic materials such as plastics and fiberglass. Plastic tanks are available in various shapes and sizes, from small bowls to aquaria up to several hundred liters capacity. Most plastics are nontoxic, but some, like polyethylene, are initially toxic and should be conditioned with clean, running water for at least two weeks prior to use, particularly for marine culture. Initial toxicity often is a result of the hardner used in the plastic manufacturing proces or mold release compounds or die lubricants used during manufacture.

Fiberglass

Fiberglass is one of most popular materials for tank construction. It is light, strong, and relatively inexpensive. It can be molded into most desired shapes. It is inert in both fresh and salt water. It can withstand the effects of ultraviolet rays if used in sun. Fiberglass tanks are normally gel-coated on the inner surfaces to provide a smooth surface that may be easily cleaned and disinfected. The fiberglass tanks can be easily drilled for installation of drains and other plumbing fixtures. A variety of tank sizes are available, from a few hundred liters up to several hundred cubic meters. Some models are available with legs, skirts, or stands so that they can be elevated above the floor or ground. Tanks smaller than 1.0 to 2.4 m in diameter are usually available in a single molded unit, but larger tanks may come unassembled in two or more sections that must be bolted or glued together [3]. Some manufacturers will assemble tanks on-site for an additional fee. Fiberglass tanks are available from numerous manufacturers worldwide. Fiberglass is strongest in tension loading, and this is the stress experienced in circular tank walls. Molded fiberglass may contain fiberglass matting. In this case the mat material must be completely covered with resin [1]. Fiberglass mat consists of very thin, sharp pieces of glass. If this material gets into the culture water, it may enter the cultured animals' gills or digestive structures, causing extensive tissue damage or internal bleeding.

Vinyl

Vinyl is a flexible plastic widely used for swimming-pool liners and other applications requiring flexibility and essentially zero permeability to water. In aquaculture facilities, vinyl is mostly used for lining purposes. In fish culture, children's wading and swimming pools are used to culture or hold fish. Most of these pools are constructed of some type of vinyl. Vinyl is flexible, and thus support must be provided for the vertical walls. In small pools it is done with inflated circular sections of vinyl. Larger pools require more rigid support, which in commercially available units usually is some type of coated steel. The flexible properties of vinyl allow the liner to conform to nearly any shape of external support as long as the support material is smooth. Because vinyl is easily punctured and cut, care must be exercised to prevent liner contact with sharp or pointed surfaces. As a lining material for fish culture ponds, vinyl has several advantages. It is relatively inexpensive, smooth, and easily cleaned. However, it is not very durable. Vinyl liners have also been used to seal ponds constructed in permeable soils. With a good sand base under the vinyl and an overlaying layer of sand for protection, vinyl will provide a long service life.

Polyethylene

Polyethylene is similar in properties to vinyl but does not elongate as much. It is also widely used for making impermeable liners for low-cost, rigid structures. It is available in rolls up to 6 m wide and 75 m long. Polyethylene can be purchased in many thicknesses based upon the requirement and is quite inexpensive. However, polyethylene does not have a long life, can be easily punctured, and sometimes may be difficult to join.

Acrylic

Acrylic materials are widely used for experimental purposes in fish culture. Plexiglas acrylic plastic has limited flexibility in thickness below about 0.35 cm. At greater thickness it is essentially rigid. Lexan is another acrylic trade name. For experimental work, transparent acrylics provide easy observation of the desired operation. Acrylics are available in a variety of colors, as well. Acrylics can be drilled, taped, and worked with machine tools into desired shapes and dimensions. In aquariums this means that plumbing can be threaded directly into the acrylic with drills and taps, which is impossible with glass. Acrylics will not shatter like glass, but being much softer, they scratch relatively easily. Solvent bonding can be done, or special glues designed for use on acrylics can be used.

Acrylics are inert to both salt and fresh water. However, they swell slightly if exposed to high humidity. This becomes a problem if one side of a large acrylic piece is subjected to air of high humidity and the other side is exposed to air of much lower humidity. The humidity difference causes the sheet to warp, and if the material is restrained, fasteners or the acrylic itself may be severely stressed or broken. Some acrylics warp more easily than others, and manufacturers' specifications should be consulted before acrylics are used in this type of service. Immersion of one side of the plastic sheet in water does not appear to cause a similar problem. Except for small tanks, acrylics are not suitable for tank construction because of cost, but they can have other applications in aquacultural facilities.

PVC

PVC can be used for tank construction, but structural support must be provided for all but small tanks. PVC can be glued or heat-welded. Primary use of PVC is not for tanks but for piping and fittings. It is inert in both fresh and salt water. As pipe and fittings it is inexpensive, light, and quite durable if not exposed to too much solar radiation or industrial pollution. PVC pipe and fittings are readily available in a wide range of sizes and wall thickness. Sheet PVC is more expensive.

Polypropylene

Polypropylene tanks are available from various suppliers in stock sizes up to at least 1500 L. Larger tanks can be purchased on special order. Larger tanks tend to be more expensive than similar-size fiberglass tanks. Another problem with polypropylene is that it is very difficult to glue anything to it. However, special heat-bonding equipment is available for bonding polypropylene to itself. Mechanical fittings are generally used with polypropylene components. Polypropylene is also inert in both sea water and fresh water.

Butyl Rubber

Butyl rubber is suitable for lining purposes. It can be joined or patched by special cements. In ponds, butyl-rubber linings may be used without a covering of soil except in places in which livestock, people, or equipment will be moving. Minimum film thickness for placement over clean, silty, or clayey gravels is 0.76 mm [1]. Soil sterilization is unnecessary if the butyl-rubber lining is thicker than 0.5 mm is used.

10.3.4 Wood

Inexpensive fish-culture tanks can easily be fabricated from wood. Wheaton [1] describes a simple method to construct a wooden tank for aquaculture. Besides being light in weight, wood is typically less expensive than most other construction materials and easy to work with. Marine plywood is often used to construct wooden tanks, but other grades can be used as well. Plywood tanks should be designed to prevent excessive flexure of the walls when the tank is filled. The tank sides should be braced to resist the static forces exerted by the water. Treated woods should not be used in contact with the culture water, because many contain substances that are toxic to fish. All exposed surfaces should be painted to seal the wood against rot. Paints containing lead or other heavy metals that may leach into the water should never be used for aquaculture purposes.

Tank interiors can be sealed with nontoxic materials such as epoxy or fiberglass resin paint. These materials take one to two days to cure, and, once cured, they form a hard, smooth surface. In place of paints or sealers, waterproof liners can be used in wooden tanks. Wooden-tank fabrication is very site-specific; therefore, no design concepts are presented here. Design of epoxy-coated plywood tanks cannot be based solely on allowable stress. Strain must be minimized because many coating materials are highly brittle. Strain or flexure of the tank walls under load will crack many epoxy coatings. Thus, selection of epoxy coatings and hardners should be done with care to ensure they have some elasticity and are nontoxic.

10.3.5 Others

Glass

Glass culture units are found almost exclusively in the aquarium trade or for public-display aquaria. Aquaria range in size from small fish bowls to over 400 L in volume. Glass is not practical for use in the fabrication of large tanks because of high cost, excessive weight, and the ease with which it can be damaged. In the aquarium-manufacturing industry acrylic is rapidly replacing glass for tank construction, but acrylics are too expensive for large-tank construction [3].

Gunite

Gunite is a strong, durable, concrete-like material that has an indefinite life. It is often used to construct small spas, swimming pools, and ornamental fish ponds. Gunite material is very compact and can be blown under high pressure over a support framework fabricated from steel reinforcing rods and small-mesh poultry fencing. The material is usually applied in a 5 to 10 cm thick layer [3]. Gunite is more expensive than liners but has indefinite life, whereas liners must be replaced every few years. Gunite has rough surface and porous texture, which are harmful to fish and the fish environment. The utility of gunite as a construction material for production tanks has yet to be proven.

Natural Fabrics

Natural fabrics such as cotton and hemp are suitable for fish netting and cage cultures [2]. These materials have been used historically. But, with the advances in plastics and other synthetic compounds, their use is now confined to subsistence farming. Most net materials today are some type of plastic, such as polypropylene, nylon, or PVC.

References

1. Wheaton, F. W. 1977. *Aquacultural Engineering*. John Wiley and Sons, New York.
2. Beveridge, M. 1996. *Cage Culture*. Fishing News Books, Oxford, England.
3. Lawson, T. B. 1995. *Fundamentals of Aquacultural Engineering*. Chapman and Hall, New York.
4. Hugugenin, J. E., S. C. Fuller, F. J. Ansuini, and W. T. Dodge. 1981. Experiences with a fouling-resistant modular marine fish culture system. In: L. J. Allen and E. C. Kinney (eds.), *Proceedings of the Bio-Engineering Symposium for Fish Culture*. American Fisheries Society, Bethesda, MD.

5. Shimozaki, Y. 1964. Production and characteristics of synthetic nets and ropes in Japan. In: *Modern Fishing Gear of the World*, vol. 2. Fishing News Books, London. [cited by Wheaton [1]].
6. Hochheimer, J. N. and F. W. Wheaton. 1995. Ozone Use in Aquaculture. Paper presented at Aquaculture '95, San Diego.
7. Damez, F. 1982. Materials resistant to corrosion and degradation in contact with ozone. In: W. E. Masschelein (ed.), *Ozonation Manual for Water and Wastewater Treatment*. John Wiley and Sons, New York.
8. Robson, C. M. 1982. Design engineering aspects of ozonation systems. In: R. G. Rice and A. Netzer (eds.), *Handbook of Ozone Technology and Applications*, vol. I. Ann Arbor Science Publishers, Ann Arbor, MI.
9. Piper, R. G., I. B. McElwain, L. E. Orme, J. O. McCraran, L. G. Fowler, and J. R. Leonard. 1982. *Fish Hatchery Management*. U.S. Fish and Wildlife Service, Washington, DC.

11 Facilities Design

M. B. Timmons, J. Riley, D. Brune, and O.-I. Lekang

11.1 Introduction

Culturing fish requires that the animals be confined in some controllable volume such as a pond, raceway, net pen, or tank. An important goal with the culturing system is to provide the fish a satisfactory culture environment so that growth and conversion of feed is as efficient as possible. Choice of the culture system is partly dependent upon the species of fish being cultured. Salmonids require very high water quality in comparison with carp and catfish, which have less-stringent water-quality requirements. The largest-scale production systems in the world have generally been constrained to ponds. As an example, the U.S. catfish industry produces 200 million kg per year from pond systems that consist of units that are typically between 2.5 to 5 ha in a rectangular shape. Raceway construction typifies the production systems used to produce salmonids, and in particular trout, and in both Europe and the United States the majority of production occurs for these species in raceways. Tanks, round, hexagonal, or octagonal, are commonly used for a variety of species, but generally for smaller scales of production. This trend is changing, however, with some of the largest farms concentrating on the use of tank structures. The advantage of tank culture is that it lends itself to easy maintenance because tanks can be designed to be relatively self-cleaning of fish waste and uneaten feed. For indoor fish farms, tank culture is by far the most dominant method of culture. In Norway, all juvenile salmon are produced in tanks using a range of tank sizes from 3 to 15 m in diameter and up to 4 m in depth. Recent advances in indoor culture technology indicate that indoor fish farming and tank culture may become the standard of the future, particularly in the United States, where environmental considerations and constraints may force tank culture in which waste streams can be controlled. Net pens are used mainly for larger fish and on sites that have good natural water currents. Net pens are generally considered cheaper per unit of fish-carrying capacity than land-based units. Most of the adult salmon production in the world uses net-pen culture systems. Each of these culture-system types is discussed in the following sections.

A key factor in selecting the appropriate culture system is the scale of operation intended. Small-scale operations of 12,000 to 20,000 kg per year often are economically competitive only because of local market opportunities and the use of family (subsidized or noncosted) labor [1, 2, 3]. Sedgwich [1] found that 20,000 kg of trout production was the smallest farm that could provide a viable income if fish production were the sole source of income. Intermediate-sized operations do not appear to offer much advantage. Easley [4] found only a 4% decrease in costs of production among farms between 12,000 and 60,000 kg. Keenum and Waldrop [5] reported the costs of production for a catfish pond operation as $1.50 U.S., $1.39, and $1.32 per kilogram at yearly production levels of 286,000, 580,000, and 1.2 million kg. In terms of commodity levels of production, scale of operation has distinct advantages. These data support sizing the fish-production system carefully to the management and labor resources that are to be dedicated to the operation. These data do not address the impact of productivity from a fixed facility, previously mentioned by Losordo and Westerman [6].

One should bear the above consideration in mind closely when selecting a culture system. The least cost is generally associated with tank culture systems, because they can be made very small with size and cost basically proportional. Once experience is gained with the particulars for management of a specific species, then selection of the culture system can be reevaluated.

One of the major considerations for intensive aquaculture operations is the required water-flow rates to maintain water-quality conditions that are conducive to high productivity. Generally, required flow rates are dictated by oxygen considerations, although ammonia loading or carbon dioxide build-up can also dictate the maximum required flow rates. These calculations are identical for tank or raceway considerations and are presented in the section on raceways, in Eqs. 11.2 to 11.5.

11.2 Ponds

Ponds are clearly the dominant design used for aquaculture accounting for roughly 40% of the total aquaculture production [7]. Worldwide the average production from ponds is 1040 kg/ha, but the range of productivity varies by a factor of 10. High-density shrimp ponds and carp ponds may produce in excess of 12,000 kg/ha. In such systems, artificial aeration and supplemental feeding are required.

Ponds are typically greater than 0.9 m and less than 2.4 m in depth. Pond depth is selected based on a compromise between being deep enough that light does not penetrate to the pond bottom (this limits rooted plant growth) and shallow enough to limit temperature stratification, reduce overall water usage, and to minimize harvesting efforts. Pond areas range from 0.3 to 7 ha for catfish and shrimp farms; size is less important than maintaining a regular rectangular shape to facilitate net harvesting and efficient usage of land space. More details are given later in this section on physical design parameters.

As biomass densities and feeding increase, generally water exchange through the pond is increased to maintain water quality. At some high rate of water exchange, the pond becomes a raceway. Ponds are dominated by algal photosynthesis. Stable algal populations in high-rate algal production ponds have been maintained at hydraulic detention times as short as 2 days, or a water exchange of 50% per day [8]. Typical ponds, because of self-shading and limited mixing, should be operated on no less than a 3- to 4-day detention time (25% to 33% water exchange rates per day). Once this exchange rate is exceeded, the algal cell concentration is washed out of the pond. Water make-up for evaporation is typically between 0.25 and 1 cm per day, which for a 1.5-m pond depth amounts to less than 1% water exchange per day.

11.2.1 Pond Photosynthesis

Ponds, being photosynthetic systems, exhibit a diurnal water quality fluctuation. The algal photosynthesis occurring in the pond water column has been represented by Stumm and Morgan [9].

$$106CO_2 + 16NO_3 + 122H_2O + HPO_4 \rightarrow C_{106}H_{263}P_{110}N_{16}P + 138O_2 \qquad (11.1)$$

This equation suggests that for every unit of oxygen added to the pond water column by photosynthesis, an equal unit of oxygen demand will be produced by algal biomass, amounting to 0.80 g VS/g O_2 produced, or 3.47 g O_2/g carbon fixed. This leads to a number of important implications with regard to pond aquaculture. First, because photosynthesis occurs only in the light, the pond environment experiences a daily or diurnal peak in oxygen production followed by a nighttime minimum of oxygen owing to respiration. The daily cycle of oxygen becomes a management problem for the fish farmer as mechanical aeration is used to prevent oxygen levels in the pond from going below lethal levels.

11.2.2 Diurnal Limits

The situation of the catfish pond can be used to illustrate this. If a pond contains no fish and no nutrients to drive algal production, the oxygen concentration in the pond water will reach the equilibrium dissolved oxygen level as defined by temperature, 8.4 mg/L at 25°C (Fig. 11.1a). If nitrogen and phosphorus are added to the pond at a rate equal to the N and P in catfish feed sufficient to support 3000 kg/ha, the pond oxygen levels will become polarized as a result of photosynthesis with a daytime high of 10.4 mg/L and a nighttime low of 6.4 mg/L (at typical pond photosynthesis rates of 2.0 g carbon/m^2/d, see Fig. 11.1b). If a minimum desirable oxygen concentration in the pond is 3 mg/L, and surface re-aeration can supply 3.6 g O_2/m^2/d at a typical wind speed of 3 m/s, approximately 590 kg of catfish respiratory demand can be sustained (see Fig. 11.1c). The catfish farmer is able to increase the carrying capacity of the pond by eliminating the bottom of the nighttime diurnal curve by applying mechanical aeration when needed. Under these conditions, unless the algal crop is being harvested, photosynthesis is not supplying oxygen to the pond. In practice, mass balances on the growing algal crop consistently fail to account for approximately 25% of the algal cell production. This algal biomass is likely leaving the water column by entering the catfish biomass through the food chain or exiting the pond by decomposition and seepage to the groundwater. This fortuitous algal harvest combined with early-morning mechanical aeration to increase pond O_2 from a low of 1.3 mg/L to 3.0 mg/L allows the catfish farmer to push the carrying capacity to 1400 kg/ha (see Fig. 11.1d).

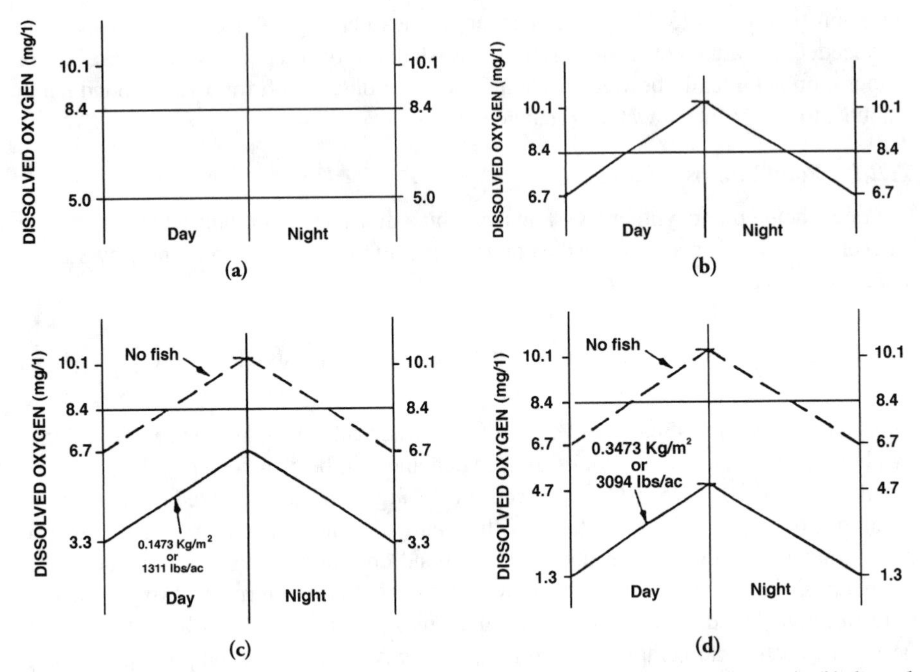

Figure 11.1. (a) Oxygen concentration at equilibrium (8.4 mg/1) with no algal growth; (b) diurnal oxygen fluctuation due to algal growth at a rate of 2 gm carbon/m^2/day; (c) suppression of oxygen concentration due to 1131 lbs/acre of catfish; and (d) pond oxygen concentration with 3094 lb/acre of catfish at 25% algal harvest.

11.2.3 Productive Systems and Polyculture

The above analysis also explains how shrimp and carp farmers can increase production to in excess of 10,000 kg/ha. In the case of carp culture, the algal biomass is the feed source to the expanding fish biomass. In effect, the oxygen demand associated with the algal biomass is stored in the fish flesh. Consequently, the fish are producing their own oxygen supply by consuming the oxygen demand produced from the algal photosynthesis. The shrimp farmer, on the other hand, produces a net oxygen supply in the pond by discharging the algal oxygen demand in the effluent. This allows the oxygen to be supplied to the crop by indirectly using the oxygen supply of the surrounding environment into which the pond farmer discharges the pond effluent. The environment must be capable of sustaining the load imposed for this practice to work; political constraints may eliminate this option.

The use of polyculture is another technique of expanding the productivity of pond aquaculture, by pond algal production or reducing decomposition of oxygen-demanding organics within the pond water column. By adding different trophic levels to the pond, such as oysters or clams, the farmer reduces the oxygen demand to the pond by storing organic oxygen demand in other animal biomass (from consumption of fecal or algal solids). Smith [10] discussed the use of polyculture techniques to shift algal production into animal biomass, which represents harvestable oxygen demand.

Drapcho and Brune [11] presented a theoretical analysis of pond aquaculture suggesting that if algal photosynthesis could be sustained at maximum levels of 6 to 12 g carbon/m^2/d, and harvested by filter-feeding fish, sustained yields in excess of 26,000 kg/ha may be possible in non-aerated ponds with limited water-exchange rates.

11.2.4 pH Limits

As oxygen cycles on a diurnal basis, so too the pH is cycling. As inorganic carbon is extracted from the water for algal carbon fixation, the pH of the water column will experience a daily rise and then, in turn, experience a nighttime decline from CO_2 release. Boyd [12] recommends a pH range of 6.5 to 9.0 as desirable range for fish production, with 4.0 to 5.0 being the acid death point and 11.0 the alkaline death point. In many cases, pH fluctuations in organically fed ponds are not a problem because CO_2 is removed from the pond as side benefit of the gas exchange that must occur to prevent death from oxygen depletion. However, if the pond water has an alkalinity of less than 25 mg/L and a production rate of greater than 3 g carbon/m^2/d, it is possible to exceed the high end of the desirable pH scale. This is particularly true if the pond is fertilized by inorganic fertilizer so that there is no organic carbon input to the pond. Under these conditions, the pond may not have sufficient buffering capacity to prevent excessive pH shifts. Under such conditions, production must be limited, or additional alkalinity must be supplied to the pond, usually in the form of lime.

11.2.5 Nitrogen Control

There is some basis for thinking that the algal crop in an aquaculture pond plays an additional important role in preventing excessive levels of toxic NH_3 and NO_2 accumulation in the pond water. However, the relative rates of pond nitrification (bacteria-mediated transformation of $NH_3 \rightarrow NO_2 \rightarrow NO_3$) compared with algal photosynthesis are not well understood. Toxicity owing to NH_3 and NO_2 build-up in ponds has been documented. However, frequent periods of cloudiness (and reduced photosynthesis) can often be sustained with little build-up of toxic ammonia, even in ponds with carrying capacity in excess of 3000 kg/ha [13]. In general, if net oxygen production rates are maintained by algal photosynthesis followed by algal harvest, one should expect that nitrogen accumulation will also be controlled by algal nitrogen fixation, combined with algal removal.

11.2.6 Pond Types

There are three common types of ponds: watershed, excavated, and embankment ponds. Watershed ponds take advantage of natural topography that includes a ravine of natural depression and simply require a dam across the lower end of the natural drainage outlet to catch seasonal rainfall. This type of pond is not well suited to aquaculture because the harvesting of fish from such a system is not practical. Control of water supply is undependable and unpredictable. Excavated ponds also are generally not used for aquaculture because they are simply holes dug into the ground and then allowed to fill from rainfall or spring flow. These ponds often result from situations in which landfill is needed and a pond depression results from removing soil for these other purposes. At best, these ponds represent opportunities for cage culture or sport-fishing opportunities.

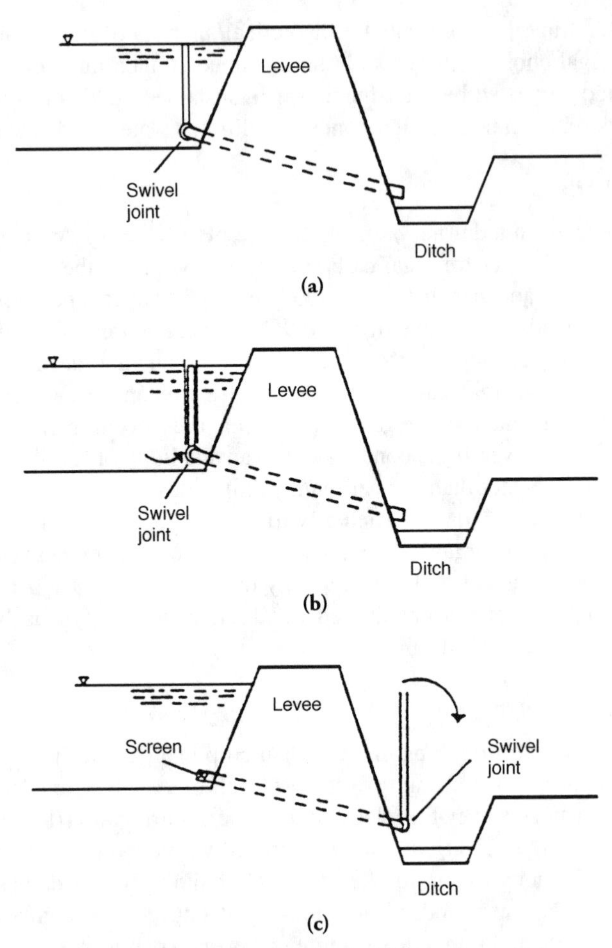

Figure 11.2. Drainage methods: (a) Turn-down drain; (b) double-sleeve turndown drain; and (c) swivel joint on the outside of the pond.

The levee pond or embankment pond is the pond type most clearly associated with aquaculture. The pond is constructed by large dirt-moving equipment that creates a rectangular structure of 0.5 to 8 hectares generally 0.9 m to 2.4 m in depth and is constructed to minimize the difficulties associated with net harvesting and truck loading of live product. Design should also incorporate provisions for draining the embankment type pond (Fig. 11.2). Figure 11.3 provides further detail on normal drainage control from such ponds. Although not shown in these figures, some provision should be made for emergency flow across the top of the levee during flooding; generally these channel depressions might be placed on either end of the levee. One must remember also to include some provision to retain the fish from swimming out of the pond, or at the conclusion of the flood situation the pond owner may be short of fish! The geometry of the levee/embankment ponds (see Fig. 11.4) is dictated by soil materials and the actual

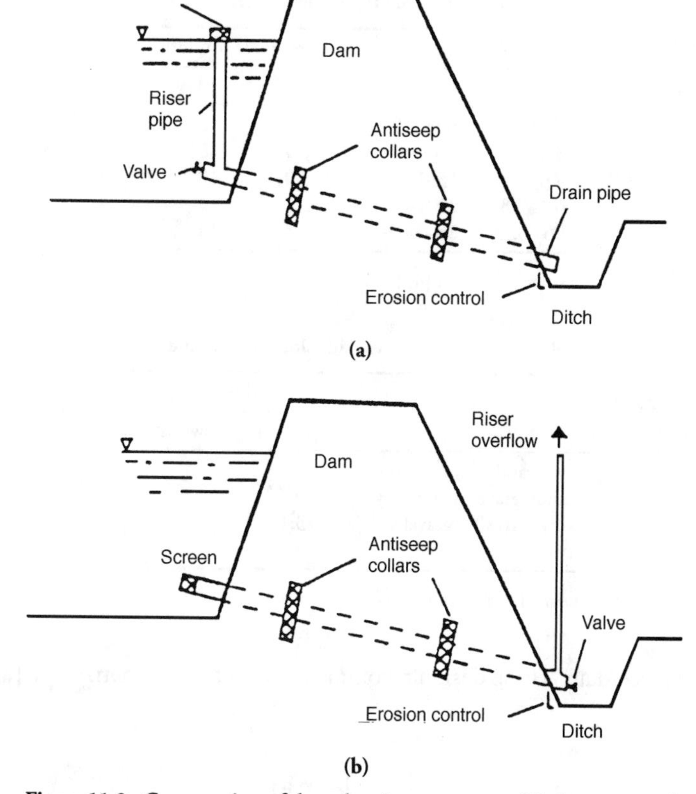

Figure 11.3. Cross section of dam showing two types of drains commonly used: (a) hooded inlet pipe; and (b) outside drain with valve. Also note use of antiseep collars.

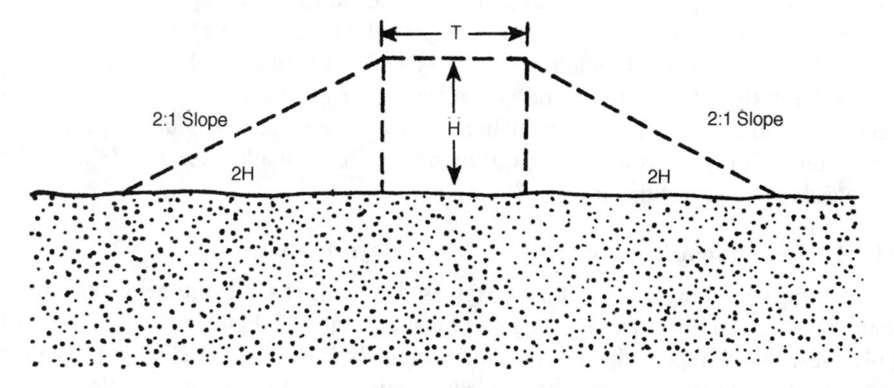

Figure 11.4. Levee showing 2:1 slope. *T* = top width. *H* = height.

Table 11.1. Dam height and minimum recommended top width of levee

Height of Dam		Minimum Top Width	
(m)	(ft)	(m)	(ft)
Under 3	Under 10	1.8	6
3.4–4.3	11–14	2.4	8
4.6–5.8	15–19	3.0	10
6.1–7.3	20–24	3.7	12
7.6–10.4	25–34	4.3	14

Source: Lawson [49].

Table 11.2. Recommended side slopes for earth dams

	Side Slope	
Fill Material	Upstream	Downstream
Clayey sand, clayey gravel, sandy gravel, sandy clay, silty clay, silty gravel	3:1	2:1
Silty clay, clayey silt	3:1	3:1

Source: Lawson [49].

height of the levee/dam. These design recommendations are summarized in Tables 11.1 and 11.2.

11.3 Raceways

Raceways have been primarily associated with the culture of salmonids, and information about that group of fishes provides the basis for this section. Information of a biological nature about the species being grown is necessary for effective design. Behavior, swimming speed, oxygen requirement, rate of fecal settling, responses to temperature, and tolerance of pollutants are examples of information useful or necessary for design. Such information is empirically derived and less readily available but critical to successful design. There are several guidelines that may be used as rough approximations useful in checking design calculations. The total amount of production of trout in a flow-through system is about 6.0 kg/L/min/y, e.g. one should expect 2273 kg to be produced in one year if the flow is 378 L/min.

11.3.1 Design Densities and Loadings (Raceways or Tanks)

Stocking or loading density of trout expressed in terms of kilograms per cubic meters should not exceed 3.16 times the length of the fish, (e.g., 64.2 kg/m^3 for a 20.3-cm fish) [14]. Westers [15] reports that salmonids are routinely maintained from 60 to 120 kg/m^3 in Michigan fish hatcheries. Recommended densities in practical farming in Norway are from 5 to 100 kg/m^3. Small fingerlings have the lowest density. For fish above 50 g, up to

120 kg/m^3 have been reported from land-based salmon farms. Arctic charr can tolerate a much higher density than Atlantic salmon. Average densities of up to 150 kg/m^3 can be tolerated. In experimental conditions, densities between 200 to 300 kg/m^3 have been attempted with good results. Poston [16] has reviewed the effects of density and oxygen consumption under high-density conditions. Under high-density carrying capacities, it is important to be sensitive to CO_2 control, because elevated levels often result. Carrying capacity is apparently affected by more than either oxygen or ammonia constraints [17].

The required flow rates or allowable loading levels for oxygen supply are based upon the available oxygen (difference between inlet and outlet concentrations in milligrams per liter), the percentage feeding rate, and the fact that roughly 220 g of oxygen are required to oxidize each kg of feed (100 g/lb of feed) [18]. After some iteration and simplification, allowable loading levels subject to oxygen constraints can be expressed as (assuming feeding activity is uniformly distributed over a 24-hour period):

$$Ld_{\text{oxygen}} = \frac{1.44 \times \Delta O_2}{2.20 \times \%_{\text{feeding rate}}} \tag{11.2}$$

where

Ld_{oxygen} = loading level in kg fish per liter per minute of flow (multiply by 8.33 to convert to lb of fish per gallon per minute)
ΔO_2 = change in oxygen level, mg/L
$\%_{\text{feeding rate}}$ = rate of feeding in percent of body weight.

For example, if there were 5 mg/L available oxygen, and the feeding rate was 2%, then the loading level would be $Ld_{\text{oxygen}} = \frac{1.44 \times 5}{2.20 \times 2} = 1.14$ kg of fish per L/min (9.5 lb/gpm). Figure 11.5 provides an illustration of loading levels based upon feeding rates and change in oxygen.

Ammonia is the second primary consideration in designing allowable fish loading densities. Westers [15] has simplified the calculation into the following form:

$$Ld_{\text{ammonia}} = \frac{UA_{\text{allowable}} \times 1000}{\Delta O_2 \times \%_{\text{UA of TAN}}} \tag{11.3}$$

where

Ld_{ammonia} = allowable fish weight per unit of flow to maintain unionized ammonia levels below some allowable level, kg fish per liter per minute of flow (multiply by 8.33 to convert to pounds of fish per gpm)
$UA_{\text{allowable}}$ = allowable or upper design limit for unionized ammonia levels, mg/L
$\%_{\text{UA of TAN}}$ = the percent of un-ionized ammonia of the total ammonia nitrogen (TAN), %, function of temperature and pH.

A design level for un-ionized ammonia of 0.0125 to 0.020 mg/L can be used [15]. Generally, the allowable loadings based upon ammonia control will be much larger than that required for supplying oxygen. After calculating both oxygen and ammonia allowable loading levels, one should utilize the smaller to calculate the required flow to

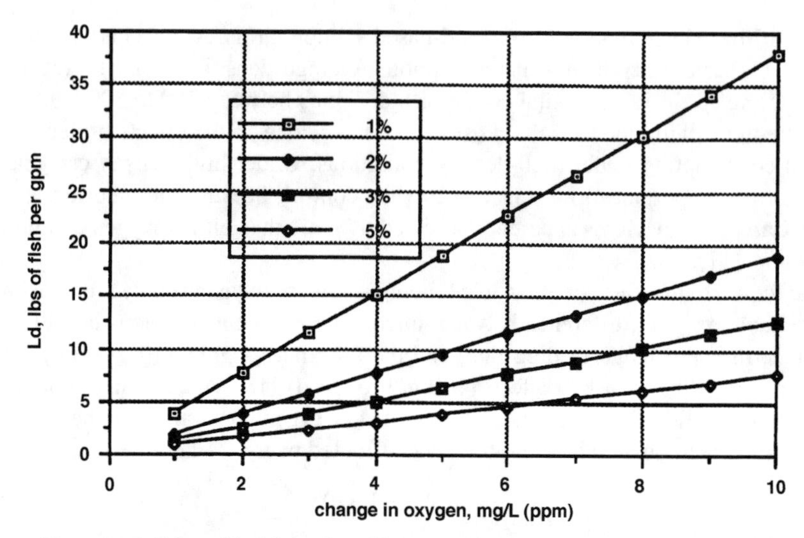

Figure 11.5. Allowable fish loading (lbs per gpm; divide by 8.33 to obtain kgs per L/min) as a function of feeding rate (% of body weight) and change in oxygen content from inlet to outlet (5 ppm of oxygen at the outlet is recommended for salmonids).

support the intended level of fish biomass:

$$Q = \frac{\text{Weight of Fish}}{Ld} \tag{11.4}$$

where Q = required flow rate, L/min (× 3.785 to convert to gpm).

Water-exchange rate can be calculated from the fish density and loading levels (0.06 constant in the following equation is converting L/min to m^3/hr):

$$R = \frac{D \times 0.06}{Ld} \tag{11.5}$$

$$R = \frac{D \times 8.02}{Ld} \tag{11.5.1}$$

where

R = water volume exchanges per hour
Ld = loading rate, kg/L/min (lb/gpm in English units)
D = fish density, kg/m^3 (lb/gpm in English units).

11.3.2 Raceway Length

Raceway length can be calculated based upon R and a minimum required average velocity for the raceway:

$$Lm = \frac{3600 \times V}{R} \tag{11.6}$$

where

Lm = required raceway length, m
V = minimum required raceway velocity, m/s.

Acceptable values for raceway velocity are discussed in detail elsewhere in this chapter.

11.3.3 Fish Growth

Water temperature is used in estimating, along with fish size, the amount of food required and associated increases in length and weight of salmonids [19, 20]. Haskell [21] wrote that trout increased 1 cm for each 4.59 temperature units (°C); a temperature unit was the average monthly water temperature minus 3.67°C:

$$G = \frac{T - 3.67}{4.59} \tag{11.7}$$

where

G = fish growth per month, cm
T = average monthly water temperature, °C.

Speece (1973) summarizes and interprets information useful in design related to nitrification and waste management control.

Example. Suppose the objective is to produce 455 kg of 20.3-cm brook trout in a flow-through, rectangular raceway. Because feeding rate and dissolved oxygen are functions of temperature, knowledge of temperature is required. We will assume a constant temperature of 10°C. One should always use the maximum expected temperature, if known, for design purposes. Feeding charts and tables listing dissolved oxygen are given by Leitritz [22] and Piper et al. [14], for example. Brook trout 20.3 cm in length and held in 10°C water require a daily amount of food equal to 1.15% of their body weight, assuming dry pelleted food is being used. The 455 kg of trout will therefore be fed 5.23 kg of feed per day. Because 220 g of oxygen are required to metabolize 1.0 kg of pelleted food [18], 5.23 kg of pellets would cause a removal of 1.15 kg of oxygen from the in-flowing water each day.

The amount of oxygen available to the raceway is dependent on temperature and flow of incoming water, and on the altitude of the raceway above sea level. Assuming an altitude of 305 m and a temperature of 10°C, the oxygen content of water in equilibrium with air is 10.9 parts per million (ppm). Not all of this oxygen is available for fish because there is a minimum level necessary for satisfactory growth. Currently, an oxygen concentration of 7 ppm is an accepted lower limit in Norwegian farms or 70% of oxygen saturation [23, 24]. In this example there are 3.9 kg (10.9 − 7.0 = 3.9) of oxygen available for each million kilograms of water (ppm). Thus, 294,872 kg (or liters) of water per day provides the 1.15 kg of oxygen required each day. A flow of 205 L/min is required for this example. Any change in atmospheric pressure or temperature causes a change in the exact flow required. Increases in water temperature and associated decreases in incoming oxygen levels can dramatically increase water-flow requirements.

Table 11.3. Maximum allowable velocity to prevent erosion

Soil Type	n^a	Clear Water (ft/s)	Colloidal Transport (ft/s)
Sandy	0.020	1.5–1.7	2.5
Loam	0.020	1.7–2.5	2.5–3.5
Clay	0.025	3.7	5.0
Shale/hardpan	0.025	6.0	6.0

[a] Manning roughness coefficient.

Equation (11.2) can also be used also to calculate the required flow for the previous example. Using the specified feeding rate, 1.15%, and the available oxygen, 3.9 mg/L, the allowable loading level for oxygen is:

$$Ld_{\text{oxygen}}\ (\text{kg/L/min}) = (3.9 \times 1.44)/(2.20 \times 1.15)$$
$$= 2.22\ \text{kg/L/min}$$

Required flow rate to support the intended fish biomass (Eq. (11.4)) is

$$Q = 454\ \text{kg of fish}/2.22\ \text{kg per L/min}$$
$$= 205\ \text{L/min}$$

This is the same flow rate as calculated before.

11.3.4 Design Principles and Considerations

Raceway design should be based upon the physical principles that control open channel flow. Wheaton [25] provides information related to physical design of raceways. Several textbooks are available that review design of open-channel flows [26]. In designing an open channel, at least two factors should be considered related to average channel velocity, V_{average}: minimum cleaning velocity and maximum allowable velocities for fish performance.

Erosion

Considerations related to erosion are generally not important because most raceways are constructed of concrete, metal, or wood. If a raceway is formed using natural soils (cuts), then one should refer to Table 11.3 for allowable velocities.

Cleaning Velocities

Minimum cleaning velocities can be estimated based on the following expression [26]

$$V_{\text{clean}} = \frac{1}{2} d^{4/9}(G - 1)^{1/2} \tag{11.8}$$

where

V_{clean} = cleaning velocity, ft/s
G = specific gravity of the material
d = particle diameter in mm.

Table 11.4. Minimum required velocity for channel cleaning

Material	Particle Size	V_{clean}, cm/s (ft/s)
Feed/feces ($G = 1.19$)	100 μm	2.4 (0.08)
Feed/feces	1/16 in	3.7 (0.12)
Silt	0.002 mm	0.9 (0.03)
Fine sand	0.05 mm	4.0 (0.13)

This expression was specifically developed from data for materials with specific gravities from 1.83 to 2.64. It is assumed to be valid as a first approximation to predict cleaning velocities for fish feces and uneaten feed because the specific gravity of fish feces is more near 1.2 for salmonid diets [27]. Ketola [28] determined specific gravities of commercial trout diets to be 1.20 for #3 crumbles and 1.13 for .25-in pellets. Table 11.4 gives minimum velocities for cleaning.

Westers [15] recommends a value of 3 cm/s (0.10 ft/s) as an effective compromise to allow heavy solids to settle rapidly yet create "good" hydraulics.

Fish Fatigue

Design of raceways (or round tanks) should also be done so as not to create excessive water velocities that would fatigue the raceway fish. Beamish [29] summarizes a considerable volume of data related to burst speeds, critical speeds, prolonged speeds, and sustained speeds for various species. From a review of the data comparing sustained speeds at various fish lengths and critical speeds, a conservative estimate of safe raceway speeds could be made of one half the critical speed (there is considerable more data on critical speeds than any other data, and no data on sustained speeds for fresh-water fishes). There is a definite relationship between critical speed and fish length. When swimming speed is expressed as fish lengths per second, the critical speeds for sockeye salmon decreased from 4.5 to 2.0 lengths/s as length increased from 10 to 90 cm. These data along with the assumed relationship between safe speed and critical speed were used to develop the following relationship to calculate allowable safe velocities in raceways:

$$V_{safe} = \frac{5.25}{\text{length}^{0.37}} \tag{11.9}$$

where

V_{safe} = maximum design velocity for raceway, which is assigned to be one half the critical swimming speed, length/s

length = fish length, cm.

For example, using the above equation for a 10-cm fish, the allowable design velocity would be 2.24 lengths/s × 10 cm/lengths = 22.4 cm/s (0.73 ft/s). Figure 11.6 illustrates the relationship between fish length and safe swimming speed. Usually, this constraint will not be a problem; however, it should always be checked.

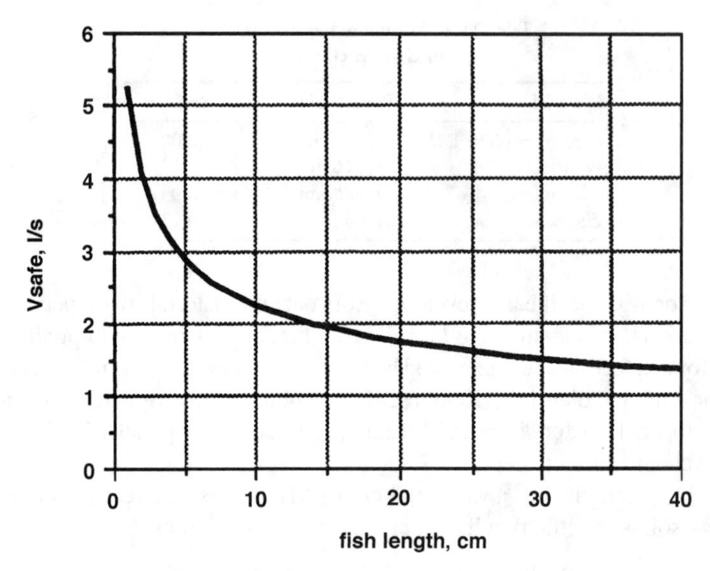

Figure 11.6. Safe raceway velocities to prevent fish fatigue as a function of fish length; expressed as fish lengths per sec (l/s).

Channel Geometry

Most raceways will be constructed from concrete or other man-made materials. The bottom width for most efficient cross section and minimum volume of excavation is given as follows [26].

$$W = 2d \ \tan\{\emptyset/2\} \tag{11.10}$$

where

W = width, m
d = depth, m
$\emptyset$ = side slope angle (vertical sides are 90 degrees).

For the typical raceway which has vertical sidewalls, then

$$W = 2d$$

It can also be shown that if equation (10) is used to calculate channel width, then the hydraulic radius, R, as used in the Manning equation, is equal to one half of the depth.

If raceways are cut into the earth, allowable side slopes are given in Table 11.5.

Channel Velocity

If a flow rate in a channel is known, for example from a pumping rate or a weir determination, then the average channel velocity can be calculated from the continuity equation,

$$V = \frac{Q}{A} \tag{11.11}$$

Table 11.5. Allowable side slope for soil constructed channels up to 1.2 m

Soil Type	Side Slope
Peat, muck	Vertical
Stiff/heavy clay	0.5:1
Clay silt loam	1:1
Sandy loam	1.5:1
Loose sand	2:1

Note. For channels wider than 1.2 m, increase the horizontal component by 50%; e.g., loose sandy soils should use a 3:1 side slope.

Table 11.6. Roughness coefficient (*n*) for different materials

Material	*n*	Range
Concrete	0.015	
Metal, smooth	0.013	
Metal, corrugated	0.024	
Plastic	0.013	
Wood	0.012	
Earth channel, rubble sides	0.032	0.025 to 0.040
Natural streams, clean, straight		0.025 to 0.033

where

Q = flow, m^3/sec
A = cross sectional area in m^2.

If the potential flow rate is needed for a specific channel, with a known bottom slope and flow depth, the average velocity can be calculated using Manning's formula (English units; [26]):

$$V_{\text{average}} = \frac{1.49\ R^{2/3}\ s^{1/2}}{n} \tag{11.12}$$

where

V_{average} = average channel velocity, ft/s
R = hydraulic radius (area of cross sectional flow/wetted perimeter), ft
s = channel slope, ft/ft decimal
n = roughness coefficient.

Values for roughness coefficients are given in Table 11.6 [26].

Having calculated the average channel velocity, flow rates are calculated from the continuity equation, $Q = VA$ (Eq. (11.11)). Solutions to the Manning equation (Eq. (11.12)) can be found by using Fig. 11.7. Flow rates for typically sized channels at different slopes are given in Fig. 11.8 (0.305-m channel) and Fig. 11.9 (1.8-m channel).

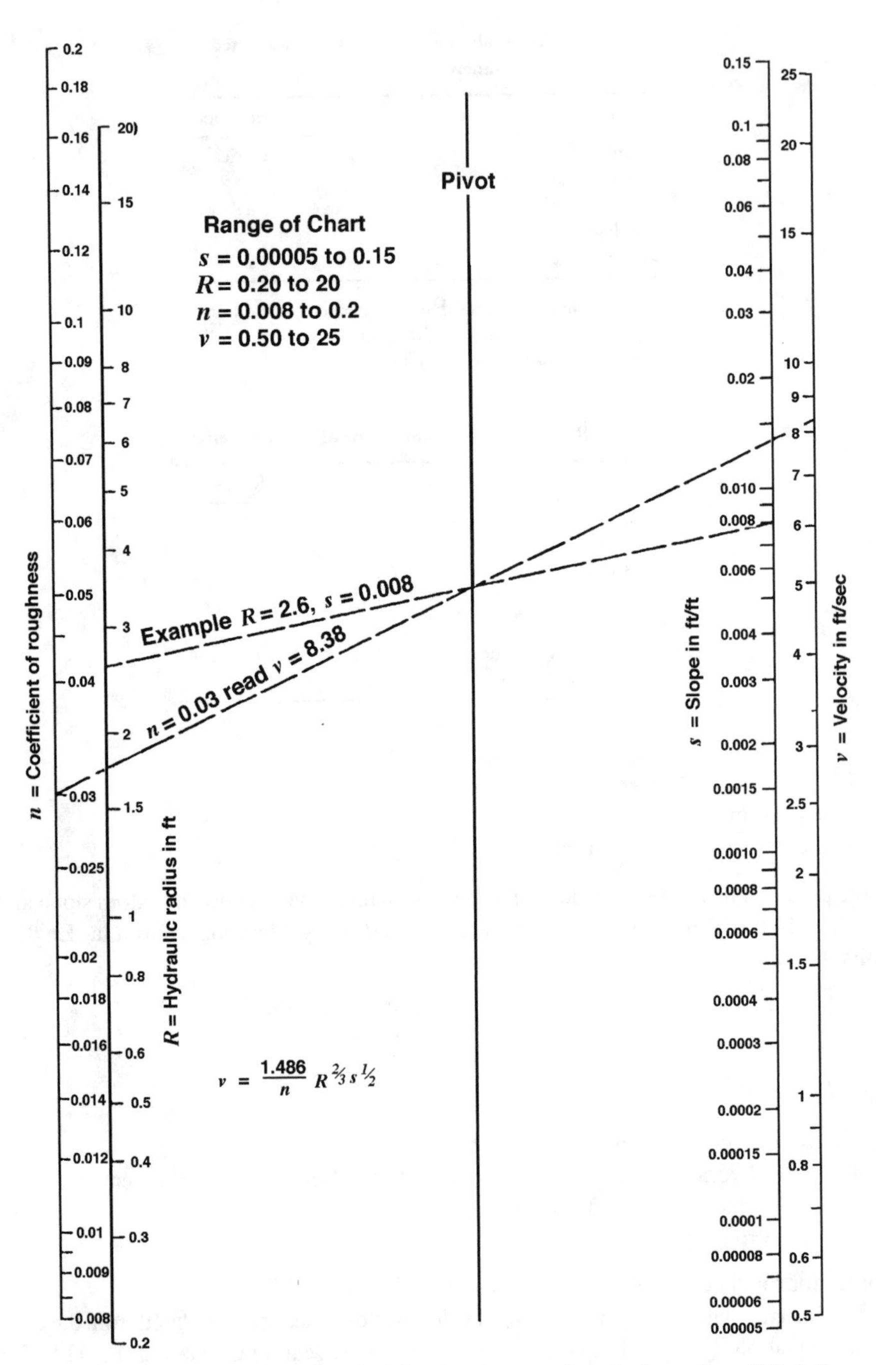

Figure 11.7. Graphical solutions of the Manning equation. (Redrawn from U.S. Soil Conservation Service, *Engineering Handbook*, Hydraulics Section 5, 1951.)

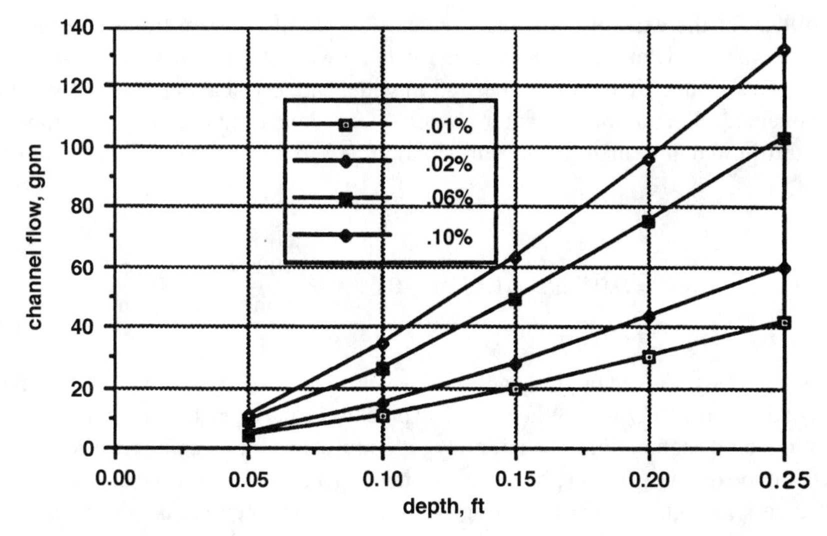

Figure 11.8. Channel flow rates for a 0.305 m (1 ft) wide, rectangular, wood channel at depths between 1.5 cm and 7.6 cm (0.05 and 0.25 feet) at slopes of 0.01, 0.02, 0.06 and 0.10%.

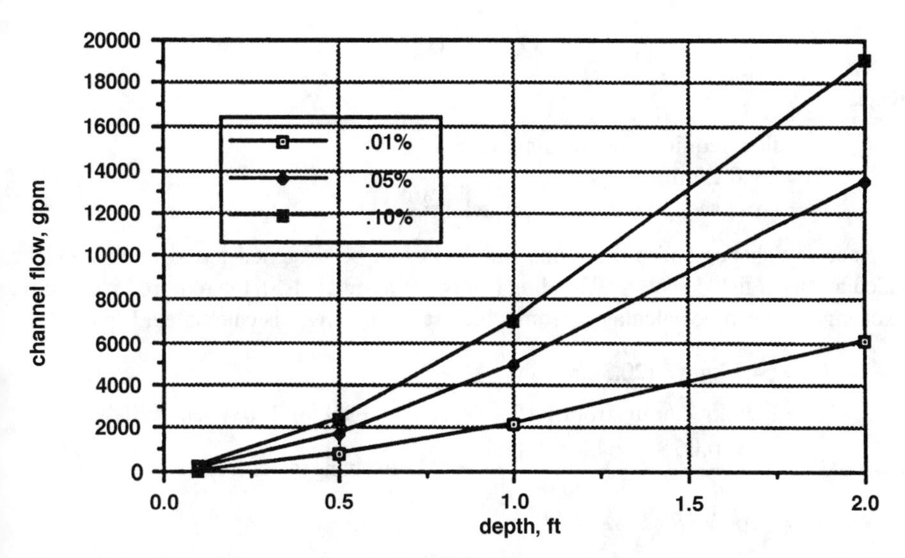

Figure 11.9. Channel flow rates for a 1.83 m (6 ft) wide, rectangular, concrete raceway at depths between 3 cm and 61 cm (0.1 and 2.0 ft) at channel slopes of 0.01, 0.05 and 0.10%.

Example. For the example previously given (455 kg of 20.3-cm trout), the required flow rate was 135 L/min and the required volume was 7.09 m^3. A typical raceway in the northeast United States would be 1.83 m wide, and a typical depth of 0.6 m would be maintained. The minimum cleaning velocity of 4 cm/s can be used to determine the minimum flow requirement to prevent siltation and to clean feed and feces from the raceway:

$$\begin{aligned} Q_{\text{clean}} &= V_{\text{clean}} A \\ &= 0.04 \text{ m/s} \times 1.83 \text{ m} \times 0.61 \text{ m} \times \frac{1000 \text{ L}}{\text{m}^3} \times 60 \frac{\text{sec}}{\text{min}} \\ &= 2679 \text{ L/min} \end{aligned} \tag{11.13}$$

This is obviously a much larger quantity of water than is required to supply the oxygen needs of the fish (136 L/min). What is seen in practice is that raceways are managed much closer to their design requirements for oxygen supply than for cleaning requirements. In fact, common opinion among practicing hatchery operators is that raceways do not work well. Clearly, a significant contributing factor is that the raceways are designed below their required cleaning velocities. In general, raceways should be designed to minimize cross-sectional area so as to promote maximum velocity. Also, where multiple races are needed, raceways should be run in series flows as opposed to parallel. This approach may require re-aeration between raceways.

If 4 cm/s is used as the minimum design velocity, then the required flow for cleaning becomes:

$$Q_{\text{clean}} = C \times A \tag{11.14}$$

where

Q_{clean} = flow required for cleaning, L/min (gpm)
C = 2360 L/min
A = cross sectional area of channel, m^2.

Raceway shapes and dimensions should in fact be designed based upon previous calculations of fish density and loading levels. Using Eq. (11.5), the required volumetric exchange rate can be calculated, from which raceway length is calculated, Eq. (11.6):

$$D = 64.2 \text{ kg/m}^3 \text{ (20.3-cm fish)}$$

$$Ld = 3.36 \text{ kg/L/min (from 1.15\% feeding and 5.9 mg/L oxygen available)}$$

$$R = \frac{D \times 0.06}{Ld} = \frac{64.2 \times 0.06}{3.36 \text{ kg/L/min}} = 1.15 \text{ exchanges per hour}$$

$$Lm = 36 \times V/R$$

Utilizing a design velocity of 4 cm/s, Lm is

$$Lm = \frac{36 \times 4}{1.15} = 126 \text{ m}$$

Returning to our design volume of 7.09 m^3, the cross-sectional area of the raceway would be only 0.056 m^2. This of course is impractical.

Because typical raceways are a fifth to a tenth of the calculated design length, it can be seen that required flow rates must also be five to ten times the flows required for oxygen considerations. Equations (11.5) and (11.6) are useful in showing the inter relationships between these factors.

Practical Considerations

Size of raceways in practice is determined by a number of other factors such as topography, soil type, construction material, and total water available. Because raceways are generally outdoor structures, ease of access and protection of fish from predators must also be considered. This may restrict the maximum depth or width of the raceway. The size and species of fish could have requirements, for example maximum velocity, that necessitate design or operational modification. Depth of water is one dimension easily controlled by addition or removal of dam boards. Changing depth changes both volume and velocity.

Raceways are normally operated in series with the discharge of the upstream raceway serving as the inflow water of the next one downstream; this is done to maintain maximum velocities through the races as previously discussed. The quality of water obviously deteriorates through the accumulation of waste products and depletion of oxygen.

Separate inlets and outlets should be designed for each raceway to allow for independent operation. This arrangement allows for isolation, which is desirable in treating disease and cleaning waste from the system. Supply and discharge pipes need to be of sufficient size to account for maximum flows expected during any phase of operation. Keyways are provided for screens and dam boards.

11.4 Net Pens

Floating net pens or cages have been used for fish culture in Southeast Asia for over 100 years. In the Western hemisphere net-pen culture of Atlantic salmon began in Norway and Scotland in the early 1960s [30]. Since that time the technology has been transferred to several countries including Canada, the United States, and Chile. The industry has progressed from small home made pens of a few cubic meters to factory-built pens with volumes over 1000 m^3. At the same time, there has been a transition from one-person or family operations with a few pens to large arrays of linked pens covering several acres. Many of these are owned and operated by multinational corporations and include floating feed storage and computerized feed distribution, on-shore feed production, fish health laboratories, net-washing facilities, and fish-processing plants.

Until quite recently, establishment of an ocean-based fish farm has been less a matter of design than an exercise in dealing with lease applications and environmental regulations, obtaining financing, establishing a supply of smolt and feed, finding suitable personnel, addressing fish health concerns, and marketing the product. Design, either of the overall system or of the components within it, has not represented a high priority. However, in the 1980s this situation changed. As the industry grew, so did concerns over the environmental effects of intensive fish culture in protected in-shore waters: the fjords of Norway, the Scottish locks, the relatively shallow waters of Washington's Puget Sound, and the coastal sites in New Brunswick and Maine. This prompted a move to deeper-water sites with better water flow but greater exposure to harsh open-ocean conditions. Major equipment failures occurred, resulting in some large insurance claims, and attention was focused on the need for proper design using sound engineering practices. The process of establishing a net-pen system can be considered in three steps, namely, site selection, component selection and/or fabrication, and system operation.

11.4.1 Site Selection

Selection of a suitable location for a net-pen system is the single most important factor in its potential for success or failure [31, 32, 33, 34]. Several factors need to be considered.

Current Speed

Production of fish under intensive conditions requires a constant supply of clean water. Net-pen systems rely on continuous adoption to ensure regular water changes within the pens. This is needed to provide a constant supply of oxygen to the fish and to remove their metabolic products and any uneaten feed. Several reports have been published studying the effects of different current speeds on the fish and the environment [32, 35], but site specificity makes it difficult to propose uniform figures. A minimum peak tidal current speed of 0.5 m/s has been proposed for the state of Maine, although several farms have been established at sites with much slower current speeds with no apparent ill effects. In Norway, it is recommended that the current speed inside the net pen should not be below 0.1 m/s. Maximum desirable current speed is limited initially by the energy wasted by the fish in its swimming to maintain position, and ultimately by its inability to swim continuously at this speed. This results in crowding of the fish to the downstream end of a pen. In addition, excessively high current speeds induce high stresses in the equipment, increasing the possibility for net failure and movement of moorings with potentially catastrophic results. Sites with current speeds in excess of 200 cm/s are not considered desirable. It is suggested that prior to purchase or lease of a site, local data be collected for that site.

Water Depth and Wave Action

Minimum water depth at low tide below the pens is a matter of some concern and much controversy. Excess feed and fish feces falling through the bottom of a net pen should ideally be flushed away by the tides to prevent accumulation and potential anoxic conditions with negative impacts on the local benthic community of organisms. To some extent, high current speeds can offset the effects of too shallow a site, but in general it is felt that a minimum water depth of 2 to 5 m under pens is required. These environmental effects have been studied [36, 37] and shown to be controllable by good site selection and avoiding excessive biomass densities. Sites are generally selected so as to offer protection from heavy wave action. Such action induces stresses on the pen systems and their moorings, as well as making daily operations difficult, dangerous, and uncomfortable. Pens and pen systems designed for exposed offshore sites need to consider the greater wave heights likely to be encountered.

Water Temperature

Salmon and trout in ocean net pens utilize feed most efficiently at temperatures from 12°C to 15°C. Higher temperatures, which can result in low oxygen and subsequent fish mortality, are generally only found in shallow bays and inlets with poor water circulation. These sites should be avoided. In winter, water temperatures can fall below 0°C, which, when combined with extremely cold winds, can result in superchill effects. Significant mortality under such conditions can result. Other species, particularly cod and haddock, have greater tolerance than salmonids for low temperatures, and attention is now turning to culture of these alternative species. An additional problem resulting from these extremely low winter water temperatures is the formation of ice over the pens. Breakup of the ice by hand has been used to prevent ice formation. Structural damage to the net pens has also occurred due to ice buildup and movement, particularly in the areas prone to ice-cold waters.

Wind

Winds result in increased lateral forces on the superstructure of a net pen and, as with wave action, this results in increased stresses on the structural members and on the moorings and makes operation of the system (feeding, inspection, harvesting) uncomfortable and dangerous. Normally a net pen is installed in such a way that the part above the water surface is small, to minimize the effect of wind on the net pen.

Aesthetics

Although lobster boats and buoys, herring weirs, and other traditional fishing gear are accepted as part of the coastal seascape and are in fact considered a positive feature by tourists and summer people, the appearance of ocean net pens has unleashed a fury of objections from these same groups. The unpredictable nature of the permitting process and the length of time required to obtain a site permit should not be underestimated.

11.4.2 Net-Pen Design and Construction

An ocean fish farm can be anything from a single homemade net pen, an outboard skiff, and an onshore feed storage shed, to a complex of pens covering several acres, with floating feed-storage and -distribution facilities and, in the case of offshore systems, accommodation for the workers.

Modern net pens are available in a wide range of sizes, materials, and designs, but all consist basically of a net, a framework or lines to give shape to the net, some means of access to the fish, flotation to support these components, and a mooring system (Fig. 11.10).

Nets

Most nets in use in commercial pen cultures today are of nylon mesh. Smaller pens have been constructed using plastic-coated wire mesh; this material is more resistant to fouling and predator penetration and is more easily cleaned, but the greater cost and

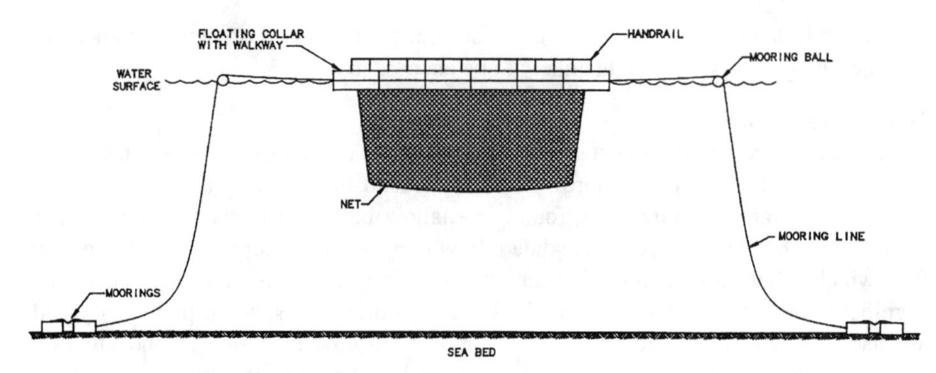

Figure 11.10. Typical arrangement for a floating collar net-pen.

difficulty of handling in large sizes make it unattractive. Semirigid plastic mesh has been developed recently and provides many of the advantages of rigid mesh while being lighter and less expensive.

Nylon nets are either custom-fabricated locally or commercially available in sizes to fit the popular pen types. A popular net size for salmon and trout farming is 13 m × 13 m × 6 m deep. However, some recent experiments on larger sizes (with circumferences 90 m and more) have demonstrated increased growth and feed conversion factors [38]. Sizing of the mesh is an important consideration. It must be small enough to retain the smallest fish to be put in the net, but use of larger meshes reduces net fouling by seaweeds and marine organisms; water circulation is improved and they are lighter and more easily handled.

Mesh size can be measured by the dimension between opposite corners of the mesh as it is stretched diagonally. For a smolt, 24- to 28-mm stretch is typically used, and for market fish, a 42- to 56-mm stretch is preferred. If two nets are used, an inner net keeps the fish in, and an outer net, typically 200-mm mesh, keeps predators out. Seals and sea lions represent a serious predation problem not only by killing the fish but by making holes in the net. Fish that escape through damaged nets represent an economic loss and also cause concern, as yet unfounded, regarding genetic contamination of the wild stocks. In order to prove effective as a predator barrier the outer net must be rigged to provide a gap between the two nets. This is not a simple problem, especially in high-current areas. Even with predator nets in place, seals have been observed to push up against both nets and bite at the fish. There are active attempts and research directed at creating effective seal deterrents, for example using sound. Also, predation from birds is a problem and requires placement of overhead netting.

Net Support and Access

Most net pens in use today provide some type of rigid structural framework on which to hang the net. Although some wood framed net pens are still in use steel, aluminum, rubber, and polyethylene are now the materials of choice. These supporting structures may be round, octagonal, hexagonal, square, or rectangular, and they are deployed as individual units or as an array with hinged connection between units (Fig. 11.11). These

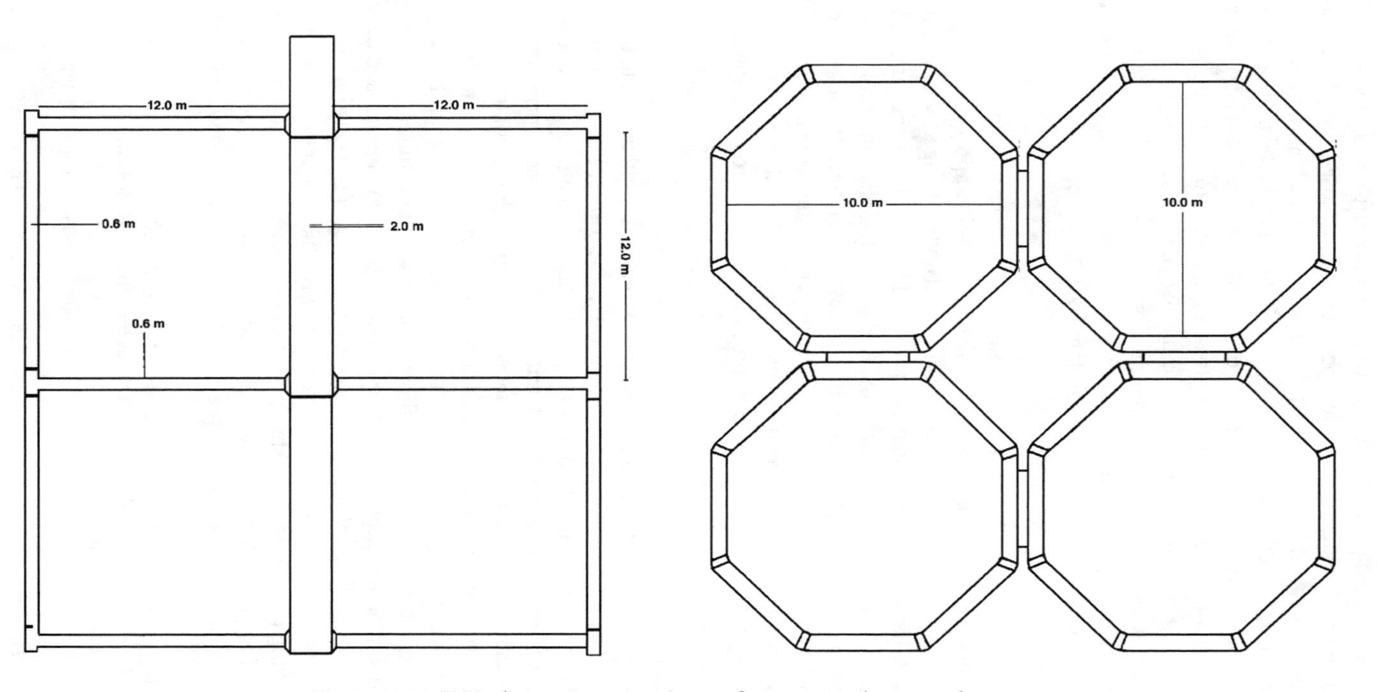

Figure 11.11. Typical grouping arrangements for square and octagonal net-pens.

hinged connections have proved to be the weak point of net-pen systems, especially if systems designed for sheltered sites have been used in exposed, off-shore conditions. The individual units can either be located as single units or anchored to a walkway (Fig. 11.11). The mooring system is more complicated if there is a walkway. For exposed sites single units are advantageous. The absence of a rigid framework tends to make the pen "invisible" to the waves, resulting in less stress on the system including the moorings. This design, however, provides no access to the fish and all servicing, feeding, and so forth must be carried out from a vessel brought alongside the net. Single units are usually kept 50 to 100 m apart [39]. Recently moorings for circular single net pens have been developed. Each pen is connected to the system by a "hen foot."

Flotation

A floating "collar" around the net is incorporated into the net-support framework. On circular pens this can be one or two circles of styrofoam-filled polyethylene pipe, up to 0.5 m in diameter with pen diameters over 30 m. Rigid bolted, flanged, watertight sections of galvanized steel pipe (up to 1 m diameter) are used for circular, octagonal, and hexagonal pens up to 15 m across. In these units the flotation collar also supports the walkways. Square or rectangular pens generally utilize a galvanized steel framework enclosing plastic-coated styrofoam, with expanded metal walkways and tubular steel handrails.

In addition to these relatively conventional designs there have been several innovative solutions proposed but not in widespread use, such as rotating spherical cages, submersible cages, and tension leg cages.

Forces on Net Pens

Two approaches are used in design of floating structures: rigid design to oppose environmental forces and flexible structures to absorb these forces. Design of an individual net pen for in-shore deployment does not represent a major engineering undertaking; it is the grouping of pens into a combined unit that causes problems, principally with the pen-to-pen linkages and the mooring systems needed in high sea conditions.

Design of large, floating, moored, ocean-based systems is a formidable challenge. Detailed engineering design of these ocean cage systems is beyond the scope of this chapter. The interest in deep-water, off-shore aquaculture has resulted in several international conferences on the engineering of systems for these conditions, and the reader is referred to the proceedings of these conferences for in-depth information on the topic.[1]

Engineers involved in net-pen design seem to agree that net pens are not being overdesigned. Moving pens further off-shore increases the stresses greatly because of great wave heights. Fatigue failures because of cyclic loading over a long period and the occurrence of peak storm conditions are factors that may be underestimated. Ideally,

[1] Conference on Engineering for Offshore Fish Farming. Thomas Telford Press, London, England, 1990.
National Science Foundation Engineering Research Needs for Offshore Mariculture Systems Workshop. University of Hawaii, Honolulu, 1991.
Conference on Open-Ocean Aquaculture. University of Maine/University of New Hampshire Sea Grant, Portland, Maine, 1996.
Second International Conference on Open Ocean Aquaculture. University of Hawaii Sea Grant Program, Maui, Hawaii, 1997.

long-term data on environmental conditions should be obtained. Barker [40] suggests using 100-year conditions in design, but if good environmental data for a particular site is available, then 50-year conditions can be sufficient. To avoid joining of front and back walls of a net, the depth-to-width ratio should not be much above 1. Use of large weight or other stretching systems can also maintain the shape of a net pen; however, they will exert considerable forces on the net.

Mooring Design

Design of mooring requires the accurate prediction of the forces acting on these floating structures. Forces acting on the fish cages are generated from three sources: wind, current, and waves. For a conservative design, it should be assumed that the maximum values of force due to each of these components act simultaneously. In addition, it is prudent to assume that all of these components may act in the same direction. This assumption simplifies the analysis, and although it may yield a very conservative estimate of the forces on the fish pen, it is not necessarily economically extravagant because the mooring system is a relatively inexpensive part of the total production system.

Riley and Mannuzza [41] describe an analytical method for calculating these forces. Wind forces act only on the portion of the structure above the water surface and may be determined as a function of maximum wind velocity, the characteristic area of the superstructure, and a form factor based on the shape of the structural members. Current and waves work together to exert a significant force on the portion of the structure below the water surface. This part of the structure is composed mostly of the nets used to retain the fish and the predator barriers. The frictional resistance of the water on the base of the superstructure is small in comparison to the drag forces on nets and may be neglected. It is assumed that the nets hang vertically from the superstructure and the current is perfectly horizontal.

As a body of water experiences the passage of a wave, the individual fluid particles travel in orbital paths at the rate of one revolution per wave period. For conditions commonly encountered in aquaculture, these orbital paths are generally circular. The diameters of these circular paths decrease with increase in depth. The horizontal component of velocity may be calculated as a function of wave height, period and length, and water depth. In the presence of a current the total velocity of the water is approximated by adding current velocity to wave velocity. Forces on the netting may then be computed as a function of absolute water velocity, characteristic area of the net, and drag coefficient for the net. Drag coefficients for various types and sizes of net have been calculated (Milne [42]) together with the effect of fouling on these numbers.

With the forces on all submerged sides of the net added to the wind force on the superstructure, a reasonable conservative figure for total horizontal force on the system is obtained. This information represents a beginning point in determining the size and number of moorings required to immobilize the structure. Different mooring and anchor shapes and sizes have different holding characteristics, some of which are well documented for the various bottom conditions likely to be encountered at the site in question, such as mud, silt, gravel, and rock. Local fishermen are undoubtedly the best consultants as to how a particular type of mooring will hold at a particular site.

The existing methods for calculating forces on net pens owing to wind, current, and waves, including that developed by Riley and Mannuzza [41], consider the net pen as a single isolated unit. In real-world situations, however, this is rarely the case. It is necessary to develop the analysis further to calculate forces on groups of net pens, evaluating the shielding effect on current forces of downstream net pens. Aarsnes, Rudi, and Loland [43] have developed prediction equations for this shielding effect. Current velocity is reduced 50% to 80% in the downstream pens, so that in a long line of pens parallel to the current direction, water exchange may be greatly reduced in the pens furthest downstream. Fouling of the nets will exacerbate this problem and can have a significant impact on fish health and growth rate by reducing available oxygen and affecting waste removal.

Operational Aspects

Feeding

Many net pen operations still rely upon hand feeding; others have gone to timed feeding systems or demand feeders; and some operators use a combination of both. Computer-controlled pneumatic feeding from a central location is now in use on some of the larger farms. Ideally, fish are fed to satiation, which requires great skill in observing feeding behavior. Some commercial systems employ feed wastage monitoring devices at the bottoms of the net pens to assist in feeding management. Feed represents approximately 40% of total production costs for net-pen salmon operations and requires careful management.

Net Cleaning

Fouling of nets by silt, marine animals, and a wide variety of marine algae can result in flow restriction to the captive fish. In addition, the stresses on the nets and superstructure are increased, both from the increase in weight and greater horizontal current-induced forces. Net pens should be cleaned regularly at least once per year, and more often if possible. Cleaning is done by divers equipped with brushes and scrapers and by surface personnel if the net can be partially raised from the water. Under the worst conditions there is a need for complete removal of the net from the structure, transport to shore, and thorough cleaning by pressurized sprays and rotary net washers. In a stocked pen, a newly cleaned net pen is carefully put in place, and then the fouled one carefully removed. Control of fouling by use of companion species in the net pen, such as winkles, crabs, and flatfish, has been tried but without any great success.

Removal of Dead Fish

Fish that die in the net pens, for whatever reason, need to be removed as soon as possible. Many fish farmers hire divers to do this "mort" removal, but some systems utilize mechanical retrieval, either with an airlift pump inside the pen or a collection device in the center of the net bottom panel emptied from outside the pen.

11.4.3 Net-Pen Summary

Net-pen technology is constantly changing; this development is driven by the need to reduce capital and operating costs and the move to less-sheltered off-shore sites with harsher environmental conditions. Well-designed, large floating collar net pens have been tested in production situations in off-shore locations in Europe and Australia and are proving to be reliable and cost effective. If present trends continue we will see larger and larger net-pen systems, moving further off-shore; greater automation and computer control of feeding, stock assessment, and inspection using underwater video cameras and image analysis; and mechanization of mort removal, net cleaning and other necessary operations.

11.5 Tanks

11.5.1 Biomass Loading

A key aspect to economical productivity for recirculating systems is to maintain the system biomass as near as possible to the biomass carrying capacity of the system. Flow requirements can use the same calculation techniques as presented in Section 11.3 Raceways to determine flow rates through the tank for oxygen and ammonia control; see Eqs. 11.2 and 11.3.

Continuous biomass loading is a technique that carries a mixture of fish sizes within the same tank and removes fish by selectively sorting entire tank populations. Small fish replace the number of fish removed, with adjustments in stocking numbers made to account for expected mortality losses. The obvious advantage is that a tank receives approximately the same amount of fish feed on a continuous basis, which maximizes the potential output from a fixed physical resource. From an operational perspective, such a system is easier to manage because the water quality demands also remain fixed, increasing reliability and robustness of the system. A primary disadvantage is increased sorting requirements, potentially causing additional stress and possible fish-behavioral problems. This technique has been employed successfully for trout [44, 45]. Cornell University has employed this technique successfully, maintaining tilapia biomass densities exceeding 200 kg/m^3; in the Cornell system, a minimum size fish for mixed rearing was 90 g at tank introduction. Continuous biomass loading is similar in principle to raising fish in a series of stages within each of which the fish biomass is increased to the maximum biomass carrying capacity at each independent stage. McNown and Seireg [46] found that fish production costs were reduced by 30% to 40% when the number of stages was increased from one to eight, with the largest reductions being obtained over the first three stage additions.

11.5.2 Labor Requirements

Ley et al. [47] reported a range of labor requirements from 0.5 to 2 man-hours per day to manage a single 4.2-m (12-ft) diameter tank of trout; similar time requirements existed to manage a 2.7-m (9-ft) diameter tank. In general, it can be assumed that time-management requirements are independent of tank size. In effect, it is the number of tanks and not the size of tanks that is important in determining management hours required. Current experience suggests that a series of tanks can be managed averaging 20 min per day per tank; labor includes daily water chemistry measurements, fish feeding, filter maintenance, and tank cleaning. A weekly maintenance of two to three additional man-hours per tank system for major cleaning activities and preventative maintenance is also necessary. Assuming a 40-hour work week, this suggests one person could manage eight tank systems. Because many operations on a farm require two people, a facility should be designed assuming two full-time employees/owners. Hourly or contract labor can be employed for special tasks, such as harvesting, hauling, processing. This then defines the size of a basic production unit as a 16-tank system. Some adjustments in labor requirements might be allowed depending upon the tank size. Round tanks are assumed because of their self-cleaning attributes and uniform water quality. Losordo and Westerman [6] assigned 8 hours per day to manage an eight-tank growout facility with a three-tank nursery.

11.5.3 Tank Shapes and Sizes

Tanks are designed in a variety of shapes and sizes. Normally, they are designed with the objective of achieving uniformity of water recirculation within the entire tank water column. Design of the inlet and outlet structures has the major impact on water-flow uniformity. Tank geometry and tank diameter-to-depth ratios also impact water-recirculation uniformity. With uniform conditions in the tank, fish will also occupy all portions of the tank water column, maximizing efficient use of the rearing space. Design of the inlet and outlet structure is also the primary factor controlling settleable solids and algae removal from tank culture systems.

The first basic principle of tank design is the flushing rate of the tank based upon the introduction of new water. Mathematically, the water-exchange rate can be expressed as follows:

$$F(\%) = (1 - e^{-t/th}) * 100 \tag{11.15}$$

where

F = water exchange rate in percent
th = theoretical holding time (time necessary to fill one tank volume with a given water flow)
t = time.

Inserting numbers into Eq. (11.15) shows that by adding a water amount equal to the tank volume during a given time t ($t = th$), 63.2% of the old water volume in the tank is exchanged with new water provided that water mixing is uniform.

Self-Cleaning

Self-cleaning is one of the most important attributes of any tank design in order to minimize the potential for large labor requirements for manual cleaning. With proper design of both the inlet and outlet structures, two flows will be created within the tank. Figure 11.12 shows these two flow patterns, with a primary flow to ensure good water distribution of the inlet water throughout the tank and a secondary flow that ensures cleaning of the walls and floor of the tank. Experiments have shown that effective floor cleaning requires a bottom velocity of at least 6 to 8 cm/s [48]. Such floor velocities are generally created if the water velocities in the tank are between 12 and 15 cm/s. High fish densities promote self cleaning of the floor by resuspending settled solids so that the secondary flow pattern within the tank moves solids to a center drain.

Attention by the culturist should be given to measuring water velocities within the water column. A small propeller type velometer can be used and placed at differen

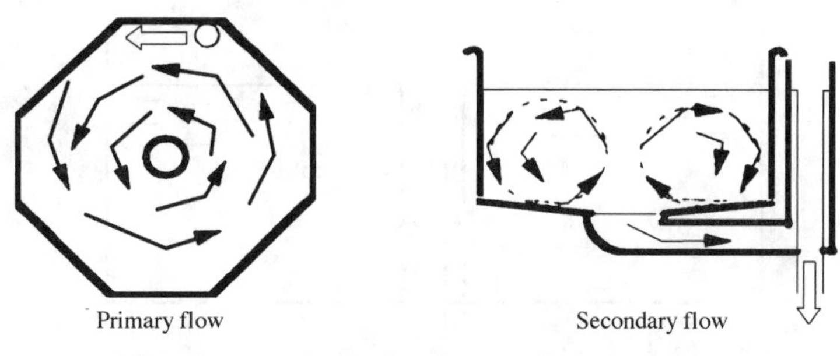

Figure 11.12. Primary and secondary flow patterns with the primary flow to ensure good water distribution of the inlet water and the secondary flow contributing to effective solids removal.

locations within the tank both horizontally and vertically to obtain a clear picture of the velocity gradients within the tank. Every effort should be made to create as equal as possible velocities and to ensure that floor velocities are not below 6 cm/s. Water exchange rates also affect the overall self cleaning of the tank. Tvinneriem [48] demonstrated that tank water-retention times for flowthrough should be no more than 100 min (flow >10 L/min/m^3 of water tank volume). However, tank flowthroughs that are less than 30 minutes can create strong vortexes and whirlpools around the center drain, causing management problems. Special designs of both inlets and outlets are required if these flow through times are not within the recommended ranges.

Shape of Tanks

Round, octagonal, and hexagonal tanks can all provide uniform flow conditions with the tank water column. The shape of the tank is important in order to avoid dead areas within the tank. Square or rectangular tanks create dead areas in the corners and require manual cleaning. Generally, such tanks are to be avoided. However, rectangular tanks are still in common use, particularly for fingerlings for which feeding rates are low and labor maintenance is not extreme. The major reason for the use of rectangular tanks is their advantage of allowing the fish to be more easily crowded and sorted; the rectangular tank also allows better usage of floor space. Because fingerling sorting is frequent during the early growth periods for food fish, the rectangular tank is still preferred by some culturists. Figure 11.13 shows an outlet design that can be used to enhance removal of solids from rectangular tanks. Another disadvantage of rectangular tanks is that they result in water quality gradients from inlet to outlet. There is not a unanimous voice on round versus rectangular tank selection, with some large commercial operations still using rectangular tanks even for large food-fish rearing. We clearly prefer the round tank.

Tank diameter-to-depth ratio also has a strong impact on the self cleaning attribute of round tanks. Tvinnereim [48] recommends that diameter-to-depth ratios should be maintained between 2 and 5; Chenoweth et al. [52] recommend between 5 and 10. For example if the diameter of a tank were 10 m, Tvinnereim would suggest that the depth should be between 2 and 5 m and Chenoweth would recommend between 1 and 2 m. The clear advantage of the smaller diameter-to-depth ratio is that more water volume per unit floor area can be maintained. If smaller diameter-to-depth ratios are employed,

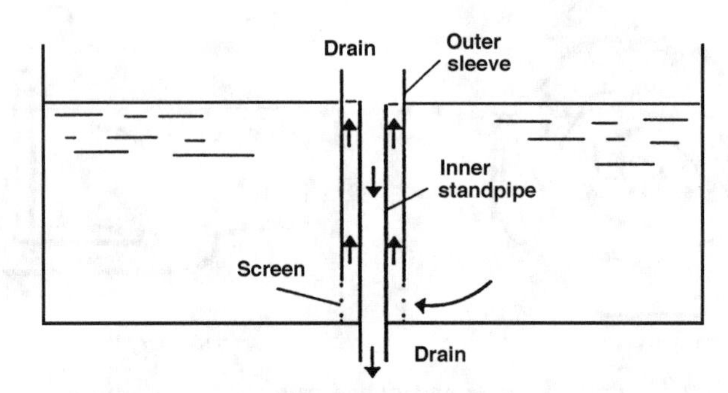

Figure 11.13. A central drain with outer sleeve for removing wastes from the bottom of a round tank.

meaning deeper tanks, then greater attention should be paid to both the inlet and outlet structures to enhance cleaning. Also, the tank throughput flow (exchange rate) will have to be on the "faster" side, if the deeper tank design is used. Cornell University has had good success following the Chenoweth or "shallower"-tank recommendation.

There is probably equal controversy surrounding the slope of the floor in round tanks. Lekang and Fjaere [23] have demonstrated that the use of flat to minimal floor slopes can be maintained without affecting the self-cleaning attribute of the round tank. Floor slopes should simply be considered in terms of draining a tank between stockings. The movement of solids from the outer areas of the tank and along the floor to the center drain depend upon the inlet–outlet structures and the fish resuspending settled solids, and not upon the slope of the floor.

11.5.4 Water Inlet

Proper design of the inlet is necessary to ensure uniform distribution and good mixing of the inlet water. Inlet design also affects the self-cleaning effectiveness of the round tank. The force of the inlet water must be sufficient to create the hydraulic characteristics desired within the tank. The most effective usage of the forces created by the inlet pipe is to fully or partially submerge the inlet pipe. Then, a series of holes or slits are made in the submerged inlet pipe to ensure that inlet water forces are imposed over the entire water column depth and not just the surface.

The impulse force caused by the inlet water flow can be expressed as follows:

$$F = \rho Q(V_2 - V_1) \tag{11.16}$$

where

F = impulse force
ρ = water density
Q = water flow
V_2 = speed of water in the tank
V_1 = speed of water from the inlet pipe orifices.

Excessive energy losses are avoided by designing the inlet pipe with velocities not to exceed 1.5 m/s and the inlet orifice velocity less than 1.2 m/s. A complete pressure-loss

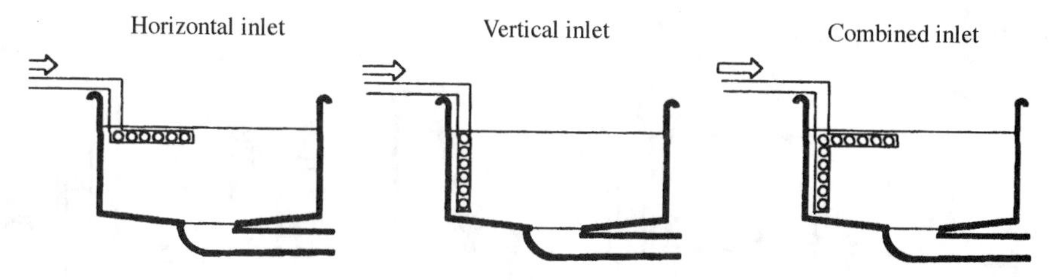

Figure 11.14. Various inlet pipe arrangements.

analysis should be performed on all the pipe systems supplying the tank to ensure proper flow delivery and to properly match the pump system with the water-delivery system.

The inlet pipe system can be arranged in several ways (Fig. 11.14). A horizontal-spray inlet pipe has an advantage of creating an effective water-distribution entry, but the secondary flow is generally weak. The vertical submerged inlet pipe creates both effective primary and secondary flows. A combined vertical and horizontal inlet arrangement is also possible and is recommended if possible (Fig. 11.14). Normally, the vertical inlet is placed near the tank wall about a fish's width away. However, as the tank diameter-to-depth ratio reduces (deeper tanks), the submerged inlet can be moved farther away from the outer tank wall towards the center drain. Compared with the current speed from the inlet orifices, the current speed in the tank should be targeted towards 0.15 to 0.25 cm/s [48]. Of course, the higher the tank current speed, the higher the floor velocities and improved solids removal. Care should be taken, however, not to overexercise the fish. Refer to a previous section for swimming speed allowances.

11.5.5 Water Outlet

The outlet function is to remove settled waste from the tank water column and to maintain water-column levels within the tank. The outlet design also affects the efficiency of the removal of settled solids. An undersized outlet or improper design can require higher water-column elevations to achieve a balance between inlet and outlet flow. A restricted outlet will result in a tank overflowing and losing water and possibly fish. Velocity in the outlet pipe will affect the sedimentation rate of settleable solids in the outlet flow. Properly designed, the outlet pipe will collect minimal solids and ideally none. This can be checked quite easily by removing the outside standpipe that controls the water level in the tank. Removing this standpipe will result in a large increase in flow which should result in any settled solids being flushed, which can be easily observed. If solids flushing is observed in this case, then either the outlet pipe must be redesigned (made smaller) or flushing of the outlet pipe as part of a daily maintenance schedule (or more often) must be implemented. Retention of solids in the outlet pipe contributes to an overall degradation in water quality for the tank and therefore should be avoided if possible. The outlet pipe should be sized to create a velocity of at least 0.3 m/s but less than 1.5 m/s, which can result in break-up of solids, making their removal more difficult.

Various outlet-structure arrangements are shown in Fig. 11.15. The most typical arrangement is the flat outlet pot in the tank with a screen covering the pot to prevent fish escaping. The outlet pot connects to an outlet pipe to a standpipe where the water level in the tank is controlled. It is important to increase the speed through the holes of

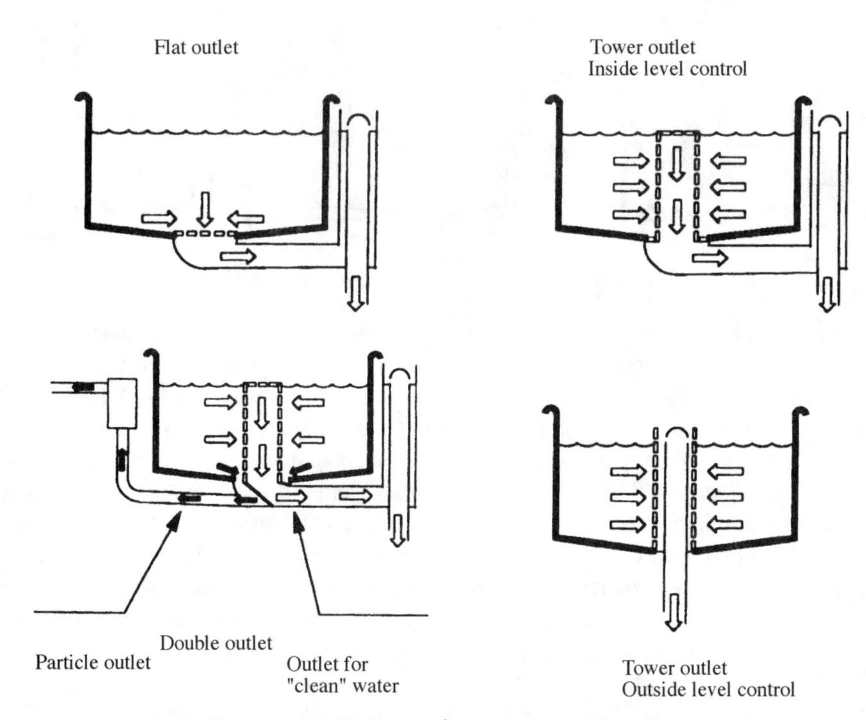

Figure 11.15. Various outlet structure arrangements.

the pot screen to prevent clogging and trapping of solids; speeds of at least 0.4 m/s are recommended [48]. The outlet pot is a critical part of the sedimentation collection system. It is very important not to reduce the water speed in this pot, because solids sedimentation will occur, which will eventually result in anaerobic conditions and havens for bacterial growth and other undesirable results. Improved cleaning of the pot can be achieved by placing the outlet pipe eccentric to the outlet pipe to create a swirling flow within the pot.

Figure 11.15 shows two types of outlet standpipes or towers, one being inside the tank and one being outside the tank. For maximum security of extremely precious animals a double standpipe might be employed, for example one inside and one outside at a lower level as an insurance against water-column loss. A disadvantage of the simpler approach of using the inside standpipe is that it makes harvesting and sorting of the tank more difficult; also, an inside standpipe is more subject to failure because of large fish bumping into it and possibly dislodging it from its normal position.

Current tank designs have begun to emphasize a double-drain approach, in which 10% to 20% of the total flow is removed via the center bottom drain and the rest of the flow is removed from either the side wall or from a center drain that removes water from top of the water column (see Fig. 11.16). Also, the removal of waste from the center drain, whether through a double-drain or single-drain approach, is enhanced by use of a double-walled standpipe where the outer pipe has screen mesh on the lower end and the inner pipe controls water level in the tank (see Fig. 11.17). Note also that the top of the

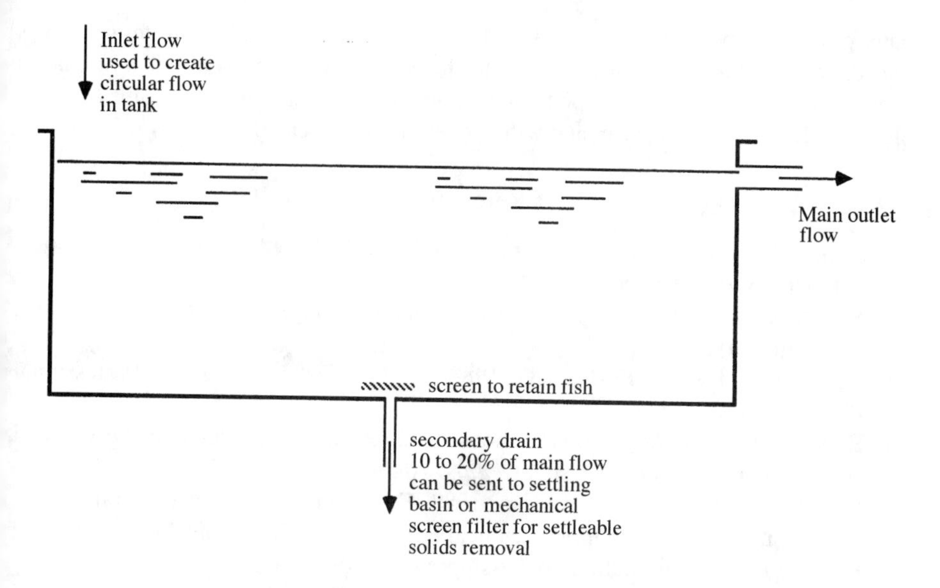

Figure 11.16. Double drain approach for enhancing solids removal while reducing tank effluent for solids removal to 10 to 20% of total flow.

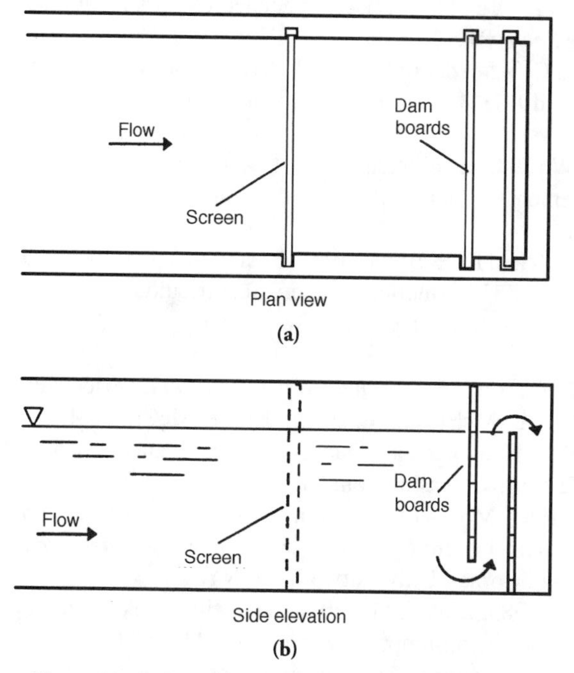

Figure 11.17. Dam boards for drawing fish wastes from bottom of raceway.

outer pipe is below the lip of the tank walls so that if the screen plugs, water overflows into the center drain pipe, instead of onto the floor. Screening should also be placed over the top of the outer standpipe to prevent fish jumping into the center pipe area and plugging the exit flow; this can and will happen if not prevented.

References

1. Sedgwick, S. D. 1985. *Trout Farming Handbook*, 4th ed. Fishing News Books, Farnham, Surrey, 160 pp.
2. Stevenson, J. P. 1987. *Trout Farming Manual*, 2nd ed. Fishing News Books, Farnham, Surrey, 259 pp.
3. Shepherd, C. J. and N. R. Bromage. 1988. *Intensive Fish Farming*. BSP Professional Books, Oxford, 404 pp.
4. Easley, J. E., Jr. 1977. Response of costs and returns to alternative feed prices and conversion in aquaculture systems. *Mar. Fish. Rev.* 39(5):15–17.
5. Keenum, M. E. and J. E. Waldrop. 1988. Economic analysis of farm-raised catfish production in Mississippi. Technical Bulletin 155, Mississippi Agricultural and Forestry Experiment Station, Mississippi State, MS 39762.
6. Losordo, T. M. and P. W. Westerman. 1994. An analysis of biological, economic, and engineering factors affecting the cost of fish production in recirculating aquaculture systems. *Journal of the World Aquaculture Society* 25:193–203.
7. Nash, C. E. 1988. A global overview of aquaculture production. *Journal of the World Aquaculture Society* 19(2).
8. Shelf, G. and C. J. Soeder (eds.). 1980. *Algae Biomass*. Elsevier Press, New York.
9. Stumm, W. and J. J. Morgan. 1981. Aquatic Chemistry. John Wiley and Sons, New York.
10. Smith, D. W. 1985. Biological control of excessive phytoplankton growth and the enhancement of aquaculture production. *Can J. Fish Aquat. Science* 42: 1940–1945.
11. Drapcho, C. M. and D. E. Brune. 1989. Design of a partitional Aquaculture System. ASAE paper 89-7527. American Society of Agricultural Engineers, St. Joseph, MI.
12. Boyd, C. E. 1979. *Water Quality in Warm-Water Fish Ponds*. Auburn University, Auburn, AL.
13. Tucker, C. S. (ed.). 1985. *Channel Catfish Culture*. Elsevier Press, New York.
14. Piper, R. G., I. B. McElwain, L. E. Orme, J. P. McCraren, L. G. Fowler, and J. R. Leonard. 1982. Fish Hatchery Management. U.S. Department of the Interior, Fish and Wildlife Service, Washington, DC.
15. Westers, H. 1991. Modes of operation and design relative to carrying capacities of flow-through systems. In: *Proceedings: Engineering Aspects of Intensive Aquaculture*, April 4–6. Cornell University, Ithaca, NY.
16. Poston, H. A. 1983. Effect of population density of lake trout in cylindrical jars on growth and oxygen consumption. *Progressive Fish-Culturist* 45(1):8–13.
17. Meade, J. W. 1991. Intensity of aquaculture production: definitions, meanings, and measures. In: *Proceedings: Engineering Aspects of Intensive Aquaculture*, April 4–6. Cornell University, Ithaca, NY.

18. Willoughby, H. 1969. A method for calculating carrying capacities of hatchery troughs and ponds. *Progressive Fish-Culturist* 30:173–174.
19. Austreng, E., T. Storbakken, and T. Asgaard. 1987. Growth rate estimate of cultured Atlantic salmon and rainbow trout. *Aquaculture* 60:157–160.
20. Jobling, M., E. H. Jorgensen, A. M. Arnesen, and E. Ringo. 1993. Feeding, growth and environmental requirements of arctic charr: a review of the aquaculture potential. *Aquaculture International* 1:20–46.
21. Haskell, D. C. 1959. Trout growth in hatcheries. New York Fish and Game Journal 6(2):204–237.
22. Leitritz, E. 1959. Trout and salmon culture. California Department of Fish and Game Fish Bulletin no. 107.
23. Lekang, O. I. and S. O. Fjaera. 1997. *Technology for Aquaculture* [in Norwegian]. Landbruksforlaget, Oslo, Norway.
24. Christiansen, J. S., M. Jobling, and E. H. Jorgensen. 1990. Oxygen and water requirements of salmonids [in Norwegian]. *Norsk Fiskeoppdrett* 10:28–29.
25. Wheaton, F. W. 1977. *Aquacultural Engineering*. John Wiley and Sons, New York.
26. Schwab, G. O., R. K. Frevert, T. W. Edminster, and K. K. Barnes. 1966. *Soil and Water Conservation Engineering*. (2nd ed.). John Wiley and Sons, New York.
27. Chen, S. 1991. Theoretical and experimental investigation of foam separation applied to aquaculture. Ph.D. thesis, Cornell University, Ithaca, NY.
28. Ketola, G. 1991. Personal communication (unpublished data). Research Scientist, U.S. Fish and Wildlife Service, Cortland, NY.
29. Beamish, F. W. H. 1978. Swimming capacity. In: W. S. Hoar and D. J. Randall (eds.), *Fish Physiology*, vol. 7. New York, Academic Press, pp. 101–187.
30. Beveridge, M. 1987. *Cage Aquaculture*. Fishing News Books Ltd. Farnham, England.
31. Edwards, D. J. 1978. *Salmon and Trout Farming in Norway*. Fishing News Books Limited, Farnham, Surrey, England.
32. Gowne, R. J. and A. Edwards. 1990. Interaction between physical and biological processes in coastal and offshore fish farming: an overview. In: *Proceedings of the Conference on Engineering for Offshore Fish Farming*. Thomas Telford, London.
33. Kennedy, W. A. 1978. Handbook of Rearing Pan-Sized Pacific Salmon Using Floating Sea-Pens. Canadian Fisheries and Marine Science Report no. 107.
34. Rudi, H. and E. Dragesund. 1993. Localization strategies. In: Reinertsen et al. (eds.), *Fish Farming Technology*. A. A. Balkema, Rotterdam.
35. Washington Department of Fisheries. 1990. Fish Culture in Floating Net-Pens. Final Programmatic Environmental Impact Statement. Olympia, WA.
36. Findlay, R. and L. Watling. 1990. Effects of salmon net pen aquaculture on microbenthic and macrobenthic communities. Maine Aquaculture Innovation Center Report no. 90-6. Bower, Maine.
37. Bayer, R. and R. Hamill. 1990. A study of fishpen aquaculture and the local lobster harvest. Maine Aquaculture Innovation Center Report.
38. Guldberg, B., A. Kittelsen, M. Rye, and T. Asgard. 1993. Improved salmon production in large cage systems. In Reinertsen et al. (eds.), *Fish Farming Technology*. A. A. Balkema, Rotterdam.

39. Karlsen, L. 1993. Development in salmon aquaculture technology. In: K. Heen, R. L. Monhan, and F. Utter (eds.), *Salmon Aquaculture*. Fishing News Books, Oxford.
40. Barker, C. J. 1990. Classification society rules for fish farms. *Proceedings of the Conference on Engineering for Offshore Fish Farming*. London: Thomas Telford.
41. Riley, J. and M. Mannuzza. 1989. Hydrodynamic Forces on Floating Fish Cages. Paper no. 89-7558. American Society of Agricultural Engineers, St. Joseph, MI.
42. Milne, P. H. 1967. Interim report of water forces on fish netting. Department of Civil Engineering, Univ. Strathclyde, Glasgow. 8 pp.
43. Aarsnes, J. V., H. Rudi, and G. Loland. 1990. Current forces on cages and net deflection. In: *Proceedings of the Conference on Engineering for Offshore Fish Farming*. Thomas Telford, London.
44. Summerfelt, S. T., J. A. Hankins, S. R. Summerfelt, and J. M. Heinen. 1993. Modeling continuous culture with periodic stocking and selective harvesting to measure the effect on productivity and biomass capacity of fish culture systems. In: *Techniques for Modern Aquaculture, Proceedings of an Aquacultural Engineering Conference*, June 21–23, 1993. ASAE Publication 02-93. American Society of Agricultural Engineers, St. Joseph, MI.
45. Hickling, C. F. 1960. *The Malacca Tilapia hybrids*. J. Genet. 57:1–10.
46. McNown, W., and Seireg, A. 1983. Computer aided optimum design and control of staged aquaculture systems. *J. World Aquaculture* 14:417–433.
47. Ley, D. R., M. B. Timmons, and W. D. Youngs. 1993. The Northern Fresh Fish Cooperative. Bulletin 465. Department of Agricultural and Biological Engineering, Cornell University, Ithaca, NY.
48. Tvinnereim, K. 1994. Hydraulic Shaping of Tanks for Aquaculture [in Norwegian]. SINTEF Report STF60 A94046, Trondheim, Norway.
49. Lawson, T. B. 1995. *Fundamentals of Aquacultural Engineering*. Chapman and Hall.
50. Maine Aquaculture Innovation Center. 1990. Development of a Benthic Threshold Environmental Impact Model for Net-pen Salmon Culture Mutually Beneficial to Aquaculture. National Marine Fisheries Service.
51. Speece, R. E. 1973. Trout metabolism characteristics and the rational design of nitrification facilities for water reuse in hatcheries. Trans. Am. Fish. Soc. 102:323–334.
52. Chenoweth, H. H., J. D. Larmoyevx, and R. G. Piper, 1973. Evaluation of circular tanks for salmonid production. *Prog. Fish-cult. 35*, 122–131.

12 Equipment and Controls

J. Hochheimer

12.1 Feeding Equipment

Feeding is usually the largest operational cost of growing fish in aquaculture. The feed can represent greater than 50% of the variable costs in growing fish. Labor and feeding equipment must also be considered as contributors to the total cost of feeding the aquacultural crop. Delivering the feed to fish at the right time, in the correct form, and in the right amount is necessary for optimal growth. The choice of feeding equipment depends on factors such as type and life stage of fish, type and size of feed, size of the operation, available labor, and the type of culture system. Therefore, a careful analysis of these factors is essential for successful and profitable fish culture.

Feeding regimes in aquaculture can be classified in one of the following three categories [1]:

1. Natural food produced in the growing system: Growing larval stages of many carnivorous fish (for example striped bass) to a size that will accept a formulated diet is most successfully done in ponds in which the fish are presented with natural food. See Geiger and Turner [2] for a more in-depth discussion of pond management techniques for producing live foods. When fish reach the fingerling size (about 2.5 cm), they are trained to feed on a formulated diet for grow-out. If omnivores or herbivores are cultured, the plankton bloom in a pond is manipulated with fertilizer additions similarly to that for carnivores. However, this bloom is continued as the fish grow to provide food for the larger fish.
2. A combination of natural and formulated food: In some extensive or semiextensive systems, fish are allowed to graze on natural foods available in the water and are also fed formulated feed to supplement their diet.
3. All formulated food: Fish in intensive systems are fed their complete nutritional requirements in a formulated diet that has been found to meet desired growth requirements.

Proper nutrition is not only a function of feed composition—the fish will not grow unless they get and consume the feed [3]. For many cultured species, dry, pelleted formulated diets have been developed to meet total or supplemental dietary requirements. There are some special cases in which feeds other than natural or formulated dry diets are used, but these are beyond the scope of this review. This section focuses on the delivery of dry diets.

Specific feeding-rate requirements depend on the size of fish being cultured and water temperatures. For example, feeding rates vary from 7% to 10% of body weight for first feedings of fry to 1% to 6% for fingerlings and larger fish [4]. However, both underfeeding and overfeeding are common in aquaculture, and care must be taken to feed the proper amount. Overfeeding wastes feed and degrades water quality; underfeeding leads to lost production.

Particle size is another important factor in feeding. Smaller fish require smaller feed particles, whereas larger fish generally require larger particle sizes. Dry feeds are commercially available in particle sizes ranging from less than 200 μm for larval fish to 1 cm or larger for bigger fish. These particles can also be made to sink or float, depending on the pelleting process used. Often floating diets are preferred by culturists so that they can observe feeding behaviors, but the choice of floating versus sinking pellets should be made with feeding preferences of the fish in mind.

Feeding frequency also depends on life stage and temperature. Larval fish generally require many feedings per day so that they can be trained to accept the dry diet. Additionally, larval fish have smaller gastrointestinal tracts and require frequent feedings to maintain satiation. Failure to feed frequently enough in the larval stages typically results in cannibalism, poor growth, and low survival rates. As fish grow, the frequency of feeding can be reduced, with the subsequent frequencies determined by species-specific requirements.

The feeding frequency also affects the operation of some aquacultural systems, especially the more intensive ones. Oxygen requirements may become critically high in highly intensified systems that are fed only once or twice per day. Around feeding time and up to several hours after, activity and metabolic rates increase dramatically, which leads to increased oxygen consumption. If the total biomass in a system is near the system carrying capacity, oxygen levels can sag below critical levels and stress the fish. Ammonia production may also become a critical factor if large, infrequent feedings are offered, especially in recirculating systems.

There are three groups of feeders and feeding methods—hand, automatic, and demand—used to deliver formulated diets. The type of feeding system depends on the life stage of fish being cultured, size of operation, type of diet, available resources, and personal preferences. Because larval and smaller fish require frequent feedings, they are often fed with automated feeders that can be set to feed throughout the day at regular intervals (Fig. 12.1). Smaller aquacultural operations may have sufficient labor to feed all stocks by hand. Most larger operations feed so much feed that hand feeding becomes a materials-handling problem unless larger, automated equipment is used. Often larger-scale fish farmers use demand or automatic feeders for feeding about 70% to 80% of the daily ration and then hand feed the remaining ration. This allows for the farmer to observe daily feeding patterns and to feed close to satiation [5]. Regardless of which method is used to facilitate feeding, the operator must periodically check for feeding effectiveness by sampling and weighing fish, calculating feed-conversion ratios, and adjusting feeding rates to reflect newly estimated biomass determinations.

When feeding ponds, distributing the feeds to as much of the pond surface as possible enables more fish to get the feed. For shrimp, feeding as much of the entire pond as possible is essential because the shrimp are less mobile and may not get to the feed. Under windy conditions, or if using floating pellets, feed on at least two sides with the

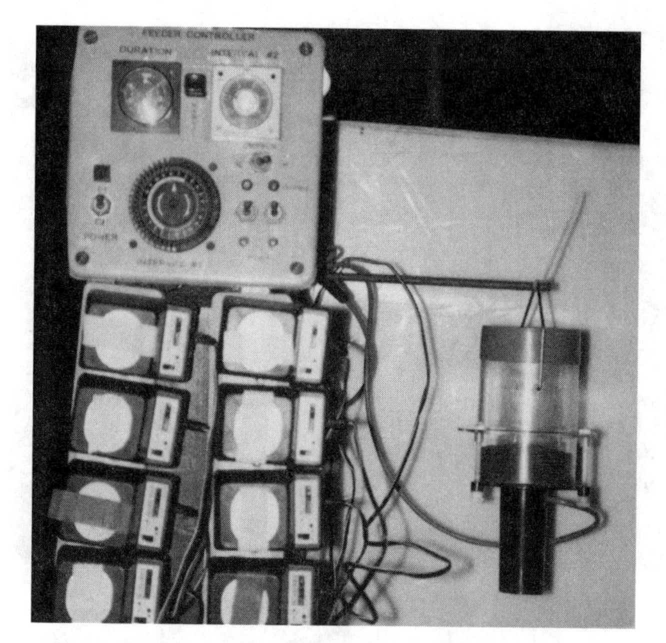

Figure 12.1. Automatic feeder (right) and electronic controller (upper left) for feeding small fish.

wind blowing away from shore. Some fish farmers use a feeding ring, especially for floating feeds, to keep the feed concentrated in one spot. When feeding a cage or net pen, a feed ring keeps floating feeds inside of the structure, thus available only for the cultured fish and not "wild" fish that may congregate around the structure.

12.1.1 Hand Feeding

Manual or hand feeding can be used in all types of aquacultural systems—cage, net, pond, flowthrough, and recirculating—and is often the method of choice for small systems [6]. Hand feeding can reduce capital expenditures for equipment, provided sufficient labor is available to perform the feeding duties. There are other advantages to hand feeding. The fish can be observed at each feeding for feeding behavior to determine if they are actively eating all of the food presented to them. Floating pellets or food platforms (flat-surface mounted about 30 cm below the surface) make observations easier. Fish can only be routinely fed to satiation by hand feeding and observation of the feeding behaviors. Observing the fish during feeding can also be critical in the early determination of disease or parasite problems.

An adaption of hand feeding is a mechanical blower system (Fig. 12.2) used in feeding larger systems, especially ponds and raceways. Feed is loaded into a hopper fixed on top of a blower. When the operator desires to feed, the feed pellets are allowed to drop into the path of the exhausting air stream and are blown out to the desired target. Mechanical blowers and hoppers are fixed to an axle and pulled around ponds or along raceways by a tractor. The power for the blower can be from either power take-off of the tractor or a

Figure 12.2. Mechanical blower feeder for pond feeding.

self-contained motor on the blower. In either case, the operator controls the flow of feed into the air path to regulate the amount of feed delivered.

The catfish aquaculture industry in the United States uses a variety of mechanical blower systems for feeding. These are commercially available or, more often, are fabricated on-site by the farmer. Care must be taken to construct the unit with sufficient integrity to withstand the weight of feed desired in the hopper and to deliver the feed to as much of the culture unit as possible. Sometimes the frequency of feeding in larger operations is dictated by labor availability and not necessarily ideal frequencies for the fish. In larger-channel catfish pond operations, the amount of feed may be as high as 224 kg/ha/d and is fed only once per day [7].

12.1.2 Automatic Feeders

An automatic feeder is basically a mechanism that delivers a prescribed amount of feed to the fish at desired time intervals. The major advantage of automatic feeders is the reduction in labor requirements necessitated by hand feeding, especially in larger operations. Automated feeders have been adapted from other livestock operations to provide cost-effective means to feed fish. There are a wide variety of automated feeders commercially available to meet the array of needs for aquaculture (Fig. 12.3). The most important considerations in selecting an automated system are (Goddard [5]; New [8]):

1. Is the system compatible with the size and form of feed being used?
2. Can the system deliver feed at the desired time?
3. Can the system deliver varying amounts of feed that reflect the changing demands of the growing fish in the culture system?

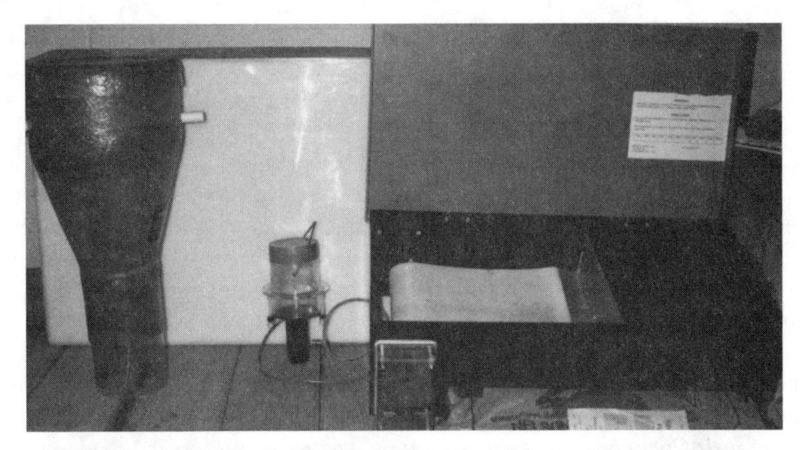

Figure 12.3. Examples of automatic feeders—from left, demand feeder, vibrating feeder, belt feeder.

4. Does the system protect the feed from degradation and spoilage due to rain, excessive heat, and moisture?
5. Are the power source, timer, and drive mechanism reliable under adverse environmental conditions?
6. Does the delivery system prevent grinding and crushing of pellets?
7. Do the capital and operational costs of the system justify the expense over hand feeding?

Automatic feeders can be powered by a variety of means, including mechanical clock mechanisms, electricity (line-powered or battery-powered), pneumatic, and hydraulic (water-powered). The choice of power source depends on availability and reliability. In remote areas in which line-powered electricity would be expensive, mechanical clock mechanisms and battery-powered systems may be the appropriate choice. Belt-type feeders with clock mechanisms as power sources are ideal for feeding continuously. Battery-powered systems can also be fitted with solar panels to maintain battery charges and increase reliability. Line-powered systems should be the most reliable, given that incoming power is consistent and that proper back-up power generation is operational. Other systems, with primary power sources such as pneumatic or hydraulic, are not as common but can be adapted to certain configurations to provide an economical system.

Some innovative feeders have been reported in the literature that are adapted to remote or environmentally harsh conditions. For example, Baldwin [9] reported on a novel feeder for minnow production in Hawaii. The automatic feeder used closed compartments arranged around a stationary axle and used a weighted float located inside of a slowly draining reservoir. As the water drains from the reservoir, the weighted float drops and pulls a cable attached to the compartments. As the compartments rotate past a tripping mechanism, a bottom trap door opens and dumps feed into the fish system. Baldwin [9] reports over three of years of operational experience with this device in often harsh conditions.

One key feature of automatic feeders is some kind of programmable device that allows the operator to vary the frequency and amount of feeding. This may be a combination of electrical and mechanical components that broadcast feed into the culture system. Some systems use electrical timers to determine the frequency, which are available from most aquacultural equipment suppliers (see Fig. 12.1). Examples can also be found in the literature—see Mortensen and Vahl [10] for an easy-to-construct electronic controller. Other aquaculturists may use some kind of mechanical clock system. Ideally, the settings for frequency will allow enough flexibility to meet the needs of the fish. For smaller fish, the frequency may need to be several times per hour, and for larger fish the frequency could be hourly or longer.

Automatic feeders use some kind of mechanical or pneumatic delivery system to present feed to the fish. Mechanical systems use vibration, sliding mechanisms, augers, rotating plate, belts, tipping buckets, or controlled orifices to determine the amount and delivery of feed. Pneumatic systems use compressed air or blowers to broadcast feed to the fish. There are several systems that use a combination of mechanical and pneumatic delivery systems.

Vibrating systems use either motor-driven or electromagnetic sources. The simplest form of vibrating system uses a hopper with a plate fixed to the bottom (see Fig. 12.1). An electromagnet is attached to the bottom plate and vibrates at prescribed intervals. The gap between the plate and hopper sides roughly determines the amount of feed, based on the duration of vibration. Feed amounts are regulated by increasing the gap and duration of vibration. This type of vibrating system can be small, holding less than 1 kg of feed for larvae or small fingerlings, or larger, designed to hold several kilograms of feed.

Several versions of sliding mechanisms are available [8]. One system uses a slider inside of a larger tube. The entire mechanism is fixed to a hopper that holds feed. The slider has a chamber, which is typically fixed in size but can be variable. The feed chamber normally rests below the hopper and is gravity-filled. When feeding is desired, the chamber slides inside of the outer tube and moves the prescribed amount of feed from the hopper opening to an orifice that allows the feed to drop into the culture system. The feeding cycle is complete when the slider returns to its resting position with the chamber under the hopper. Power for the system is used to move the slider and can be either electromagnetic or electromechanical.

Augers or endless screws can be used to deliver feed directly to fish or to pneumatic blowers for broadcast to fish in larger systems. Screw-type augers connected to a motor can be turned on and off to present a prescribed amount of feed to the fish. Feed is usually gravity-fed to the auger. Some auger systems use disks attached to an endless cable that is pulled through a pipe or tube. An advantage of the cable and disk system can be less grinding of feed as it is conveyed through the tube. For larger feeds, pellet integrity becomes a major consideration, as excessively fine pellets are wasteful when uneaten. Some augers deliver feed to a blower or compressed air tube for increased broadcasting of feed to larger areas.

Automatic feeders that use belts driven by mechanical clock mechanisms (Fig. 12.3) are well-suited for feeding fry and fingerlings, as they can provide continuous or frequent feedings. Belt feeders are ideal for training swim-up fry and first-feeding fish to accept dry, formulated diets. By providing small amounts of food over the major portion of a day, the fish are more readily trained to accept the dry food and less likely to cannibalize each other. Typically an 18- to 20-hour photo period is provided to these early life stages to maximize feed adaptation [4].

Larger belt systems can be attached to a hopper and motor driven to deliver larger amounts of feed. Feed falls from a hopper to the belt and is carried over the tank and drops off into the water. The frequency and duration of belt operation determines the frequency and amount fed.

Another feeder that is particularly well suited for smaller feeds is the rotating-plate feeder [5]. This type uses a disk or plate on which feed is placed. The plate slowly rotates, and feed is pushed into an opening with a stationary arm. The amount placed on the plate and the speed of rotation determines the amount delivered to the fish. Because the plate is constantly moving, this feeder can provide continuous feeding. An adaption of this method uses a rotating arm instead of plate. For feeds that are excessively oily or sticky, this method may present problems.

Tipping buckets can also be used to deliver prescribed amounts of feed to a culture system. A series of buckets are hinged and placed under a revolving spindle with a series of spikes used to tip the buckets. The spindle is rotated at a frequency that determines the timing of tipping. Each bucket is filled with the desired amount of feed for a given feeding. Winfree and Stickney [11] report on the design and operation of a small tipping bucket-type feeder.

12.1.3 Demand Feeders

The cultured fish can be allowed to determine for themselves how much feed is made available when demand feeders are used. Demand feeders can be used to make food continuously available and allow for the fish to feed to satiation. Typically, fish fed with demand feeders consume amounts of feed similar to what they would eat *ad libitum* by hand [6]. The demand feeders must, however, be readily accessible to the fish, and sufficient numbers of feeders must be strategically located around larger systems.

Most demand feeders consist of a feed hopper with an opening controlled by a trigger that extends into the water (Fig. 12.3). The hopper is designed to hold from about 1 kg for smaller models to over 450 kg in larger models [12]. Demand feeders are commercially available from a variety of aquacultural suppliers (see *Aquaculture Magazine's Annual Buyer's Guide* for sources) and many have been constructed from materials found around most fish farms. Materials such as plastic jugs, 2-L soft drink bottles, plexiglass disks, plastic balls, and welding rods have been fabricated into demand feeders.

Fish are trained to hit the trigger of a demand feeder when hungry and then become self-feeding. Some demand systems use a touch-sensitive trigger to activate some kind of mechanical delivery system (such as those described in the section on automatic feeders). Other modifications of the demand feeder use an in-water feed tray that activates feed release as feed is eaten from the tray. Typically all of the demand feeders are adjustable for food size and amount released per trigger activation. Goddard [5] reported that it typically takes about 7 to 10 days for rainbow trout to train and feeding to stabilize on demand feeders.

Avault [13] summarized the advantages of demand feeders for rainbow trout. Demand feeders make feed available 24 hours per day and result in less size variability in the harvested fish. Trout fed with demand feeders can have up to 10% better feed conversion than in hand-fed systems. Other advantages from demand feeders are lower production costs from reduced labor requirements and increased feeding efficiency. Trout on demand feeders showed fewer disease problems and fewer problems with dissolved oxygen sags and ammonia spikes from heavy feedings. One other significant finding was a shorter growout—8 to 9 months as compared with 10 to 12 months using hand-feeding methods.

McElwain [12] reported that in tests at the Hagerman National Fish Hatchery (Hagerman, ID, U.S. Fish and Wildlife Service) demand-fed fish showed weight gains 30% greater and lower food conversions than those fed by hand. Additionally the Hagerman tests showed that over 45,000 kg of steelhead trout were raised for $4200 (1981 dollars) less than hand-fed fish. At the Lamar Hatchery, (Lamar, PA, U.S. Fish and Wildlife Service), demand-fed brook trout were twice the length of trout hand-fed under similar conditions. Stickney [6] reported that *Tilapia aurea* gained about 72% more weight when fed by demand feeders than when fed by hand.

Some of the disadvantages reported by Avault [13] include a build-up of wastes under the demand feeders, with some increased incidence of bacterial gill disease. It was postulated that other feeding methods presented feed over a wider area and stimulated the fish to keep the water stirred to reduce build-up of wastes in the tanks. Conventional demand feeders are suitable for dry feeds only. Adjusting the feed output from demand feeders requires constant attention, as does adjusting filling rates (McElwain [12]). Excessive moisture around the feed opening can cause crusting and prevent food from dropping out of the feeder. Wind and waves can trigger the demand feeder and release unwanted food. Many commercial feeders have wind screens around the feeder outlet to help reduce wind-induced triggering.

12.2 Pumps

Moving water is an essential operation in most aquacultural operations. Some aquaculturists have the luxury of using gravity to move essentially all of their water, but these facilities are rare. Pumps are pieces of mechanical equipment that add energy to fluids. Pumps in aquacultural facilities are primarily used to move water from one point to another (e.g., draining ponds and tanks, circulating water in tanks, flowing water through raceways, or filtering water to remove ammonia or solids). Like many other mechanical systems, proper design and selection of pumping systems is essential for the desired outcome. Improperly selected pumps can lead to increased costs and, in some cases, catastrophic failures in part or all of a system.

There are several definitions that are important in the understanding of pumps and pump selection (Wheaton [14]; Lawson [15]).

1. *Head* (or *pressure head*) is the total pressure that the pump must work against. Often expressed as a height of water, head includes the sum of discharge pressure (the vertical distance from the pump to the point of discharge), static pressure at the discharge (any pressure on the fluid as it leaves the pumping system), velocity

head (pressure in the system from movement of the fluid), friction head (pressure losses due to friction in the components of the system), and net positive-suction head (the absolute total head at the inlet to the pump, which can be positive if below atmospheric pressure or negative if above atmospheric pressure).

2. *Brake power* is the power available at the output of the unit powering the pump.
3. *Water horsepower* is the power imparted to the fluid.
4. *Efficiency* is the ratio of the water horsepower to the brake power; this provides a value to compare pumps for an application.
5. *Flow* is the volume output of the pumping system.

Selecting the proper pump requires knowledge about total system head, desired flow rates, suction lift, and properties of the fluid [16]. The piping components of the pumping system are also included in a proper evaluation. Water is the primary fluid found in aquacultural operations, although some other fluid applications may be present, notably air and hydraulic oil. This section focuses on pumping water.

A wealth of information is available from pump manufacturers about their pumps and how to select the proper pump. Descriptions about the pump and its operational requirements are found in manufacturers' literature. Pressure and flow-rate information is presented in tabular form or as a pump curve. Additionally, pump efficiencies are generally presented as a part of the pump data. Net positive-suction head information is provided by the manufacturer for each pump.

The properties of water that are important include temperature, salinity, and the amount of solids present in the fluid. As temperature changes, so do flow and net positive-suction head characteristics of the water. Water temperature is also important for compatibility with the pump's component materials. Salinity (or other corrosive factors) can play an important role in determining compatibility with pump materials. For example, in salt-water systems, wetted pump materials must be plastic, stainless steel, fiberglass, or other noncorrosive materials. Pump specifications will generally indicate compatibility with corrosive fluids. Some pumps are well suited to pumping solids; others are not. Again, the manufacturers' data will provide some useful information related to solids-handling capabilities.

12.2.1 Types of Pumps

Different pumping applications require different types of pumps, depending on parameters such as the flow and pressure desired, the fluid being pumped, and the pumping environment. There are a variety of pump types used in aquacultural applications, although centrifugal and airlift pumps are the predominant types. Other major categories of pumps include rotary and reciprocating. Table 12.1 gives examples of the different types of pumps that are available. Table 12.2 compares some of the characteristics of centrifugal, rotary, and reciprocating pumps. It is important to remember that the manufacturer's specifications should be followed when deciding on a model for a particular application.

Centrifugal

Centrifugal pumps use some type of rotating vane to pressurize and move water. Figure 12.4 illustrates two types of centrifugal pump. As water is directed from the inlet

Table 12.1. Examples of pumps used in aquaculture (after Wheaton [14] and Lawson [15])

Centrifugal	Rotary	Reciprocating
Volute	Propeller	Piston
Diffuser	Regenerative turbine	Diaphragm
Deep-well turbine	Sliding vane	Rotary piston
Jet	Gear	
Mixed flow	Lobe	
	Screw	
	Flexible tube	

Sources: Lawson [15]; Wheaton [14].

Table 12.2. Comparison of pump characteristics

Characteristic	Centrifugal	Rotary	Reciprocating
Use on suction lift	Poor[a] to excellent[b]	Limited; pump must be filled with fluid	Limited; pump must be filled with fluid
Pump life	Excellent[c]	Poor for dry fluids, excellent for lubricating fluids[d]	Good
Quiet operation	Yes	No	No
Pressure relief	Unnecessary unless flow is shut off	Bypass valve required	Bypass valve required
Flow (L/min)	7 to 2400	4 to 150	7 to 50
Pressure (kPa)	0 to 1400	130 to 2100	340 to 25,000
Viscosity of fluid	Low	High at lower RPM	Low[e] to High[f]
Typical power range (watts)	3 to 12,000	180 to 3700	250 to 7500
Pipe sizes inlet/outlet (cm)	0.3 to 10	0.3 to 3.8	1.3 to 1.9
Common construction	Cast iron, aluminum, bronze; naval bronze; types 304 and 316 stainless steel; thermoplastic	Cast iron, bronze, aluminum; Ni-resist; silvercast; type 304 stainless steel	Cast iron, bronze, aluminum; polypropylene
Typical applications	General water transfer, pond draining, recirculation, tank draining and circulation	High-pressure spraying, liquid transfer, washing operations	High-pressure spraying, liquid transfer, washing operations, high solids transfers

[a] Straight centrifugal—about 2 m with suction hose completely full.
[b] Self-priming can be up to 6 m with initial fill of pump casing.
[c] When used with compatible, nonabrasive materials.
[d] For parts that touch, the better the lubricating the fluid, the longer the pump life.
[e] With twin piston or plunger pump.
[f] With diaphragm pump.

Figure 12.4. Centrifugal pumps—closed impeller (left) and open impeller (right).

to the center of the rotating vane (*impeller*), centrifugal forces move the water outward along the vanes of the impeller. The outwardly moving water picks up speed and is directed away from the impeller by the pump casing to the discharge. The dynamic head (*velocity*) is converted to static head as the water backs up in the piping system at the discharge. Water moving away from the center of the impeller creates a pressure drop and allows more water to be drawn in from the inlet. The amount of pressure developed in a centrifugal pump is a function of the pump characteristics, the inlet pressure, and the outlet configuration.

Important characteristics of a centrifugal pump for use in aquaculture include the impeller, casing, and power source. Impellers are classified as *open*, *semienclosed*, and *closed*, describing the exposure of the impeller vanes inside of the pump. Open impellers use vanes attached to a small hub and have large clearances between the impeller and pump walls. Thus, open impellers are better suited for pumping water with high solids content. The trade-off for solids-pumping ability when using an open impeller is a reduction in efficiency compared with other pump types. Semienclosed impellers have vanes that are attached to a single full face plate. A semienclosed impeller can also handle solids, but not as well as open designs. The advantage of a semienclosed impeller is the increase in efficiency over the open impeller. Closed impellers have vanes between two full face plates. Closed impellers have the highest efficiencies but lack the ability to pump solids.

The casing design of a centrifugal pump also determines its operating characteristics. A volute casing has a spiral-shaped water path that increases in area as it approaches the outlet. This allows for the velocity to decrease smoothly and for the energy to transfer smoothly from dynamic head to static head. Single volute pumps are better for use in situations in which the pump will be operated at peak efficiency. Double volute pumps perform better when the pump output is throttled back for long periods of time [15]. Diffusion-type casings have stationary vanes located radially around the impeller. The diffusion vanes reduce turbulence as the water transitions from dynamic to static head within the pump, which results in greater efficiencies [14]. Diffuser casings are better suited to water with a low solids content, and volute casings are the choice for applications with higher solids contents.

Rotary Pumps

Rotary pumps have a rotating unit within a casing that forces the fluid from the inlet or low-pressure side to the outlet or high-pressure side. Most rotary pumps have positive displacement—the discharge is directly related to the speed of the pump. Two exceptions to this are propeller and regenerative turbines [14]. All rotary pumps should be operated at designed output flows and pressures to avoid inefficiencies and possible damage to the pump. All positive-displacement pumps should have pressure relief in the outlet to avoid damage to the pump in the event of discharge throttling or shut off.

The most common rotary pump used in aquaculture is the propeller pump, which is aptly suited for applications with low head and high volume outputs [15]. A simple propeller pump consists of a pipe with an internally mounted propeller that draws water from one end and discharges it out the other. The power required for the propeller pump is directly related to the desired output. Propeller pumps have been used for moving water from ponds and for circulating water within larger tank systems. Care must be taken to protect the fish in the system from the moving propeller blades. Screened inlets are often used.

Some other rotary pumps that may be used in aquacultural applications include sliding-vane, gear, and flexible-tube pumps. Sliding-vane pumps consist of an eccentric rotor with radially mounted vanes that slide. As the rotor moves, the vanes are pushed outward to contact the casing. The volume between two vanes is pressurized as it flows from the inlet to outlet port by decreasing the area between the vanes and casing because of the eccentricity. A modification of this technique is the flexible-vane pump. Gear pumps trap fluids between rotating gear teeth and pressurize the fluid. The flexible-tube pump (or *peristaltic pump*) can be used to deliver smaller volumes of liquids at constant rates (Fig. 12.5). Rotating rollers pinch flexible tubing against a casing wall to move the fluid through the tubing. Because the fluid only contacts the tubing,

Figure 12.5. Peristaltic or tubing pump and controller.

flexible-tubing pumps are ideal for corrosive fluids or in applications that require the pumping system to be sterilized.

Reciprocating Pumps

One type of reciprocating pump that can be very useful in some aquacultural applications is the diaphragm pump. A flexible membrane is fitted into a cavity and flexed back and forth. The vacuum created on the up stroke of the diaphragm draws the fluid through a valve and into the chamber. The pressurized downstroke forces the fluid out of the chamber and through a second valve. Diaphragm pumps are bulky and rather heavy but have several desirable operational characteristics. For applications that require a self-priming pump, have high levels of solids, or may result in the pump running dry for more than a few moments, the diaphragm pump is ideal.

Airlift Pumps

An airlift pump is simply a rising column of air bubbles in a tube submerged in the water (Fig. 12.6). An airlift pump can be an efficient means to aerate and circulate water in a variety of aquacultural systems, provided that the water does not have to be pumped very high above the surface. The bubble/water mixture inside of the airlift creates a fluid of lower specific gravity then the water alone. The amount of lift is a function of the difference in specific gravity between the fluid inside of the tube and that outside. As long as sufficient air is injected into the submerged tube, water will flow.

The submergence ratio of the lift tube is the primary factor in determining the amount of lift available. Submergence ratio is defined as the ratio of the total length of the lift tube to the length that is below the water surface [15]. As the submergence ratio increases, more of the lift tube is below the surface and the airlift pump is more efficient. See Wheaton [14] and Reinemann and Timmons [17] for more complete information on the design and operation of airlift pumps.

Airlift pumps have been used in many applications, including circulating water in ponds and tanks, pumping water from tanks, and enhancing aeration with air or pure oxygen. In fact, airlifts are excellent for circulating water in tank and pond systems if placed at 100% submergence. Turk and Lee [18] found a $200 labor savings and $2000 difference in yearly electrical costs in a comparison of hatchery and grow-out production of squids and cuttlefish using airlift pumps and centrifugal pumps. The airlift systems performed as good or better than the traditional pumped systems in the areas of water circulation, water quality, animal performance, labor for maintenance, equipment and material replacement, and energy usage.

12.2.2 Power Source

Selecting the power source for a pump is also an important consideration when designing a pumping system for aquaculture. Power to the pump shaft must be efficient and at the correct speed for the operating characteristics of the pump. Typical power sources used in aquaculture include electric motors, internal combustion engines, and air (for airlift pumps). The selection of a power source will require an evaluation of the following [14]:

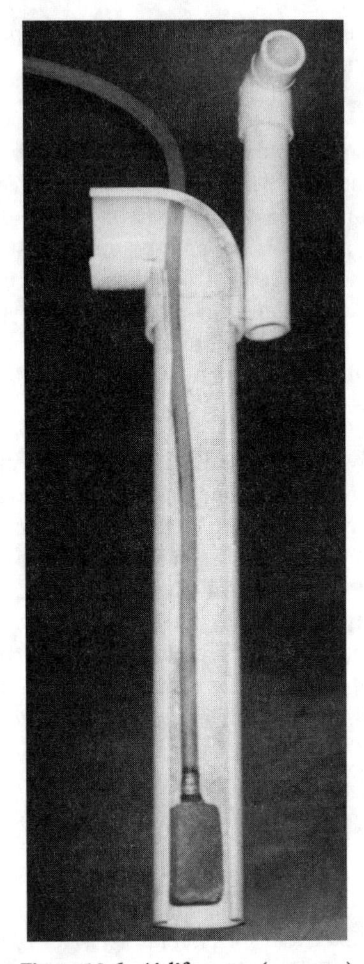

Figure 12.6. Airlift pump (cut-away).

1. The amount of power required
2. Capital cost of the power unit
3. Operating costs of the power unit
4. The frequency and duration of pump use
5. Availability of energy

Another important consideration in selecting the proper power source is the power-output rating of the power unit. Continuous-duty electrical motors are rated to operate at 100% of their rated output. Gasoline and diesel engines can only be loaded to about 60% for air-cooled gasoline, 70% for water-cooled gasoline, and about 80% for diesel [14]. Larger electrical motors need starter circuits to handle the increased electrical load during initial start-up of the motor. Adding loss of phase and overload protection in the starter circuit is often cost-effective through preventing costly repairs caused by weather-related electricity problems.

12.3 Harvest Equipment

Harvesting fish from aquacultural systems represents the most labor-intensive part of the operation [19]. Harvesting of the crop requires specialized equipment and site-specific modifications to concentrate the fish, remove them from the water, and then sort the crop according to market requirements [20]. Throughout a typical production cycle, fish are harvested several times for restocking, sorting and grading, health and feeding assessment, and shipment to markets. Many food fish and all fish being sold for grow-out or stocking purposes must be alive and relatively stress-free to be able to survive after harvest. Different species and life stages may also require unique harvesting methods. Larger operations, such as ponds, raceways, or tank systems, require the operator to handle many kilograms of fish at a time. Safe and efficient harvesting equipment is needed to handle the large quantities of fish produced in these large systems. In all harvest operations, minimizing stress and mechanical damage to the fish is essential. Therefore a harvesting method must allow easy and safe handling for both the harvest crew and the fish.

Harvesting and handling fish causes stress to the animals. Too much stress to the fish weakens their immune systems and can lead to poor performance and even death. Typical harvest operations crowd fish into confined spaces and remove them from water for a period of time. Care must be taken to provide adequate water quality during the crowding. Additionally, the time out of water must be minimized. It is also very important to prevent mechanical damage to the fish during the harvest. Fish with bruises, cuts, punctures from spines and fins, and missing scales look bad and are less desirable for live- and whole-fish markets.

Proper planning and equipment preparation is essential for an efficient and safe harvest. Ponds and culture systems should be planned and constructed with harvesting in mind. Ponds must be accessible to the harvest equipment, including tractors and trucks for larger harvests. Shuttling fish from a pond to a hauling tank and then to a hauling truck adds extra handling and stress. Make sure the pond berms are wide enough for the kinds of equipment that will be on them. Roads on berms must also be protected with gravel to prevent erosion and rutting from the heavy equipment. Tucker and Robinson [7] summarize planning considerations for successful harvesting fish:

1. Develop a plan prior to harvesting.
2. Maintain all equipment in good working order.
3. Check for off-flavor in food fish.
4. Do not harvest sick fish.
5. Stop feeding 24 hours prior to harvest.
6. Try to avoid harvesting in hot weather.
7. Do not harvest when water quality is poor.
8. Conduct the harvest as efficiently as possible.

In most larger ponds used for food-fish and shrimp production, concentrating the crop is done with seines. Matching the proper seine to the particular harvest is essential for efficient harvest of a pond. The seine must be deep enough to form a bag sufficient to carry all of the catch. Additionally, maintaining the depth of the seine in the water column by properly pulling it through the water will keep fish like carp and tilapia from burrowing under the seine and fish such as grass carp and milkfish from jumping over the net. The mesh size of the net should be as large as possible to minimize drag, but small enough to contain the fish to be harvested. Some seines also have a larger bag in the center for containing the fish or may have a smaller bag fitted with a drawstring to attach to a livecar. As the fish are corralled by the seine, they eventually end up in the bag. Jensen and Brunson [21] summarize the seining process for larger ponds.

The harvesting efficiency (measured as the percentage of harvestable fish that are harvested live) in ponds varies according to the experience of the crew, the type and condition of the pond, the condition of the equipment, and the water temperature [7]. A skilled crew is likely to harvest more fish with less stress than a novice crew. Experience at harvesting builds the skill level. Newer ponds tend to have harder bottoms and reduce the amount of mud collected in the net. Older ponds also tend to have more irregularities in the bottom from aeration equipment and other pond operations. Older, unrepaired equipment will allow more fish to escape or add to stress if components breakdown. Fish become more lethargic in colder water and are less likely to become stressed than those in warmer water. Under ideal conditions, a seining can probably harvest about 80% to 90%; three seine hauls can typically harvest up to 99% of the fish in a pond [7]. If all of the fish must be harvested the pond must be drained and fish picked out of the mud to complete the job.

12.3.1 Types of Harvesting Equipment

There are many different types of harvest equipment available for pond and tank culture systems. A variety of nets, seines, and traps can be used to harvest from ponds. Jensen and Brunson [21] provide an excellent summary of harvest equipment for pond culture. Fish pumps and piscilators are mechanical devices used to move fish from crowded areas in ponds and tanks to transport tanks. Other equipment found in harvesting operations includes livecars, boom trucks, loading baskets or nets, transport trucks and tanks, sorting and grading equipment, small boats, and tractors. Tanks and raceways may use crowders to concentrate fish for easy removal.

Nets

Netting for harvesting equipment is available in a variety of materials and sizes. Materials should be selected that are compatible with the fish to be harvested and are sufficiently durable to last many years. Most netting used in aquaculture is made from knotted polyethylene or nylon. Knotted nets are suitable for fish like catfish and carp. Woven or knotless netting is more suited for scaled fish, as it is softer and less likely to abrade fish or knock off scales. Nylon netting should be treated with a tar-based or plastic coating to prevent sun damage and deterioration and snagging of spines and fins for some species [7]. Some nets, such as cast or gill nets, are made from monofilament. The mesh size should be sufficiently small to keep the desired size of fish contained, but large enough to minimize drag as the net is pulled though the water. Smaller meshes tend to retain mud as well. Consult net manufacturers for additional information on sizing and materials for nets.

Cast Nets

Cast nets are circular nets usually made from monofilament nylon of various mesh sizes. Cast nets are relatively inexpensive and are useful for sampling fish from ponds. When samples are needed for determining feeding rates, growth rates, and fish size, cast nets are an easy and effective tool. Shrimp are also easily sampled with cast nets. In fact, shrimp broodstock are easily and less stressfully caught with cast nets [22]. Many species easily entangle in cast nets. Practice and patience are required to develop the skill to properly throw a cast net [21].

Lift Nets

Lift nets are small square nets that are placed on the bottom or suspended with a bridle below the surface. By using soft, small meshed netting, the lift net is useful in catching and sampling small fish, fingerlings and fry with minimal injury [21]. Fish are lured over the lift net with food or light and then the net is rapidly lifted out of the water. An adaptation of the lift net is the float net, which is placed on the bottom and released to float to the surface, trapping the fish.

Seines

Seines are rectangular nets with floats along the top and weights at the bottom. The weights and floats keep the net suspended through the depth of the water and the seine is pulled through the water to crowd fish. Seines work best in ponds with flat, regularly shaped bottoms, in which there are no obstructions [21]. Seines can be used in ponds that are not drained, such as many catfish ponds. For ponds that are drained or require complete harvest, seines can also be used to harvest the majority of fish prior to draining. This is considered less stressful to the fish than draining alone.

The floats are usually made from cork or foam and keep the net at the top of the water column. The weighted bottom or lead line is usually done with lead weights or lead cored ropes. Often the lead line also has a mud line attached to help the lead line move efficiently along the bottom without digging into the mud. The mud line is made from a loosely fitting water-absorbing rope line braided along the lead line. Rope mud lines are often used on regular, hard-bottomed ponds. Other designs for mud lines include rubber rollers, which have been found to be better suited for soft, irregular bottoms (Jensen and Brunson, 1992).

The recommended length of the seine used to harvest an entire pond is usually 1.5 times the width of the pond, but some farmers prefer longer seines. Shorter seines are used to sample fish from a pond by pulling the seine across a corner and then to shore. The height of the seine should be about 1.5 times the maximum depth. For most ponds greater than 0.2 ha, a mechanically drawn seine is preferred because of the drag resistance from the area of the seine in the water. Trucks, tractors, and other motorized vehicles are used to pull larger seines through ponds for harvest.

Gill Nets

Gill nets can be placed in ponds to selectively harvest larger fish. The gill net is made from monofilament and placed across the pond. The mesh size determines the size of fish captured. Some farmers drive the fish to the net with a boat, and most prefer nighttime harvesting to limit the visibility of the net in the water. Gill nets are extremely stressful to the fish and must be checked often.

Traps

Traps can be used to capture smaller fish, such as fingerlings or fry. Some of the advantages to trapping include [3]:

1. Better condition of the harvested fish, because they are not subjected to silty, poor quality water
2. Reduced injury to the fish by handling smaller numbers of fish at a time
3. That draining the pond is not necessary
4. Successful harvests in ponds with abundant vegetation
5. Avoidance of nuisance organisms such as tadpoles and crayfish
6. Reduced labor

One major disadvantage to trapping is that supplies of fish may be variable in the trapping process; also, the entire pond is not harvested at a single time.

One popular trap is the V-trap, which is placed in front of the drain. The V-trap consists of a mesh or screen box for collecting the fish and a V-shaped opening that directs the fish into the box. The V can be constructed from mesh or glass. Fish swim into the glass or mesh wing that extends from the apex of the V outward. When they encounter the wing, the fish tend to swim along it and enter the trap through a narrow opening in the apex of the V. Once inside the trap, the fish have difficulty exiting through the narrow opening. The trap is placed in the water and submerged so that about 10% of the height is above the water line. Moving current and collecting food organisms attract larval fish to the trap. For species that are attracted to cooler water, the V-trap can be placed at the outlet of a pipe flowing water. There are many modifications to the V-trap and its use in ponds; see Piper et al. [3] for other examples.

Seine Reels

Larger seines (> 60 m) are difficult to manually handle and are typically stored and transported on hydraulically operated reels (Fig. 12.7). A seine reel is primarily used to store the seine, but some are modified with second reels to assist in collecting the net during the harvest when the net is pulled in to concentrate the fish. Reels are commercially available or can be fabricated from materials found on many farms. A 1.5-m diameter seine reel can typically hold up to a 365-m seine [21].

Livecar

Once a pond is seined and fish concentrated along one berm, they are often transferred to a livecar (sometimes called a *sock*). The livecar consists of an open-topped net bag with a tunnel-shaped opening in one end terminating at a metal frame that fits into the bag end of a seine [21]. The seine has a drawstring at the terminus of the bag that allows the metal frame of the livecar to be connected. The harvested fish are crowded into the livecar for holding until they are dipped from the pond. About 2200 kg of fish can be held in each 3-m section of approximately 1-m wide livecar when water temperatures are below 27°C [21].

Figure 12.7. Seine reel with feed storage bin on right.

Loading Basket

If many fish are to be harvested, a boom with a dip net or loading basket is used. The loading basket is made of a hoop-shaped net with a bottom trap door for dumping the fish from the net. A boom truck or backhoe fitted with an extension arm is used to maneuver the loading basket. Most harvesters use an in-line scale to determine the weight of fish as they are removed from the pond.

Other Harvest Gear

In tanks or raceways, harvesting is generally easier because the structure is smaller and regularly shaped. Some systems continuously harvest by grading and selectively harvesting the fish. For fish that grade easily, bar or other adjustable graders can be pushed through the tank to crowd fish and allow smaller fish to pass through the grader. Once the fish have been graded, the desired fish are dipped from the tank for transport. Tanks can also be partially or completely drained to aid in fish removal. In net pens or cages, the fish can be crowded and dipped out of the structure.

Some fish farmers are using some mechanical equipment for harvesting from ponds and tanks, especially once the fish have been crowded. Fish pumps and piscilators can be used to lift fish from ponds or tanks for sorting and loading onto transport trucks.

Fish pumps are usually either centrifugal pumps, airlifts, or vacuum pumps. The centrifugal-type fish pump has specially designed impellers that allow the fish to pass through the pump relatively unharmed. The size of the fish will generally determine the amount of damage. Some fish pumps have the capability of dewatering some of the fish as they are moved through the pump and piping system. Airlift pumps are similar in design to those used in moving water in other aquacultural tasks, only the piping is larger to accommodate the fish. Varadi [20] shows some designs for fish airlifts. Most commercially available vacuum fish pumps operate by intermittently sucking up a batch of fish and water into a container during each duty cycle [20].

Piscilators, or *fish screws*, have been used recently to move fish from ponds and tanks. A piscilator is either a large rotating drum with internally attached vanes or a fixed drum with an moving auger. Water and fish are gently moved up through the drum. Piscilators are becoming a popular method of moving and sorting fish, especially fingerlings. The primary advantage of piscilators is the minimal physical damage done to the fish during transport. A disadvantage is the increased time required to remove a large group of fish from a pond or tank.

Grading Equipment

Grading boxes are used frequently in aquaculture to sort fish according to size. Boxes with either fixed or variable openings in the bottom are used to allow smaller fish to pass through the voids. Fish are placed in the box that is floating in a tank or pond, and smaller fish are allowed to pass into the water under the grader. Some commercially available bar graders have mechanisms that easily allow the user to vary the slot width. For smaller fish, screens with oval holes can be effective for sorting (see [23] for an easy-to-construct grader for smaller fish). In general, rounder fish tend to work better in graders than do fish that are flatter-shaped. The flat fish tend to hang up in the grader and not pass through until the slot width is large enough to allow some of the larger fish to pass as well.

In tanks and raceways, moving or fixed graders can be used to sort fish. A moving grader can be mechanically or manually pushed through a tank or raceway and smaller fish allowed to swim through the grate. Stationary graders are placed in the tanks and allow smaller fish to freely move through while confining larger fish to one side.

12.4 Monitoring Equipment

As the intensity of aquacultural systems increases, the need for monitoring of critical components and processes becomes more necessary. Many fish are lost in aquacultural systems because of the tripping of a circuit breaker due to an electrical overload, shorting of a motor starter from the loss of one phase in a three-phase system, or the inopportune failure of a relatively simple component, such as a pipe fitting or airline. The proper design of a culture system should include an analysis of the critical components and processes to determine which are essential in the monitoring plan. The critical components and processes include anything that will cause undue stress or loss of fish in the event of a system or component failure.

Oxygen levels in the culture system are usually the primary concern, because a failure of the oxygenation system could rapidly lead to mortality. Other possible critical components include temperature, water levels in tanks that could drain, water flows, electricity, and ammonia and nitrite levels.

A monitoring system may consist of a variety of components, including both automated and manually monitored parameters. The monitoring system can also be used to control equipment to provide better water quality without operator intervention. For example, oxygen monitoring can be coupled with controls to turn on and off additional oxygen injectors. Then as the need for oxygen increases and decreases, the proper amount is added. This provides better water quality and growing conditions for the fish, as well as more economical operation for the owner.

The analysis of the monitoring requirements should consider the necessity of each parameter to be monitored and the frequency of monitoring. It should look at both the direct and indirect effects of a system component failure to determine monitoring requirements. For example, clogging of a solids filter might directly affect solids removal, but could indirectly lead to overflow of a tank. Only those parameters that are essential to the operation and safety of the fish should be considered in a monitoring program. One should determine the need for automated monitoring of each parameter versus manually monitoring using test kits or portable meters.

Extra parameters add to the overall cost and labor requirements of the aquacultural system. Some parameters are also very expensive to monitor reliably in an automated system, and the need for these should be carefully justified. Maintenance of sensors in the harsh environment of an aquacultural system is time-consuming and costly in terms of replacement parts and labor. Redundancy and reengineering of some parts of the system to reduce or eliminate the need for monitoring should be considered. For example, one could put external and internal standpipes on tanks to eliminate the need for level sensors in tanks. These sensors foul easily and will give false readings if not impeccably maintained.

Kaiser and Wheaton [24] summarized the design considerations for a monitoring system:

1. Formulate clear objectives: List the desired functions of the monitoring system for both present and future.
2. Evaluate the economic feasibility of each objective: The monitoring system should help to increase profitability. Determine the capital and operational costs of each option.
3. Determine the system inputs: Assess the parameters to be monitored, including the level and frequency of monitoring.
4. Determine the hardware and software needs: Evaluate sensors and data collection equipment available for the task. Determine availability and compatibility of software for the hardware.

12.4.1 Sensors

Sensors are the part of the monitoring system that measure the critical variable (Fig. 12.8). Simply defined, a sensor converts a physical property into an electrical signal [24]. Sensors can be simple switches that open or close a circuit depending on a certain condition, or they can measure a voltage change that relates to changes in a particular parameter of concern.

Specifications provided by the manufacturer provide important information for the selection of sensors. Some of the more important considerations about a particular sensor include [24]:

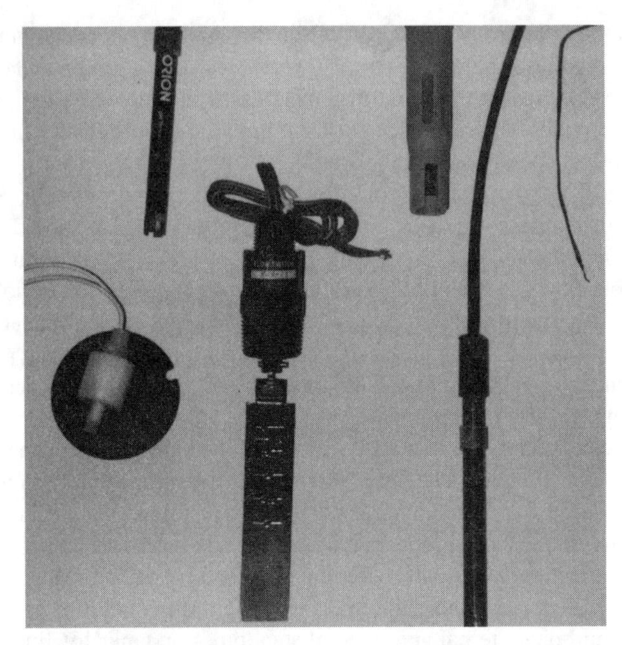

Figure 12.8. Examples of sensors—clockwise from top left, pH, dissolved oxygen, thermocouple, ammonia, flow, level.

1. Measurement range: The level of the measured variable in the culture system must fall in the range of measurements for the sensor.
2. Measurement sensitivity: The minimum detectable change of the measured variable by the sensor must be at a level sufficient to provide meaningful monitoring information.
3. Sensor drift: The frequency of changes in the output signal without changes in the input should be small.
4. Response time: How long will the sensor take to stabilize on a measurement?
5. Calibration frequency and procedure: Sensors that require frequent calibrations will need automatic calibration processes. Determine the procedure for calibration and evaluate the suitability of materials with the aquaculture system.
6. Operating conditions: Check with the manufacturer to determine if the sensor is suitable for the intended environment.
7. Interferences: Are there substances in the system water or environment that will interfere with the measurement? Check for therapeutants and any chemicals that may be added.
8. Maintenance: What steps and supplies are needed for maintenance?
9. Expected life: What is the life expectancy of the sensor?

Sensors are typically the weakest components of a monitoring system. The harsh environment of the fish tanks and system water tend to corrode or foul most sensors rather quickly. Some sensors are inherently unstable and require constant recalibration. Manufacturers of sensors are using a variety of new techniques to make the sensors more stable and reliable. Sensors available today are much more sensitive and reliable than those of a few decades ago.

The critical parameters that should have serious consideration for frequent or constant monitoring are oxygen, temperature, ammonia, and possibly electricity levels. Other parameters such as pH, water flow, or water level may in some systems require more frequent monitoring. System pH is relatively easy to monitor using a pH meter and probe or wet chemistry. Probes for pH are very reliable and can be adapted to an automated system with ease. Manufacturers of pH equipment should be consulted for details. Water level and flow are also relatively easy to monitor with readily available sensors. One should consider fouling and the need to routinely clean the sensors prior to purchasing and installing. Most water level and flow sensors can be of the simple on/off switch type and therefore can be set up on an automated system to detect an open or closed circuit.

Oxygen Sensors

Most dissolved-oxygen sensors are polarographic sensors that measure a current produced by the consumption of oxygen between two electrodes. The consumption of oxygen is proportional to the amount present, which is a function of how much oxygen diffuses across a permeable membrane. The sensor can be connected to meters specifically designed for monitoring oxygen or can be connected to a computerized data collection system. The manufacturers specifications give additional details.

Oxygen sensors can be air-calibrated and typically require calibration at long (> 24-hour) intervals. Maintenance of the membrane and filling with a potassium chloride solution are the usual maintenance procedures in clean water. For prolonged exposure to culture water, additional maintenance will require cleaning of the membrane periodically.

For many aquacultural operations, monitoring dissolved oxygen several times per day will provide sufficient information about the status of oxygen. For ponds and many tanks culture systems, the oxygen levels may not change very rapidly. Pond systems are reasonably well understood and can usually be monitored two or three times per day. Manual readings are probably cost-effective for many pond systems. For more intensive systems, a failure of an oxygenation component could mean extensive mortality in less than an hour. These very intensive systems require either redundant components or automated readings of oxygen several times per hour. Control circuits that use dissolved oxygen readings to administer the amount of oxygen in a system can be cost-effective if high densities are cultured.

Temperature

Temperature sensors come in a variety of configurations, including thermocouples, resistance temperature detectors, and thermistors. See the Omega Temperature Handbook for a more detailed discussion on temperature measurement and sensors [25].

A *thermocouple* is two dissimilar metal wires connected together. If exposed to changing temperatures, the wires generate a current that is proportional to the temperature. Thermocouples are well studied and available from a variety of sources. One should be careful to select thermocouples that measure temperatures within the desired temperature range and are made from materials compatible with the environment.

Resistance temperature detectors (RTDs) rely on the fact that resistance in metals changes with changing temperatures. Most RTDs are made from platinum, nickel, or nickel alloys. RTDs tend to be more linear than thermocouples and much more stable, but they are more fragile and require a small current to provide a voltage to be measured [25].

Thermistors are semiconductors that have an inverse relationship between temperature and resistance. Thermistors have the advantage of being much more sensitive than RTDs or thermocouples. The construction of the thermistor makes it more fragile than RTDs, and thermocouples and are less tolerant to large temperature differences, which is not a problem in aquaculture [25].

Ammonia Sensors

Ammonia levels can become critical in many aquacultural systems and need to be routinely monitored in intensive culture systems. Although ammonia is a critical variable to monitor, careful consideration should be given to the type of monitoring system. For operations that will only require routine checks on ammonia, test kits and small spectrophotometers should be sufficient. However, for heavily loaded recirculating systems, frequent ammonia monitoring may be necessary to ensure efficient biofilter operation and fish health.

Kaiser and Wheaton [24] reported on an automated system for monitoring ammonia in recirculating aquacultural systems. The sensor used in this application was an ammonia gas–sensing electrode. This electrode measures the free ammonia gas in a solution using a selective membrane. To measure total ammonia in a water sample, the sample pH is increased to greater than 11 to ensure that all ammonia in solution is in the form of ammonia and not ionized as ammonium. An automated system must take this into account and have a measurement chamber capable of automatically adjusting the sample pH for each measurement.

12.4.2 Monitoring and Control Systems

With the advent of relatively inexpensive personal computer systems and data acquisition systems, automated monitoring is becoming much more affordable than even a decade ago. Fast computers coupled with data-acquisition boards and data-monitoring software are making the task of developing a custom monitoring and control system easy and economical.

There are several data-acquisition and -control systems commercially available for aquacultural applications. Kaiser and Wheaton [24] and Ebeling [26], [27] present descriptions of data acquisition and control systems.

System Components

Parts of a monitoring system include sensors to measure the desired variables, an interface to convert the electrical information into a form readable by a computer or microprocessor, a computer, software to run the system, and displays. Matching the components is the most important step in deciding what to use in a monitoring system.

Kaiser and Wheaton [24] present several key considerations in selecting the monitoring system components:

1. Determining the inputs: Are the signals analog or digital?
2. Accuracy and resolution: What is the desired accuracy and resolution of the measurements and can the data conversion and computer provide this resolution?
3. Determining system outputs: Does the control function require analog or digital output? What displayed and stored data will be required?
4. Selecting a computer: Is the computer rugged enough for the aquacultural environment? Is it compatible with the other components?
5. Selecting software: Does the computer software easily interface with the user and components?

Alarms and Calling Systems

One of the most important functions of a monitoring system is to provide alerts to the system operator in the event of malfunctions and problems. If critical variables are sensed to be outside of acceptable limits, alarms need to be sent out for corrective action to be initiated. It is important to design and test the monitoring and alarm system so that false alerts are not sent out too frequently. Too many false alarms usually results in the operator becoming less likely to respond.

Alarms must be constructed and operated so that the proper individuals are alerted to adverse conditions. Visual and audible alarms can be placed in key areas within a facility to alert workers of problems (Fig. 12.9). Loud and obnoxious audible alarms placed outside of a facility are not likely to be very effective during hours when there are no onsite workers. Instead, remote alarms need to be employed. All alarm systems must also be capable of routine testing to ensure that the systems are working properly.

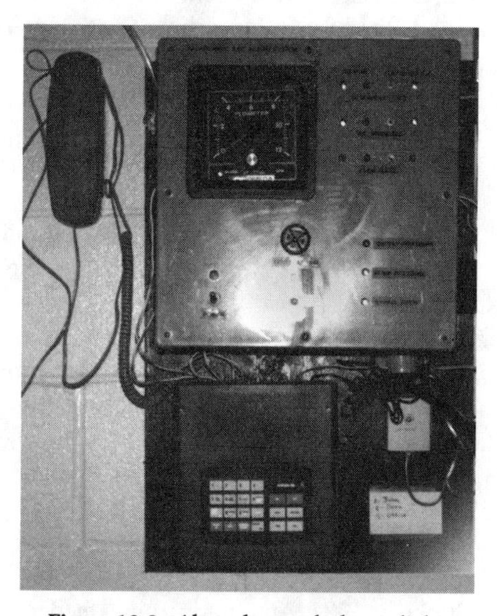

Figure 12.9. Alarm box and phone dialer.

One external alarm that has shown promise is the automatic phone dialer. This device is commercially available and quite simple to use. Most of the inexpensive models were designed for monitoring open/closed contacts, the presence of line-powered electricity, temperature, and noise levels. Basically they are inexpensive burglar alarms. The units accept either four or eight on/off inputs, which can be microswitches or analog outputs from a computer monitoring system. Each input to the dialer is then routinely scanned by the unit, and an alert conditions exists if one is triggered. Also, if temperature exceeds a preset limit, a noise above a threshold is detected, or electricity goes off, the unit goes into alert status. The unit is programmed to dial one or more phone numbers and will continue to cycle through the number list until an acknowledgment is accepted. The user can also call into the system and poll the unit for a status.

Sometimes pagers are used with the dialers to give the operator mobility with constant availability to the dialer. A variety of more complex dialers are also available with more features and capabilities. The system needs and compatibility with other system components must be evaluated prior to purchase.

References

1. Swift, D. 1993. *Aquaculture Training Manual.* Fishing News Books, Blackwell Scientific Publications, Oxford, England.
2. Geiger, J., and C. Turner. 1990. Pond fertilization and zooplankton management techniques for production of fingerling striped bass and hybrid striped bass. In: R. Harrell, J. Kerby, and V. Minton (eds.), *Culture and Propagation of Striped Bass and Its Hybrids.* Striped Bass Committee, Southern Division, American Fisheries Society, Bethesda, MD, pp. 79–98.
3. Piper, R., I. McElwain, L. Orme, J. McCraren, L. Fowler, and J. Leonard. 1982. *Fish Hatchery Management.* United States Department of the Interior, Fish and Wildlife Service, Washington, DC.
4. Shepherd, C., and N. Bromage. 1988. *Intensive Fish Farming.* Blackwell Scientific Publications Professional Books, Oxford, England.
5. Goddard, S. 1996. *Feed Management in Intensive Aquaculture.* Chapman and Hall, New York.
6. Stickney, R. 1994. *Principles of Aquaculture.* John Wiley and Sons, New York.
7. Tucker, C., and E. Robinson. 1990. *Channel Catfish Farming Handbook.* Van Nostrand Reinhold, New York.
8. New, M. 1987. *Feed and Feeding of Fish and Shrimp: A Manual on the Preparation and Presentation of Compound Feeds for Shrimp and Fish in Aquaculture.* United Nations Development Programme, Food and Agriculture Organization of the United Nations, Rome, Italy.
9. Baldwin, W. 1983. The design and operation of an automatic feed dispenser. *Aquaculture* 34:151–155.
10. Mortensen, W., and O. Vahl. 1979. A programmed controller for automatic feeders. *Aquaculture* 17:73–76.
11. Winfree, R., and R. Stickney. 1981. Automatic trough feeder developed. *Aquaculture Magazine*, May–June:18–19.

12. McElwain, I. 1981. *The Use of Self Feeders in Fish Culture*. Lamar Information Leaflet 81-01. U.S. Fish and Wildlife Service, Lamar Fish Culture Development Center, Lamar, PA.
13. Avault, J. 1981. Feeding methods affect production and profit. *Aquaculture Magazine*, May–June:38–39.
14. Wheaton, F. 1977. *Aquacultural Engineering*. John Wiley and Sons, New York.
15. Lawson, T. 1995. *Fundamentals of Aquacultural Engineering*. Chapman and Hall, New York.
16. Bankston, J., Jr., and F. Baker. 1994. Selecting the Proper Pump. SRAC Publication no. 372. Southern Regional Aquaculture Center, Stoneville, MS.
17. Reinemann, D., and M. Timmons. 1986. Airpump. Northeast Regional Agricultural Engineering Service, Cornell University, Ithaca, NY.
18. Turk, P., and P. Lee. 1991. Design and economic analysis of airlift versus electrical pump driven recirculating aquacultural systems. In: *Engineering Aspects of Intensive Aquaculture*, Proceedings from the Aquaculture Symposium, Cornell University, April 4–6, 1991, NRAES-9. Northeast Regional Agricultural Engineering Service, Cornell University, Ithaca, NY.
19. Pillay, T. 1990. *Aquaculture Principles and Practices*. Fishing News Books, Blackwell Scientific Publications, Oxford, England.
20. Varadi, L. 1985. Mechanized harvesting in fish culture. In: T.V.R. Pillay (ed.), United Nations Development Program: *Inland Aquaculture Engineering*. ADCP/REP/84/21. Food and Agriculture Organization of the United Nations, Rome.
21. Jensen, G., and M. Brunson. 1992. Harvesting Warmwater Fish. SRAC Publication no. 394. Southern Regional Aquaculture Center, Stoneville, MS.
22. Lee, J., and M. Newman. 1992. *Aquaculture: An Introduction*. Interstate Publishers, Danville, IL.
23. Schipp, G. 1995. A simple, adjustable grading box for small fish, *Progressive Fish-Culturist* 57:245–247.
24. Kaiser, G., and F. Wheaton. 1991. Engineering aspects of water quality monitoring and control. In: *Engineering Aspects of Intensive Aquaculture*, Proceedings from the Aquaculture Symposium, Cornell University, April 4–6, 1991, NRAES-9. Northeast Regional Agricultural Engineering Service, Cornell University, Ithaca, NY.
25. Omega. 1992. *The Temperature Handbook*. Omega Engineering, Stamford, CT.
26. Ebeling, J. 1991. A computer based water quality monitoring and management system for pond aquaculture. In: *Engineering Aspects of Intensive Aquaculture*, Proceedings from the Aquaculture Symposium, Cornell University, April 4–6, 1991, NRAES-9. Northeast Regional Agricultural Engineering Service, Cornell University, Ithaca, NY.
27. Ebeling, J. 1995. Engineering design and construction details of distributed monitoring and control systems for aquaculture. In: *Aquacultural Engineering and Waste Management*, NRAES-90. Northeast Regional Agricultural Engineering Service, Cornell University, Ithaca, NY.

13 Waste-Handling Systems

Steven T. Summerfelt

13.1 Introduction

In the past, intensive fish-farming practices were often developed without much consideration for either the real or perceived effect of their waste on the farm's surrounding environment. Intensive fish-farming practices used water resources for two reasons: to carry oxygen to the fish and to receive the waste produced in the system (metabolic by-products and other materials) and disperse or carry these wastes away so that they did not accumulate in or around the fish farm to harmful and undesirable levels [1]. Therefore, in an unregulated environment, the carrying capacity of a fish-farming water resource is only limited by either the amount of oxygen that can be delivered or by the rate of dispersal or removal of metabolic wastes [1]. Recently, however, the increased emphasis on reducing, managing, and controlling effluents, as well as the growing competition for and conservation of water resources, have created a more difficult regulatory, economic, and social environment for existing aquaculture facilities to operate under [2]. Also, it is clearly much more difficult to locate and permit proposed aquaculture facilities with today's environmental awareness, more stringent regulations, and scarce untapped water resources. Therefore, technologies and strategies to manage or reduce the wastes generated during aquaculture production are being developed to reduce its demand for water resources and abate its effect on the environment.

Increased aquaculture production efficiencies, reduced waste, or reduced water use have resulted from improved feed and feeding strategies, application of aeration and oxygenation technologies, water-reuse technologies, and more widely used effluent-treatment technologies.

Technological advances now make it possible to culture a wide variety of species in almost any location. Recent trends have been towards an increase in the use of technologies to intensify production within larger intensive systems, including cage culture systems, single-pass or serial reuse systems, pond systems, and enclosed recirculating systems.

More stringent water-pollution control and water-use permitting has made systems that reuse water more attractive, because these systems reduce the volume of water to be treated and thus the size of water-treatment facilities, concentrating the wastes sufficiently for their economical removal [3]. Water-reuse systems typically use clarifiers or filters to remove particulate solids, biological filters to reduce dissolved wastes, strippers or aerators to add oxygen and decrease carbon dioxide or nitrogen gas levels, and oxygenation units to increase oxygen concentrations above saturation. Depending upon the level of reuse and water-quality requirements, processes to provide pH control and advanced oxidation may also be required.

Reductions in production costs within recirculating systems have been achieved through increases in intensification and scale [4, 5]. Even so, not all water-treatment technologies scale up equally well, and the more functional and less costly technologies for treating large flows are being adopted [4, 5]. As methods are developed to reduce the higher cost of production within these systems, recirculating systems are being more widely adopted commercially. However, it is still unclear how many of the new producers using large recirculating systems will be profitable in several years.

This chapter briefly summarizes regulation governing aquaculture wastewater, reviews the amount and type of aquaculture metabolic by-products generated, and describes several procedures for collection and disposal of these by-products, reducing their environmental impact or allowing for the water to be directly reused. The information provided includes descriptions of technologies that are used for treating water for reuse or for treating aquaculture wastewater just before it is discharged.

13.1.1 Effluent Regulations

The disposal or beneficial reuse of aquacultural effluents and solids wastes is regulated at the federal level and often at the state level. The purpose of the discharge regulations is "to restore and maintain the chemical, physical, and biological integrity of the nations waters." In the U.S., federal regulations fall primarily under the Environmental Protection Agency (EPA), which, under the Clean Water Act, has the authority to regulate all discharges of pollutants, including both point-source and non–point source effluents.

The EPA, under the Clean Water Act, provides National Pollution Discharge Elimination Systems (NPDES) permits to control point-source discharges to waters of the United States, but these are often administered by individual states under the oversight of the EPA. NPDES permits broadly define pollution in wastewater discharges so that it may include suspended solids, nutrients (phosphorus and nitrogen), biochemical oxygen demand, dissolved oxygen, pH, bacteria, toxic chemicals, volatile organic molecules, metals, pesticides, and other material. In order to minimize waste output or recycle waste products, EPA policy supports the use of best available technology (BAT) and best management practices (BMPs), the use of effluent for irrigation water, and the use of organic solids for soil enrichment, fertilizer, or animal feed [6, 7].

The EPA has not formally established technical standards for discharges from aquaculture facilities, but it does provide guidelines in the Clean Water Act [7]. Regulations under section 40 of the Code of Federal Regulations require NPDES permits for farming cold-water aquatic organisms when annual harvests exceed 9090 kg and the rearing units discharge for more than 30 days annually. Similarly, NPDES permits are required for farming warm-water aquatic animals when annual harvests exceeds 45,454 kg and the rearing units discharge for more than 30 days annually.

Regulations may suggest a best available technology (BAT) or limit either the concentration or total mass of a given constituent discharged per day.

13.2 Materials to Remove

Beveridge et al. [8] define aquaculture wastes as "the materials used in aquatic animal production which are not removed during harvesting and which may eventually find their way into the environment at large." These wastes may include feed and feed fines, the fish's metabolic by-products, and chemicals used to treat fish disease, disinfect water or equipment, or control fouling, aquatic vegetation, or other nuisance organisms.

Water-quality parameters are of concern in fish culture if they cause stress, reduce growth rate, or produce mortality. Water quality is also of concern if the effluent characteristics (e.g., biochemical oxygen demand, suspended solids, phosphorus, or nitrogenous compounds) of the culture facility must be controlled to meet water-pollution requirements [1]. Water-quality limitations reduce the profitability of fish production, because the reduction in water quality that leads to stress and the deterioration of fish health reduces growth and increase the risk of disease and catastrophic loss of fish [9, 10, 11].

The water-quality criteria required for maintaining a healthy and rapid growth environment (Table 13.1) are the basis for designing and managing aquaculture systems, and the parameters of primary concern are oxygen, nitrogenous and phosphorus compounds, carbon dioxide, solids, total gas pressure, chemotherapeutants, and pathogens. Their production or reduction can lead to concentrations that affect the growth and health of the fish and that may require treatment before the water is used, reused, or discharged. Each of these parameters is discussed subsequently here.

Table 13.1. Suggested water quality criteria for optimum fish health in freshwater

Water Quality Parameter	Piper et al. [12] on Trout and Salmonids	Laird and Needham [19] on Atlantic salmon	Losordo [20] on General Guidelines for Recirculating Systems
Maximum concentration			
Carbon dioxide	10 mg/L	10 mg/L	20 mg/L
Un-ionized ammonia	0.0125 mg/L	0.0125 mg/L	0.02–0.5 mg/L
Nitrite	0.1 mg/L	0.02 mg/L	0.2–5.0 mg/L
Nitrate	3.0 mg/L		1000 mg/L
Total suspended and settleable solids	80 mg/L	80 mg/L	
Ozone	0.005 mg/L	0.005 mg/L	
Optimum range			
pH	6.5–8.0	5.5–8.5	6.0–9.0
Dissolved oxygen	5.0 mg/L to saturation	>6 mg/L	>6.0

Source: Noble and Summerfelt [11].

13.2.1 Nitrogenous Compounds

Nitrogen is present in the protein used in fish feeds. Once introduced into the fish farm, the nitrogen in the protein in the feed can be converted to ammonia, nitrite, nitrate, and nitrogen gas. Ammonia and nitrite are the two forms of nitrogen that most often present problems to the fish. However, the total nitrogen (including proteinatious nitrogen) and nitrate concentrations may be regulated in aquaculture discharges.

Ammonia

Ammonia is a metabolic by-product of protein metabolism and is the primary nitrogenous metabolite excreted by fish. Production of ammonia, however, is dependent upon factors including the type and life stage of fish, feed type, and even the feeding frequency. Approximately 30 mg of total ammonia nitrogen ($NH_3 + NH_4^+$) is produced by fish per gram of feed consumed [12, 13]. Ammonia is also produced by three additional reactions in recycle systems using biological treatment: biological deamination of organic compounds (waste feed and feces), endogenous respiration, and cell lysis [14]. At high levels, un-ionized ammonia (NH_3) is toxic to fish ($LD_{50} = 0.32$ mg/L for rainbow trout and 3.10 mg/L for channel catfish; [15]) and at lower concentrations (0.05 to 0.2 mg/L of un-ionized ammonia) it causes a significant reduction in growth [16]. For salmonid fishes, optimal conditions for growth requires an un-ionized ammonia concentration less than 0.0125 mg/L [12]. Therefore, ammonia must be removed from aquaculture systems before it can accumulate to toxic levels.

Un-ionized ammonia is the toxic form of ammonia; it associates with water to form hydroxide (OH^-) and ammonium (NH_4^+):

$$NH_3 + H_2O \Leftrightarrow NH_4^+ + OH^- \tag{13.1}$$

The equilibrium of ammonia is a function of the water's pH,

$$[NH_3]/[NH_4^+] = 10^{(pH - pK_a)} \tag{13.2}$$

where pK_a is a function of temperature (T in degrees Kelvin),

$$pK_a = 0.09018 + 2729.92/T \tag{13.3}$$

such that decreasing pH or temperature shifts the equilibrium to form more ammonium.

Nitrite

Nitrite (NO_2^-) is an intermediate formed during the biological oxidation of ammonia to nitrate (nitrification). The 96-hour LC_{50} value of the nitrite ion to rainbow trout is 0.20 to 0.40 mg/L as nitrite-N [17]. However, the toxicity depends upon the concentration of calcium or chloride ions, which can increase the tolerance of rainbow trout to nitrite by a factor of 20 to 30 times [16]. To maintain a healthy, growth-promoting environment, nitrite concentrations of less than 0.012 and 0.1 mg/L were recommended by Westin [18] and Wedemeyer [10], respectively (Table 13.1). Incomplete nitrification causes accumulations of higher nitrite concentrations in recirculating systems, and [20] provides guidelines of less than 0.2 to 5.0 mg/L nitrite, which would depend upon species and life stage. Aqueous nitrite is in equilibrium with nitrous acid, which is more toxic to aquatic organisms than nitrite [21]

$$HNO_2 + H_2O \Leftrightarrow H_3O^+ + NO_2^- \tag{13.4}$$

The equilibrium of nitrite with nitrous acid ($K_a = 4.5 \times 10^{-4}$ at 25°C) is dependent upon pH (according to an equilibrium relationship similar to Eq. (13.2)), and a shift to lower pH values shifts the equilibrium towards nitrous acid and thus increases the nitrite toxicity. At normal culture pH, however, equilibrium favors essentially only the nitrite ion.

13.2.2 Solids

Aquatic organisms are affected by solids in suspension and by settleable solids as they are deposited by sedimentation. The effect of the solids on fish is dependent on the species, age and stage of the reproductive cycle [22]. Suspended and settleable solids may affect reproductive behavior, gonad development, and the survival of the egg, embryo, and larval stages of warmwater fishes [22]. In certain intensive systems, suspended solids have been cited as one of the major factors limiting production [23]. Suspended solids have been associated with environmentally induced disease problems [24, 25, 26] and have been reported to cause sublethal effects such as fin rot [27] and direct gill damage [24, 25] in rainbow trout.

Solids are produced in aquaculture systems as uneaten feed, feed fines, fish fecal matter, algae, and sloughed biofilter cell mass (in recirculating systems). Uneaten feed and feed fines, which can represent 1% to 30% of the total feed fed [8], have a large effect on the total solids production. However, feed waste can be controlled with new feeding techniques [28, 29, 30, 31, 32], and feed fines can be reduced by careful handling and separation techniques. If waste feed is minimal, solids production is about 30% to 50% of the daily feed input [12, 13].

Shear forces (water turbulence, fish motion, pumps, etc.) can break apart fecal matter, uneaten feed, and feed fines soon after they are deposited, allowing them to disintegrate into much finer and more soluble particles [33]. For example, as much as 30% to 40% of the total suspended solids generated in a recirculating system will decay if solids are filtered and stored in a pressurized-bead filter (a type of granular-media filtration unit) between 24-hour backwash cycles [34]. The rapid dissolution and breakup of fecal matter, waste feed, and feed fines releases ammonia, phosphate, and dissolved and fine particulate organic matter, which are harder to remove (when dissolved or as a fine particulate) and deteriorate the water quality within the system and in the discharged effluent [35, 36, 37].

Because of the detrimental effects of solids on the receiving water, fish health in the culture environment, and the performance of other unit processes within recirculating systems, solids management is one of the most critical process in aquaculture systems [38, 34, 39, 23]. To improve solids management, the freshly produced solids should be rapidly flushed from the fish culture tank, without pumping, directly to a solids-removal unit that does not store solids within the recirculating flow.

Particulate Size

Solids are classified according to size as soluble (< 0.001 μm); colloidal (0.001–1.0 μm), supracolloidal (1.0–100 μm); and settleable (> 100 μm). Dissolved solids can be removed by biological conversion to cell mass or by chemical oxidation. Colloidal and supracolloidal solids can be clarified if they can be flocculated by chemical addition (aluminum and iron salts, polymers, or ozone) or if they can be floated. However, with the exception of ozone (discussed in the section on ozonation), which shows great promise, the use of flocculant aids for enabling clarification of colloids has not been common. Larger supracolloidal solids can be clarified by filtration or flotation, and their removal efficiency can be increased by flocculation. Settleable solids can be removed with or without flocculation by sedimentation or by filtration.

Shear Resistance

The shear resistance of aquaculture solids is generally better than that of flocculant particles found in water and wastewater-treatment plants. The shear resistance of aquaculture solids, however, is still low enough to cause serious consideration when selecting the best clarification process [33].

Specific Gravity

The specific gravity of fish fecal matter is likely to vary depending upon conditions. Suspended solid specific gravity values of 1.005, 1.13 to 1.20, and 1.19 were reported by Robertsen [40], Timmons and Young [41], and Chen et al. [39], respectively.

Chemical Make-up

The general chemical make-up of the particulate stream should be known to select the clarification process properly. The process selection can depend on whether the particulate matter is relatively organic or inorganic, hydrophobic or hydrophilic, protein or fat, reactive or inert, positively or negatively charged, temperature-stabilized, or pH-stabilized. Knowledge of the particulate composition can provide insight into which clarification mechanisms are beneficial (precipitation, flocculation, flotation) or warn of problems that may occur (biofouling, scale deposition, corrosion, abrasion, flotation).

13.2.3 Dissolved Matter

Dissolved organic matter can accumulate within recirculating systems, because it is not readily biodegradable owing to its size or chemical nature and because daily replacement of water is usually low (5% to 50%). Dissolved organic matter promotes growth of heterotrophic microorganisms, which increase system total oxygen demand and inhibit nitrification by competing for space and oxygen in a biofilter. Some heterotrophic (saprotrophic) microorganisms in the system may consist of facultative pathogens (i.e., *aeromonas*) that can result in epizootics if the fish are stressed by handling or sublethal water-quality conditions [11].

13.2.4 Carbon Dioxide

Fish excrete carbon dioxide through their gills as a function of both ventilation rate (which depends upon the blood's oxygen concentration [42]) and the carbon dioxide–concentration gradient across the gill [43, 10]. The volume of carbon dioxide produced during respiration is about the same as the volume of oxygen consumed [44], effectively about 0.3 to 0.4 g of carbon dioxide per gram of feed for salmon and trout [13].

Although carbon dioxide is a gas, it is not a major contributor to gas-bubble disease because it does not contribute significantly to gas supersaturation [45]. Aqueous carbon dioxide concentrations of 20 to 40 mg/L reduce the capacity of fish blood to transport oxygen [46] and contribute to the deposition of calcium in their kidneys, with higher carbon-dioxide concentrations corresponding to increasing deposition [47]. Smart et al. [47] found reduced growth and higher feed conversion ratios in rainbow trout reared at concentrations of 55 mg/L carbon dioxide. Rainbow trout seem tolerant of 30 to 35 mg/L carbon dioxide concentrations, but 55 mg/L was found to cause death [48]. However, actual toxicity levels vary according to fish species, fish age, and other water quality variables such as oxygen concentration, alkalinity, or pH [10, 48]. The recommended limit on the carbon-dioxide concentration in trout or salmon culture is 20 mg/L [49]. The 20 mg/L recommended safe level may be conservative, especially if oxygen in the water is high [45, 10].

The concentration of carbon dioxide in water depends upon the pH and the amount of inorganic carbon (carbonic acid, bicarbonate, and carbonate) present (discussed in the section on carbon-dioxide removal).

13.2.5 Pathogens and Chemicals Used in Aquaculture

Disease and the use of chemicals to treat fish disease, disinfect water or equipment, and control fouling, aquatic vegetation, and other nuisance organisms are important issues in aquaculture [8, 50, 51]. To avoid introduction or spread of fish diseases, in certain instances there are strict regulations on fish transport, fish escapement, and effluent disinfection. In addition, use of certain chemotherapeutics is restricted to protect the cultured fish or the environment that would be contacted by the chemotherapeutics.

Management practices designed to prevent water-quality degradation and the occurrence of diseases are critical to a successful aquaculture facility, especially facilities that reuse a large portion of water [52, 11]. Use of chemotherapeutants in the water or feed are methods for treating disease that act on the symptom but present special considerations if the water is to be discharged, reused serially for fish culture, or treated in biological filters. To avoid introducing pathogens in the first place, aquaculture facilities should develop protocols for disease prevention and control. A sound protocol would limit entry of disease carried on fish being introduced to the system and make provisions to exclude pathogens carried by the water supply, facility personnel, visitors, or unwanted animals [11].

13.2.6 Nutrients

Phosphorus and nitrate do not become growth-limiting in intensive systems. Phosphorus and nitrate are primarily a concern as they result in effluent discharge problems. Feed and fecal matter can rapidly leach phosphorus and disintegrate and should be removed from the flow as rapidly as possible [53, 54, 8, 55, 35, 56, 50, 57, 37].

Phosphorus and nitrogen can be either soluble or bound with particulate matter. Most of the effluent nitrogen released (75% to 80%) is in the form of dissolved ammonia or nitrate [55, 58], depending on the effective use of feed and how much nitrification has occurred. The fecal solids contain about 1% to 4% phosphorus (on a dry-weight basis) and about 2% to 5% nitrogen (on a dry-weight basis) [58]. Literature reviews indicate that the filterable or settleable solids contain most (50% to 85%) of the effluent phosphorus, but relatively little (about 15%) of total effluent nitrogen [55, 58]. The large variability in the phosphorus fractionation between dissolved and particulate matter is largely due to the variability within the type and level of phosphorus in the feed, which in turn influence the phosphorus solubility and availability (to the fish).

13.3 Methods to Remove Ammonia

In once-through or single-pass aquaculture systems, ammonia is flushed away with fresh water. In reuse systems, however, insufficient ammonia is removed through simple dilution with fresh water, and ammonia must be removed from the system in another manner [59, 13]. The primary method for treating ammonia in recirculating aquaculture systems is by bacterial oxidation [60, 21, 61].

Biofiltration, a term coined for labeling attached-growth biological-treatment processes, uses microorganisms to biologically oxidize a portion of the ammonia nitrogen and the biodegradable organic matter in a waste stream. Retention of the microbe population is obtained by its attachment to the wetted or submerged surfaces of the biofilter medium. Ammonia is oxidized by autotrophic bacteria called *nitrifying bacteria*. Biologically degradable organic compounds are oxidized by heterotrophic microorganisms (i.e., bacteria, protozoa, and micrometazoa). Typically, the autotrophic and heterotrophic microbes coexist on the media surface in the biofilter.

A wide range of biofilter types are used to treat dissolved wastes in aquaculture, but the most common include submerged filters, trickling filters, rotating biological contactors, pressurized-bead filters, and fluidized-sand biofilters (Fig. 13.1). These units are discussed after microbial action in biofilters and biofilter media are reviewed.

13.3.1 Microbial Action

The oxidation of organic compounds and ammonia in static-bed biofilters does not take place simultaneously, as it can in a suspended-growth treatment process (such as activated sludge), but successively [14, 62]. If sufficient BOD is present, organic compounds will be oxidized preferentially before ammonia is utilized as heterotrophic microorganisms outcompete autotrophic bacteria for space on the media at the inlet of the biofilter [63, 64, 62]. The larger heterotrophic microorganisms may also prey upon the population of nitrifying bacteria [62] and the carbonaceous oxygen demand exerted by the heterotrophic organisms can limit the autotrophic nitrifiers located down-filter [13, 65, 63, 64].

Bacteria convert ammonia into nitrate in a two-step process called *nitrification*. Two different groups of nitrifying bacteria, both believed to be obligate autotrophs (that is, they consume carbon dioxide as their primary carbon source) and obligate aerobes (which

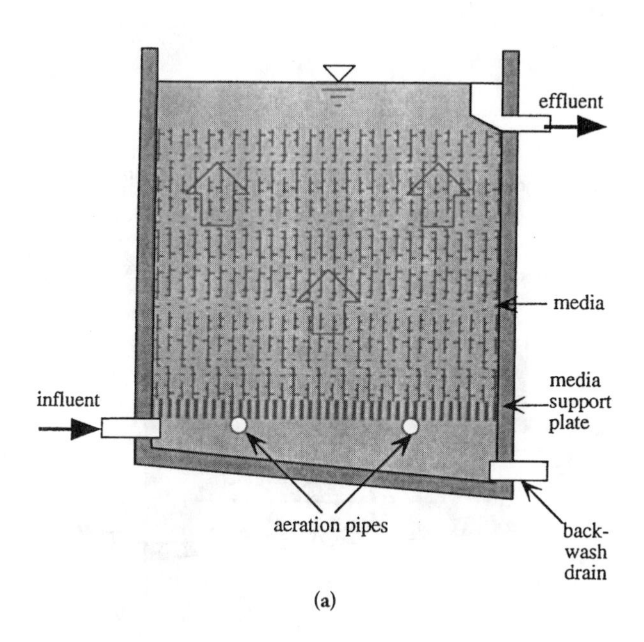

(a)

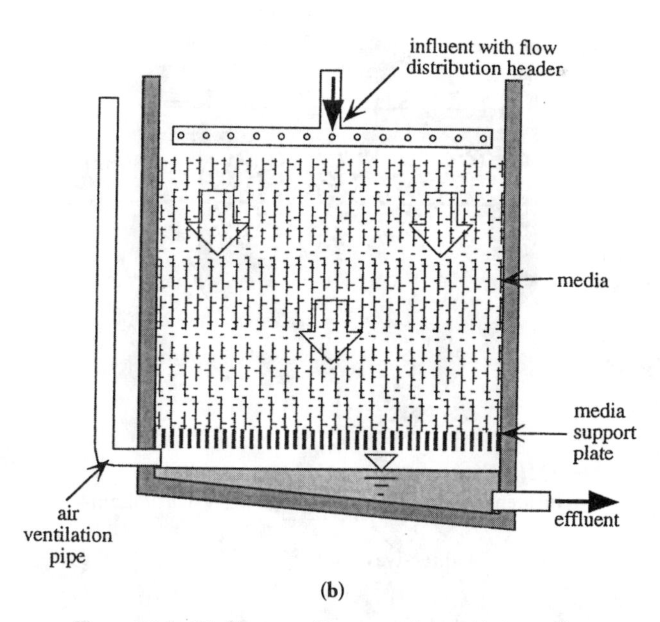

(b)

Figure 13.1. Biofilters used in aquaculture. (a) An upflow submerged filter with back-wash capabilities; (b) a trickling filter; (c) a rotating biological contactor; (d) a bead filter with propeller wash; (e) a fluidized-sand biofilter;

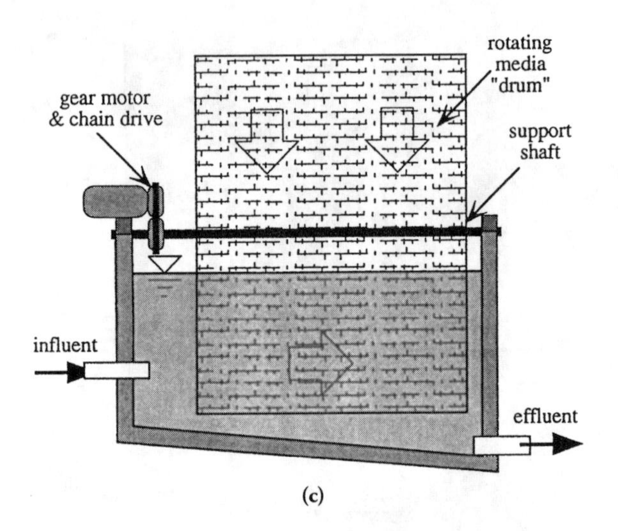

(c)

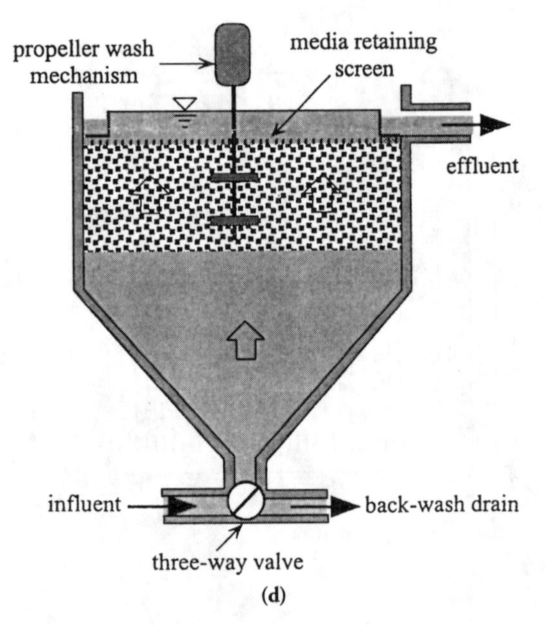

(d)

Figure 13.1. (cont.)

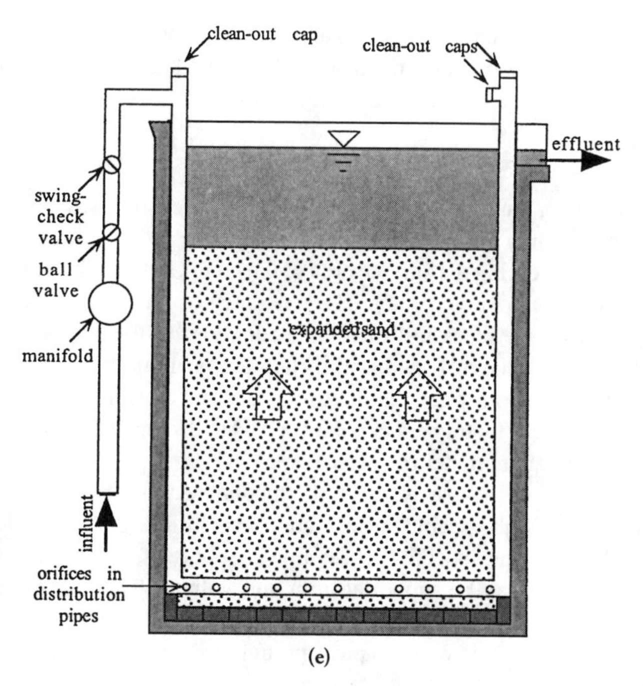

Figure 13.1. (cont.)

require oxygen), are needed for complete nitrification [66]. Although there are several species of nitrifying bacteria, Cutter and Crump [67] found that *Nitrosomonas* showed considerably higher ammonia conversion rates than the others. Consequently, it is generally assumed that *Nitrosomonas* converts ammonia into nitrite, NO_2^-, and that *Nitrobacter* converts the nitrite into nitrate, NO_3^-. The stoichiometry for these energy conversion reactions, ignoring cell synthesis, is [68]

$$NH_4^+ + 1.5\,O_2 + 2\,HCO_3^- \rightarrow NO_2^- + H_2O + H_2CO_3 \tag{13.5}$$

$$NO_2^- + 0.5\,O_2 \rightarrow NO_3^- \tag{13.6}$$

Taking nitrifier synthesis together with nitrification, the overall relationship among ammonium, bicarbonate, and oxygen consumed and cell mass, nitrate, water, and carbonic acid produced is [68]

$$\begin{aligned} NH_4^+ + 1.86\,O_2 + 1.98\,HCO_3^- \rightarrow\ & 0.0206\,C_5H_7NO_2 + 0.980\,NO_3^- \\ & + 1.041\,H_2O + 1.88\,H_2CO_3 \end{aligned} \tag{13.7}$$

As given by the stoichiometry, nitrification and nitrifier synthesis requires consumption of about 4 mg of oxygen and between 6.0 and 7.4 mg of alkalinity (as $CaCO_3$) for every mg of ammonia as nitrogen converted to nitrate [68].

13.3.2 Media

The ammonia and organic oxidation capacity of biological filters is largely dependent upon the total surface area available for biological growth and the efficiency of the area utilization. Ideally, increasing the surface area of the media will result in a corresponding increase in ammonia-removal capacity. The efficiency of nitrification per unit surface area is dependent upon the accessibility of the media surface, the mass-transfer rate into the biofilm, the growth phase of the biofilm (lag, log, stationary, or death phase), and the competition with heterotrophic microbes for space and oxygen [69, 70]. Nitrification rates reported for different filter types range from 0.04 to 0.78 g total ammonium nitrogen removed per day per square meter of surface area [71].

To work effectively in recycle systems, the media used in attached growth systems must have a relatively high specific surface area (i.e., surface per unit volume) and an appreciable voids ratio. The specific surface area is important, as it controls the amount of bacterial growth that can be supported in a unit volume. The voids ratio is critical for adequate hydraulic performance of the system. The voids ratio characterizes how much space is provided for the water to pass through the media in close contact with the biofilm. In some cases (e.g., trickling filters and rotating biological contactors) the voids must also provide space for air ventilation so that oxygen can be transferred to the flow within the bed.

The media used in the biofilters must be inert, noncompressible, and not biologically degradable. Typically, relatively large structured or randomly packed media are used in submerged filters, trickling filters, and rotating biological contactors. Randomly packed media can be sand, crushed rock or river gravel, or some form of plastic or ceramic material shaped as small beads, or larger spheres, rings, or saddles. Structured media can be crossed stacks of redwood slats, or more commonly plastic blocks composed of corrugated plates or tubes. Pressurized-bead filters and fluidized-bed filters use a finely graded plastic or sand media with average equivalent diameters generally from 1 to 3 mm and from 0.1 to 1.5 mm, respectively.

Biofilter start-up can take 4–12 weeks, depending on conditions. Biofilter function can be impaired if a biofilter has too little oxygen or is overloaded with solids, biochemical oxygen demand, or ammonia. Increased loading can reduce the capacity of the biofilter to complete the two-step bacterial conversion of ammonia to nitrate, because increasing the loading increases the competition for space and oxygen among heterotrophic organisms and *Nitrosomonas* and *Nitrobacter*. *Nitrobacter* require the nitrite produced by *Nitrosomonas* and tend to be located towards the outlet end of the biofilter. If space and oxygen become limiting when loading is increased, *Nitrobacter* are generally the first to be displaced, as they are the last in the line of microbial consumers, but *Nitrosomonas* can also be displaced or suffer from oxygen-limiting conditions. Conditions that cause a reduction in the relative amount of *Nitrobacter* compared with *Nitrosomonas* will cause the nitrite concentration across the biofilter to increase, which can become toxic within the recirculating system water. Two approaches have been applied (sometimes simultaneously) to counter the problem of insufficient ammonia and nitrite removal in large recirculating systems. Biofilters are sized with more available surface area for microbial treatment and with consideration of oxygen requirements, so that the biofilter has both the space and oxygen capacity to handle higher organic and ammonia loading rates; and ozone is added to the system to oxidize excess nitrite to nitrate (as discussed in the section on ozonation).

13.3.3 Submerged Filters

The submerged filter consists of a bed of media through which the wastewater passes in an upward (Fig. 13.1) or downward direction. The media is generally large (5–10 cm rock, or 2–5 cm random or structured plastic media) so as to provide large void spaces. Also, the upward (or submerged downward) flow allows control of the hydraulic detention time.

The submerged filter also captures some of the biological solids [72, 73, 74]. Submerged filters may have problems that result from heavy loading of organic matter coupled with improper solids flushing or underdrain design. Solids accumulate in static filter beds based on both physical and biological mechanisms. Physically, suspended or settleable solids are entrapped within the filter voids because of settling, sieving, and interception. Biologically, cell mass is produced by the microbial metabolism of substrates (primarily organics and ammonia) passing through the filter, both on the surface of the media and in the voids between the media. Cell mass resulting from nitrifying bacteria is desirable. Cell mass resulting from heterotrophic organisms, although required to some extent, increases plugging problems in static-media filters. Therefore, conditions that allow the filter to plug or to exhaust the dissolved oxygen that is available in the water must be avoided.

The characteristics of the media in a static-media filter ultimately influence the removal of suspended, settleable, and dissolved organics as well as the head loss generated and the success of flushing the filter. Media with sufficient voids for use in biofilters can be random media such as uniform crushed rock over 5 cm (2 in) in diameter, or plastic media over 2.5 cm (1 in) in diameter, or structured media such as plates or tubes that have large voids. Providing adequate flushing of solids from the filter is critical [73, 74]. Solids sometimes have to be flushed from the submerged biofilter media voids daily by reversed-flow flushing and less often by air scour coupled with reversed-flow flushing. A drainable filter, a properly designed underdrain, a high velocity of a reversed flow of water, and a means for air scouring are all required for good solids flushing. Underdrains must provide a support for the media and force the water, during regular operation or during flushing, to flow uniformly through the bed.

Two innovative modifications of submerged filters have been developed to reduce two common operational problems: low available dissolved oxygen and plugging problems. The operational problems can be alleviated with supplemental oxygenation and bubble aeration. Both modifications raise the level of available oxygen, and serve to break-up clogging to a variable extent. Vigorous aeration strips off thick growths of biofilm, thus helping to prevent biofilm growth from clogging the pores in the filter media [75]. These aerated biofilters have been effectively used to maintain good water quality in recirculating systems [71].

13.3.4 Trickling Filters

Trickling filters appear similar to the down-flow submerged biofilters, except that the water level in the trickling filter vessel is kept below the media and the downflow keeps the media wet but not submerged (Fig. 13.1). Because it is not submerged, air ventilates through the wet media, carrying oxygen to the bacteria and providing opportunity for carbon dioxide to strip.

Trickling filters have been used extensively in wastewater treatment throughout the twentieth century. The most important reason for the popularity of the trickling filters is their ease of operation, self-aerating action, and moderate capital cost. Modern trickling filters use plastic media because they are light and offer high specific surface areas (100–300 m^2/m^3) and void ratios of more than 90%. Trickling filters have been widely used to remove dissolved wastes in recirculating systems [76, 61, 77], and a range of design criteria has been reported: hydraulic-loading rates of 100 to 250 m/d, media depths of 1 to 5 m, media specific surface area of 100 to 300 m^2/m^3, and ammonia-removal rates of 0.1 to 0.9 $g/m^2/d$.

As with most of the attached growth biofilters, trickling filters with high specific surface areas can sometimes be less efficient than filters of lower specific surface areas because of channeling, regional plugging, or a hydraulic load that is not sufficient to moisten all the area of the densely packed media.

13.3.5 Rotating Biological Contactors

Rotating biological contactors function by rotating honeycombed or corrugated disks or tubes on a shaft through a tank containing the wastewater. Rotation of the media through the atmosphere provides oxygen transfer to the flow and the microbial film, which helps to maintain aerobic conditions for nitrification. Rotating biological contactors have attracted attention because of their low head requirement, ability to maintain an aerobic treatment environment, and relatively low operating costs [78]. However, the capital cost of rotating biological contactors is typically much more than that of trickling filters, submerged filters, or fluidized-sand biofilters.

Typical rotating biological contactors have disks submerged to 40% of the diameter and rotational speeds of 1.5 to 2 rpm [79]. Rotating biological contactors can treat small flows, but they also scale to lengths of 8 m (25 ft) and total surface areas of approximately 8400 m^2 (85,000 ft^2) to treat large flows [79]. Catastrophic failure of either the disks or the shaft due to heavy biomass growth were major problems in early units [79]. However, most of the problems were mechanical and recent (properly designed) rotating biological contactors have been more reliable.

13.3.6 Pressurized-Bead Filters

The pressurized-bead filter passes flow upward through floating plastic beads (Fig. 13.1). A screen at the top of the vessel prevents the beads from washing out. Operating pressures are typically 35 to 100 kPa (5–15 psi).

The pressurized-bead filter has become a popular biofilter to treat small or moderate flows (less than 1000 L/min). It is modular, compact, and straightforward and effectively removes ammonia if operated properly [80]. The pressurized-bead filter also effectively removes solids from a flow and is sometimes used only for solids removal or as a hybrid filter that combines both nitrification and solids removal. The bead filter is typically either bubble-washed or propeller-washed during the backwash procedure, which ensures that trapped solids are broken free from the beads.

Nitrifying bacteria and heterotrophic microorganisms grow on the floating plastic beads, which are cylinders about 3 mm long, and in the void spaces surrounding the beads. Because the pressurized-bead filter removes solids so effectively, large populations of heterotrophic microorganisms can grow in the bead filter and must be periodically removed with the trapped solids in order to maintain nitrification [81]. Too frequent and vigorous back-flushing, however, has been reported to remove to much of the nitrifying bacteria, so it is important that frequency and intensity of back-flushing be controlled properly [81]. Nitrification in pressurized-bead filters is still best when a majority of the solids have already been removed from the flow.

13.3.7 Fluidized-Bed Biofilters

Fluidized-sand biofilters have been shown to be reliable [82] and especially cost-effective [83, 4, 5] units for treating dissolved waste in large-scale recycle production systems; they have also found wide application in industrial and municipal wastewater treatment [84, 85, 86]. Understanding of the use of fluidized-sand biofilters to treat dissolved wastes within recirculating aquaculture systems has been improved as a result of recent studies, as reviewed by Summerfelt and Wade [5].

Fluidized-sand beds operate by injecting an equal distribution of water across the biofilter's cross-section at the bottom of the granular bed. Water flows up through void spaces between the granules in the bed. The bed expands, becoming fluidized, when the velocity of water through the bed is sufficiently large to result in a pressure loss greater than the apparent weight (actual weight less buoyancy) per unit cross-sectional area of the bed. Sands have a specific gravity 2.65 times that of water (1.0); therefore, the headloss across an expanded sand bed is about 0.9 to 1.0 m of water head for every 1.0 m of initially static sand depth (Summerfelt and Cleasby, 1996). Once the bed has been fluidized, the pressure drop across the bed remains constant at all bed expansions. During fluidization, the relative amount of bed expansion is dependent upon the media's shape (i.e., sphericity) and diameter, and the superficial velocity and temperature of the water [87].

Size grades of filter sand are available commercially and are usually specified by a sand's effective size (D_{10} = the sieve-opening size that will pass only the smallest 10%, by weight, of the granular sample) and uniformity coefficient (UC = a quantitative measure of the variation in particle size of a given media, defined as the ratio of D_{60} to D_{10}). The D_{10}, an estimate of the smallest sand in the sample, is the sand that will expand the most at a given velocity. The D_{90}, an estimate of the sieve size that will retain only the largest 10% of sand in the sample, is the sand fraction that will expand the least at a given velocity. The D_{90} can be estimated from the effective size and the uniformity coefficient [88]

$$D_{90} = D_{10} \cdot (10^{1.67 \cdot \log(UC)}) \tag{13.8}$$

Fluidized-sand biofilters in aquaculture typically use an extremely hard, whole-grain crystalline silica sand, which is finely graded and has a mean effective diameter of 0.2 to 1.0 mm and a uniformity coefficient between 1.3 to 1.8. Fluidized-sand biofilters are often operated at expansions of 20% to 100% (based on clean sand). Biofilm development increases bed expansion. Because sands are not perfectly uniform, larger sands move to the bottom of the fluidized beds, where they expand less than the smaller sands that have migrated to the top of the bed. The average expansion of a bed at a given velocity depends upon the size gradation within the bed (i.e., uniformity coefficient), which makes it important to predict expansion of both the largest and smallest fractions of sand [89]. In particular, the largest fraction of sand must expand at the velocity selected.

Filter sand is inexpensive ($40 to $70/m^3) and has a high specific surface area (4000 to 45,000 m^2 surface area per cubic meter of volume), which reduces the cost per unit of surface area ($0.02 to $0.001 per square meter surface area) and makes fluidized-sand biofilters compact (Table 13.2). The low cost and high density of sand surface area makes it relatively inexpensive to build fluidized-sand biofilters with excess surface area and, thus, excess nitrification capacity.

The design of the flow-distribution mechanism is most critical for reliable operation of fluidized-sand biofilters. The flow-distribution system must deliver an equal amount of flow across the base of the bed, prevent loss of media, operate without detrimental fouling (or have a fouling prevention system) and, in some cases, support the bed. There are a wide variety of distribution mechanisms used to inject water into the bottom of large fluidized-sand biological filters. Each mechanism differs in how flow is transported and distributed. However, most mechanisms used in recirculating aquaculture systems transport the flow through a manifold, starting at the top of the biofilter, that runs down the inside of the reactor to the base of the sand [87]. Summerfelt et al. [82] describe one method for designing flow distribution mechanisms for large-scale fluidized-sand beds.

Because of practical considerations based on biofilter geometry, pressure drop, and reactor oxygen demand, the depth of sand in aerobic fluidized-sand biofilters is generally designed to be 1 to 2 m, unexpanded. The total head pressure required across a commercial-scale fluidized-sand biofilter is moderate (generally 3.5–7.0 m of water head [5–10 psi], depending upon the height of the biofilter).

Nitrification rates and biofilter bed maintenance requirements are both dependent upon the sand size, water temperature, and ammonia and organic loading rates. Nitrification rates reported range from 0.2 to 0.4 kg/d ammonia per cubic meter of expanded bed volume in cold-water systems [90] to 0.6 and 0.8 kg per day per cubic meter expanded bed volume in warm-water systems [80, 91]. The nitrification rates

Table 13.2. Relative characteristics of two fluidized-sand biofilters constructed at different farms compared with a hypothetical plastic media trickling filter

	Fluidized-sand Biofilter #1	Fluidized-sand Biofilter #2	Plastic Media Trickling Fliter
Flow capacity (L/min)	1520	2280	2000
Media size, sand size = D_{10} (mm)	0.24	0.24	25
Media specific surface area (m^2/m^3)	11,300	11,300	180
Hydraulic loading rate (cm/s)	1.0	1.0	0.17[a]
Media depth (m)	1.00[b]	0.76[b]	2.5[a]
Design TAN removal rate (g/d/m^2 surface area)	0.06[c]	0.06[c]	0.2[c]
Design feed load[d] (kg/day)	58	64	59
Fiberglass vessel ID × height (m × m)	1.8 × 3.7	2.1 × 3.0	5.0 × 3.0
Media volume (m^3)	2.5	2.7	49.0
Media cost per unit volume including shippping (\$/m^3)	150	153[e]	420
Cost of media (\$)	380	415[e]	20,600
Cost of fiberglass vessel w/shipping (\$)	3100	2360[e]	6300
Cost of pipe laterals, manifold, valves, fittings, support frame, and bricks (\$)	1500	1685[e]	400
Cost of assembly (\$)	1100	1085	700
Total biofilter cost (\$)	6080	5545[e]	28,000[f]

Source: Summerfelt and Wade [5].

[a] The design criteria for the tickling biofilter were taken from Nijhof (1995).

[b] The depth of clean sand when unfluidized.

[c] As a safety margin for a commercial system, the design total ammonia nitrogen (TAN) removal rate listed here is at least 50% lower than the nitrification rate for (1) fluidized-sand biofilters as measured during research at the Freshwater Institute (unpublished data) and (2) trickling biofilters as extrapolated from data reported by Nijhof (1995).

[d] Assuming that 0.03 kg TAN are produced per 1.0 kg of feed fed.

[e] Costs were provided by Robert Freeman (Glacier Springs Fish Farm, Gunton, Manitoba, personal communication).

[f] This biofilter was not constructed and these costs are best estimates.

in cold-water and warm-water systems compare favorably if compensations are made for the differences in reaction rate owing to temperature. Sand size also influences biofilm growth and bed management [90]. The optimum effective sand size (D_{10}) is about 0.2 to 0.25 mm for cold-water systems, about 15°C [90], and about 0.5 to 0.7 mm (20–40 mesh sand to 16–30 mesh sand) for warm-water systems, about 25°C to 30°C [92, 91].

Plastic pellets can also be used to make fluidized-bed biofilters [93, 94]. These pellets typically come from the plastic molding industry and can be obtained from suppliers of bulk plastics. The pellets are roughly spherical or cylindrical, and range in size from 1 to 3 mm. Plastic pellets can be used unmodified, although surface-modified pellets are available. Plastic pellets can be purchased in different molecular formulations, or in formulations containing a heavy filler such as silica. Such pellets will have a specific gravity equal to, just greater than, or just less than 1.0. Using plastic beads having a specific gravity greater than or less than water allows for either up-flow expanded (specific gravity > 1) or down-flow expanded (specific gravity < 1) beds. Compared with fluidized-sand biofilters, fluidized-plastic bead filters have lower headloss (< 7 kPa [< 1 psi]) and lower requirements for expansion velocity and for delivering oxygen due to their lower specific surface area.

13.3.8 Biofilter Comparison

Each of the different biofilters has advantages and disadvantages that can affect its use. For example, trickling filters and rotating biological contactors have the advantage that they both aerate the flow during treatment, which strips some carbon dioxide and ensures that aerobic conditions are maintained for nitrification. Submerged filters (without aeration), fluidized-bed biofilters, and pressurized-bead biofilters do not provide aeration, and the only available oxygen for biofilter respiration is carried in the influent flow.

The pressurized-bead filter, if operated under the right conditions, has the advantage of filtering solids from the flow at the same time ammonia is removed [80]. However, storing solids in any biofilter may not be desired (as discussed previously). And, nitrification can be reduced by back-flushing pressurized-bead filters too frequently or vigorously [81].

An advantage of fluidized-bed biofilters is that they trap and store few solids relative to submerged static media biofilters, such as large media biofilters and pressurized sand or bead filters. Additionally, biosolids within fluidized-sand biofilters can be managed by selection of sand size or by more frequent sand replacement, so that about 50% (sometimes more) of the biofilter's oxygen demand goes towards nitrification [90; unpublished data, Michael Timmons, Cornell University, Ithaca, NY, personal communication].

Use of low–surface area media is a disadvantage in rotating biological contactors, trickling filters, and submerged filters because it results in larger and typically more costly biofilters. Conversely, pressurized-bead biofilters and fluidized-bed biofilters use small granular media with high surface areas, which allows the use of smaller and sometimes less costly biofilters to treat the same flow (compared with rotating biological contactors, trickling filters, and submerged filters). It is critical not to cut back media surface area on biofilters to reduce their cost and space requirements; biofilters with insufficient surface area to meet the ammonia removal demands will produce nitrite and fail often.

Trickling filters, rotating biological contactors, submerged biofilters, and fluidized-bed biofilters have been scaled to treat very large flows (> 4000 L/min). On the other hand, the pressurized-bead filter is typically not been scaled to treat more 1000 L/min, although a unit to treat between 1000 and 2000 L/min has been reported [95]. Because of the relatively low cost of filter sand, the fluidized-sand biofilter scales up especially cost-effectively to treat very large flows [4].

The two lowest-cost and most reliable biofilters for treating large flows (> 4000 L/min) may arguably be the fluidized-sand biofilter and the trickling filter. However, because of the lower cost and higher surface area of sand, fluidized-sand biofilters are about one fifth the cost of plastic media trickling or submerged biofilters [5]. And, although fluidized-sand biofilters do not aerate, as is the case with trickling filters, the likelihood of oxygen limitations within fluidized-sand biofilters is reduced because of their relatively large hydraulic loading rate, short hydraulic retention time, and efficient use of oxygen for ammonia removal.

13.4 Methods to Remove Solids

Solids influence the efficiency of all the other component functions as well as the potential for disease within recirculating systems [23]. Therefore, solids removal is often considered the most critical process to manage in recirculating aquaculture systems. Ideally, solids should be removed from the system as soon as possible and with a minimum exposure to turbulence, mechanical shear, or microbiological degradation.

Waste solids can be removed from aquaculture flows by either settling within the rearing tank or through the use of a solids removal unit following the rearing tank. Several unit process options are available for removing waste solids from aquaculture flows: settling basins, granular media filters, microscreen filters, and flotation or foam fractionation. Solids removal can also be enhanced by ozone addition [123].

13.4.1 Settling Basins

In sedimentation, particles having a specific gravity greater than the surrounding water settle out and are removed from the water column. As clarified water passes out of the basin the settled particles collect and form a sludge blanket on the bottom of the basin. Settling basins are typically designed with hydraulics that minimize turbulence and provide time for interception of the particle with the bottom of the clarifier (Fig. 13.2).

Unthickened solids in aquaculture usually settle as discrete particles [96]; that is, they do not change in size, shape, or density and thus settle at a constant rate. The settling velocity of very small discrete particles can be modeled by Stoke's law:

$$\nu_p = \frac{g \cdot (\rho_p - \rho_L) \cdot D^2}{18 \cdot \mu_L} \tag{13.9}$$

in which v_p is the settling velocity, ρ_p is the particle density, D is the diameter of the particle, ρ is the water density, μ is viscosity of the water, and g is acceleration due to

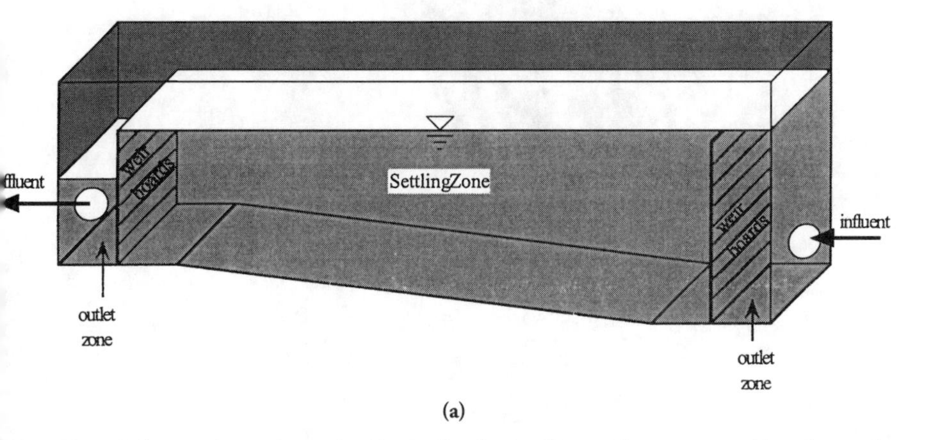

(a)

Figure 13.2. Sedimentation units can be (a) a settling basin, (b) a swirl separator, or (c) a plate settler.

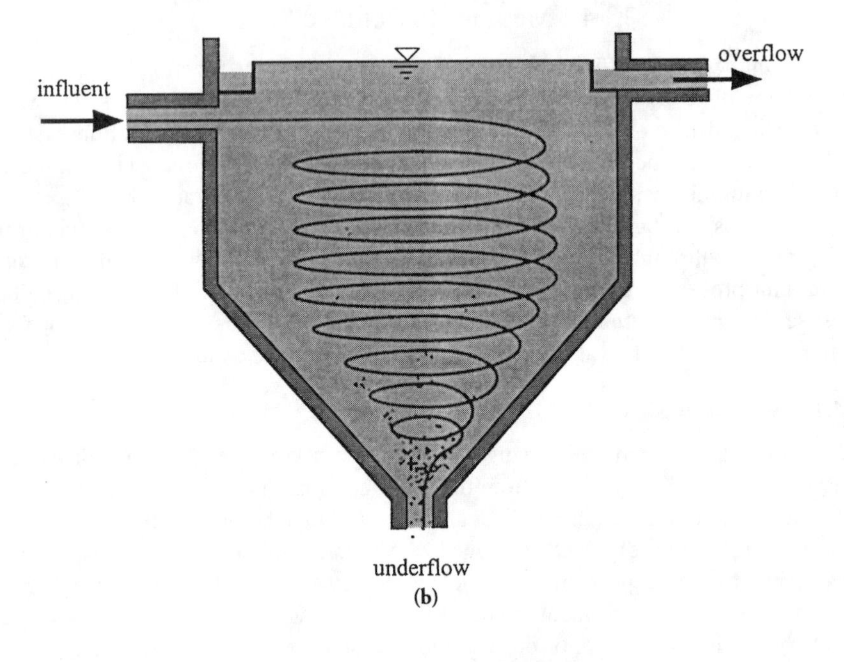

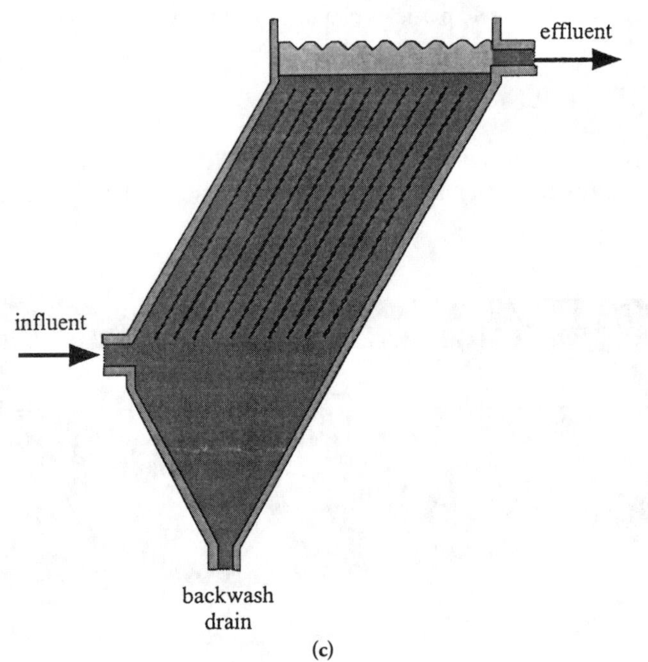

Figure 13.2. (cont.)

gravity. Stoke's law holds only if the settling velocity of the particle is in the laminar region. Intact feces of rainbow trout settle at 1 to 2.5 m/min, depending upon fish size, which is relatively fast [97].

Once the characteristic settling velocity is known, the removal of discrete particulate suspended matter in an ideal settling basin can be related to its flow rate and the horizontal plan area of the tank [98]. Fish-farm settling basins are designed using at least one (or more) of the following criteria: overflow rate (v_o), hydraulic retention time (τ), and/or mean fluid velocity (v).

$$v_o = Q/A_h \tag{13.10}$$

$$\tau = V/Q \tag{13.11}$$

$$v = L/\tau = Q/A_{cs} \tag{13.12}$$

in which Q is the volumetric flow rate of water and V, L, A_h, and A_{cs} are the settling basin volume, length, surface area, and cross-sectional area, respectively.

Overflow rates reported are less than 5 $m^3/m^2/h$ [13], 1.7 $m^3/m^2/h$ [102], 2.4 $m^3/m^2/h$ [97], and 1.7 to 3.3 $m^3/m^2/h$ [99, 100]. Hydraulic retention times measured at sixteen commercial fish farms averaged 30 to 40 min [101] and others recommended designing settling ponds with a theoretical hydraulic retention time of at least 15 to 30 min [13, 102, 103]. Henderson and Bromage [101] recommended that mean fluid velocities across settling ponds should be less than 4 m/min, and if possible less than 1 m/min; Warren-Hansen [97] recommend mean fluid velocities of less than 1.2 to 2.4 m/min.

The geometry of settling basins should be designed to promote plug flow and reduce short circuiting of flow; therefore, settling pond length-to-width ratios should be greater than 4:1 to 8:1 [104]. Liao and Mayo [13] recommended that settling basins be designed with at least 1 m water depth.

Unfortunately, dissolution of nutrients and the resuspension of solids settled and collected on the bottom of settling basins can reduce the expected performance of these clarifiers [57]. Henderson and Bromage [101] estimated that settling ponds could capture an estimated 97% of their solids loading if resuspension of settled solids was not a factor. However, due to resuspension of settleable solids, they report that settling ponds do not effectively remove solids if inlet concentrations of solids are less than 10 mg/L and that attaining effluent concentrations of less than 6 mg/L is difficult.

Two variations on the principle of clarification by sedimentation have been used in an attempt to increase the efficiency of solids removal (Fig. 13.2): hydrocylcones and tube or plate settlers.

Swirl Separator

Swirl separators (Fig. 13.2) are also called *tea-cup settlers* and *hydrocyclones* and operate by injecting the water at the outer radius of a conical tank so that the water spins around the tank's center axis [105]. The spinning creates a centrifugal force (radial acceleration) that moves the larger or denser particulates towards the wall where they settle and can be removed. The strong primary rotating flow also creates a secondary radial flow that moves along the bottom towards the tank center. To obtain strong centrifugal forces in wastewater treatment applications, water must be injected into hydrocyclones at high velocities, and this can result in high pressure losses. Hydrocyclones have been used primarily for treating municipal or industrial wastewater flows that contain particles of high specific gravity; for example, sand and grit have a specific gravity 2.65 times that of water. In aquaculture, however, swirl separators are used in low-head applications to concentrate solids in a smaller underflow (5%–10% of total flow) [97, 106, 107, 108, 53, 109, 110]. Compared with the removal efficiencies for sand and grit, the swirl separator is less effective in aquaculture applications because of the low specific gravity of aquaculture solids. However, with an upper and a lower drain, swirl separators do not store the solids in the flow, which is an advantage over a settling basin.

Tube or Plate Settlers

Tube or plate settlers consist of a sequence of inclined tubes or plates that are stacked several inches apart (Fig. 13.2). Use of tube or plate settlers increases the effective settling area per unit volume and reduces the depth to which a particle must settle to contact a surface [111, 112]. These settlers have been used to increase solids removal (particularly when space is limited) in the drinking-water treatment industry [113], and occasionally in the wastewater-treatment industry [111] and in aquaculture facilities [114, 97, 115]. Tube or plate settlers effectively remove solids from aquaculture systems; however, they are relatively expensive, and the periodic backwashing of the solids from the unit requires labor and results in lost water (sometimes a problem in warm-water recirculating systems).

13.4.2 Microscreen Filters

Microscreen filters are sieves. They strain water-bound particles larger than the filter-screen (mesh) openings. A cleaning mechanism is necessary to remove solids from the filter. Depending on the type of microscreen filter, the filter is cleaned by either hydraulic flushing, pneumatic suction, mechanical vibration, or raking. Cleaning of the sieve can occur continuously, periodically, or on demand. The drum (Fig. 13.3), Triangel, and disk microscreen filters are the three main variations used. Microscreen filters are now used by many fish farms to remove solids from aquaculture influents [116], effluents [108, 53, 103, 117, 118, 109, 119, 120, 57], and from the flows in water-reuse systems [115, 121, 58, 48, 122, 123]. Microscreen filters also require little space, have a high hydraulic capacity and an acceptable pressure drop, and are relatively easy to install.

Microscreen filters are especially attractive when used to remove solids from large flows due to economies of scale [4]. Also, microscreen filters backwash several hundred to several thousand times daily [115, 121], which makes them one of the few devices that do not store the solids removed within the flow. Frequent backwashing comes with an operating cost, a pressure-wash system (400–700 kPa [60–100 psi]) that requires maintenance and that produces a sludge water volume of between 0.2% and 2% of the total passing flow [124, 121]. In warm-water recirculating aquaculture systems, the cost of heating make-up water can be relatively high [125], and replacing water

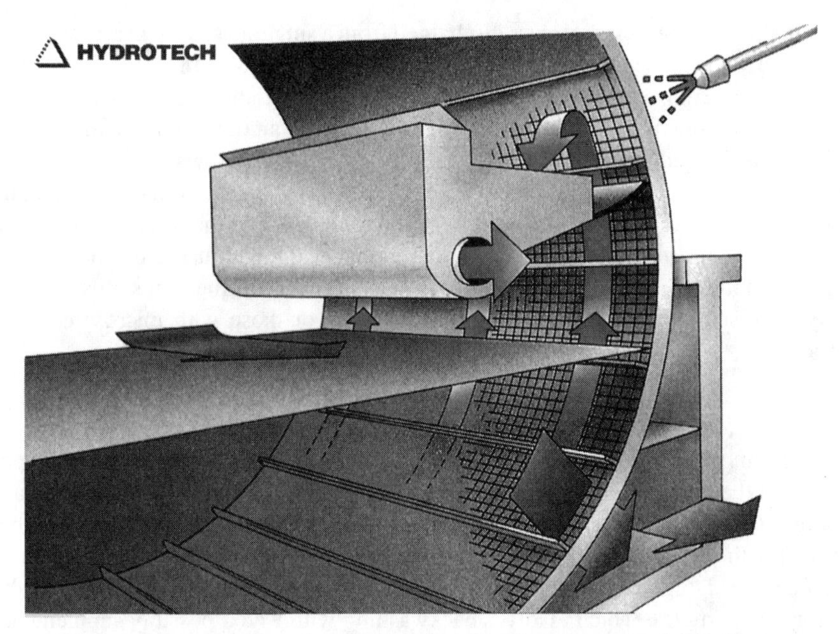

Figure 13.3. Functional mechanism of a drum filter (courtesy of Hydro Tech). Arrows indicate the path taken by the treated flow and the filterable solids as they pass through the unit.

lost from the system during filter backwash may become cost-prohibitive. Bergheim and Forsberg [124] reported using a vacuum to suck solids from the microscreens and reduced sludge water production by 50- to 100-fold compared with a typical pressurized washing mechanism.

The performance of a microscreen filter is largely dependent on the size of the microscreen panel openings, influencing the unit's hydraulic capacity, the fraction of particles removed, the sludge-water production rate and concentration, and the filter-wash frequency [121, 119, 120]. Finer openings remove more solids; however, selection of the minimum opening size through the microscreen is a compromise between treatment requirements and operational factors, such as hydraulic loading and backwash requirements [121, 119]. Summerfelt et al. [121] demonstrated that microscreen filters with panels containing 60-μm openings removed more TSS but also required much more frequent washings than filter panels having 110- and 200-μm openings. This study showed that finer microscreen openings increased backwash frequency nearly four-fold, produced ten times more backwash volume, and reduced the concentration of TSS in the backwash about 3.5-fold. The optimum microscreen opening size may be situation-specific, but microscreen panel openings of 60 to 100 and 80 to 110 μm have been recommended by Kelly et al. [120] and Summerfelt et al. [121], respectively, and the majority of panel openings sizes that have been reported are in this range: 55 to 80 μm [103], 60 μm [117], 60 μm [118], 60 μm [124], 60 μm [109], 65 μm [108], 80 μm [58], 80 μm [48], and 80 μm [123].

The majority of particles by mass discharged from a single-pass system are larger than 30 to 40 μm and can be removed by microscreen filtration [124, 118, 119]. Microscreen filters can also remove more than half of the solids concentration approaching the filter unit in recirculating fish-culture systems [48, 123], which amounts to more than 97% of the solids discharged from the system within the filter-backwash water.

Sizing of the microscreen filter unit is ultimately based on the system flow rate, fish loading, characteristics of the waste solids to be removed, and the desired microscreen opening size. Unfortunately, the aquacultural engineering literature does not provide a precise method for sizing microscreen filters, and selection of filter units must be based largely on discussions with commercial distributors or those with microscreen filter experience.

13.4.3 Granular Media Filters

Solids removal in granular media filters occurs if particles are smaller than the pore openings at the granular media surface, or are too weak to sustain the hydraulic shear at the pore entrance, and pass into the void spaces occurring within the granular media bed. Once within the bed the particles can be removed by straining, sedimentation, absorption, or chemical bridging mechanisms. Periodically, the granular media bed must be flushed of entrapped particles, which involves backwashing with water, possibly supplemented with a prior air wash.

Granular media filters include sand filters, plastic-bead filters, and diatomaceous earth filters. Plastic-bead filters are widely used in recirculating aquaculture systems to remove ammonia and/or solids from relatively small flows ($<$ 1000–2000 L/min).

Granular media filters are commonly used in drinking-water treatment plants and following the secondary clarifiers in wastewater treatment. Depth filtration is used in some fish hatcheries for reducing the entry of pathogens and often follows a chemical oxidation process used for disinfection [126]. In addition, large gravity sand filters (Fig. 13.4) are used to treat aquaculture effluents before they are discharged [53, 57]. If strict effluent discharge limits are imposed, the improved and more consistent solids-removal potential that can be achieved with granular media filters may make them obligatory to remove total phosphorus levels from fish farm discharges. However, granular media filters are a technology of last resort in aquaculture because they require a complex backwash mechanism and have relatively larger capital and operating costs per unit of flow treated.

13.4.4 Dissolved Air Flotation and Foam Fractionation

Flotation and foam fractionation are two options for removing solids. Flotation can be used as an alternative to sedimentation for removal of biosolids, grease, hydrocarbons, protein, fibers, and algae [127, 128]. In this process, the solids attach to air bubbles, float to the surface, become concentrated, and are removed. The fine bubbles, when attached to a solid particle, increase the density difference between the biosolid and water (by changing the particle's specific gravity from just over 1.0 to conserably $<$ 1.0). Flotation can produce a greater density difference between the water and rising air-entrained

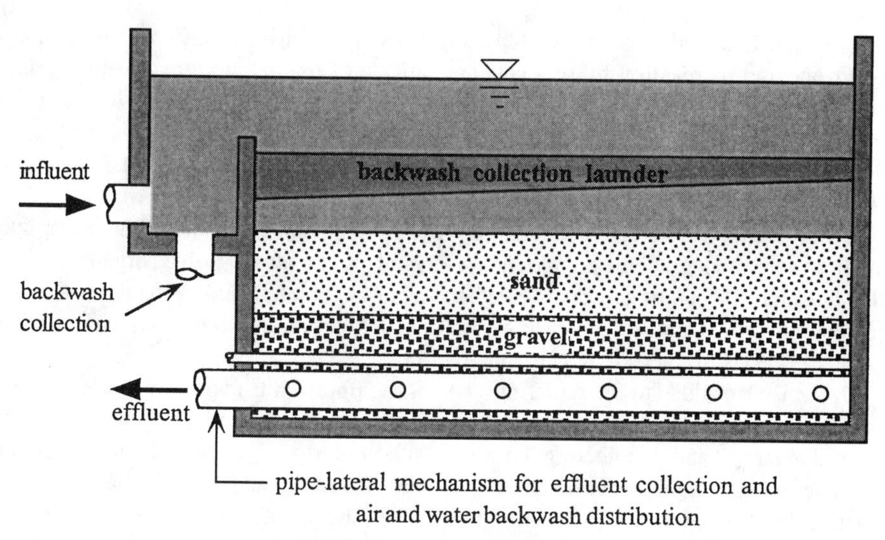

Figure 13.4. Illustration of a gravity sand filter.

sludge than what is obtained between water and sinking biosolids, which makes flotation promising.

There are three basic mechanisms used to produce bubbles for flotation: dispersed air, dissolved gas, and electrolytic flotation. The primary difference among the three flotation mechanisms is the bubble diameter produced; systems using dispersed air (often called *foam fractionation*) generate bubbles larger than 100 μm; and systems using either supersaturations of dissolved gas or electrolytic cleavage of water can precipitate bubbles in the 10- to 100-μm size range and are called *dissolved air flotation* [129]. The flotation of solids is highly dependent upon the bubble diameter, solids concentration, air-to-water ratio, surface chemistry of the solids, and surfactant concentration in the water [130]. Chemical addition (a strong oxidant such as ozone, or a salt or polymers) can aid in the formation of a cohesive froth to improve the quality of the treated water. Dispersed air flotation removes mostly particles smaller than 30 μm [131], which are the fine particles that the more standard solids-removal processes miss.

Clarification of solids with foam fractionation has been demonstrated in recirculating aquaculture systems using dispersed air [132, 131, 133, 23, 134, 135] and with a dispersed ozonated-air mix [136, 137, 138, 139]. However, because foam fractionation does not remove many of the particles greater than 30 μm, it is usually used in parallel with a settling tank or filter.

13.4.5 Ozonation

Dissolved organic matter, including refractory compounds, and fine particulate organic matter that accumulates in recirculating systems can be removed by increasing the daily water-exchange rates or by introducing oxidizing agents. To avoid high water-exchange rates, ozone is added to some recirculating systems in order to improve water quality. Ozone and its reaction by-products are capable of oxidizing a great many organic substances [140, 141]. Ozone also has a rapid reaction rate and produces few harmful reaction by-products (in freshwater), and oxygen is produced as a reaction end-product. However, ozone is toxic to both humans and aquatic organisms and must be used with care.

Ozone breaks refractory and relatively nonbiodegradable organic compounds into smaller, more biodegradable compounds [142] and precipitates dissolved organic molecules and microflocculating fine particulate matter [143]. Ozone's action on fine particulate matter improves its removal by settling, filtration, or flotation, as discussed by Summerfelt et al. [123] and Summerfelt and Hochheimer [144].

Ozone is also added to recirculating systems to improve nitrite removal and to reduce fish disease by improving water quality (which reduces or eliminates environmental sources of stress) and disinfecting the water. Although ozone can cause large reductions in microorganisms, the rapid reaction of ozone with nitrite and organic matter can, in practice, limit its effective disinfecting power (i.e., the disinfecting power is a function of the product of ozone residual concentration multiplied by its contact time). Therefore, achieving large microbial reductions in recirculating systems requires much greater ozone dosages than are required for water quality control alone and requires much higher ozone dosages than are typically required for disinfecting single-pass inflows [145a, 144].

13.4.6 Discussion of Solids-Removal Options

Traditionally, settling basins have been the most prevalent unit process used to remove solids from aquaculture effluents, if an effluent treatment unit has been used at all [145b]. Sedimentation has been an attractive effluent-treatment process because it is simple, requires little maintenance or water head, and has a moderate cost. On the other hand, treating large-volume aquaculture discharges with settling ponds requires a considerable amount of flat area (much more space than the other treatment units require), and, due to dissolution and resuspension of solids, settling basins are not as effective in achieving effluent TSS concentration of less than 6 mg/L [101]. Therefore, effluent treatment with settling basins may not meet strict regulations on the discharge of TSS. Nor is sedimentation an ideal method for removing suspended solids encountered in recirculating aquaculture systems. For these reasons, and to avoid storing solids in the flow, there has been a definite trend towards use of microscreen filters for rapid solids removal from aquaculture effluents [53, 57].

Ozonation and foam fractionation are two processes that are commonly added to supplement solids removal in recirculating systems.

13.5 Methods to Dispose of Solids

The settleable fraction of solids from clarifier backwash often is concentrated within settling basins [102, 145, 124, 117]. Thickening occurs in the bottom of the settling basin to solids concentrations of about 5% to 10%. Other sludge-thickening methods are sand

and wedgewire beds, filter presses, centrifuges, vacuum filters, and reed beds [146, 147, 148, 149].

Biosolids removed from aquaculture effluents are a residual product of wastewater treatment; therefore, many governments classify and regulate aquaculture solids as an industrial or municipal waste [7]. Other locations, however, consider aquaculture solids to be an agricultural waste, because these solids are composed of fish manure and uneaten feed and are considered a nontoxic nutrient source [7].

There are at least four options for beneficial reuse and disposal of the biosolids produced from an aquaculture facility [147, 150, 148, 149]: agricultural application, composting, vermiculture, and reed drying beds.

Land application of manure and other organic wastes (including wastewater) to fertilize agricultural crops is governed in most states by guidelines or regulations that limit the amount of pathogens, heavy metals, and other contaminants, nutrient-content, soil type, and plant nutrient-uptake characteristics to prevent run-off or groundwater contamination [151, 147, 7]. Applying the waste solids to on-site agricultural fields is almost always the cheapest method of solids disposal if adequate agricultural land is available on-site [150, 57].

Composting uses thermophilic bacteria to stabilize waste solids while producing a valuable soil amendment [150, 149]. Aerated static piles are the most common method used to compost biosolids [150, 149].

Vermiculture is an alternative for disposing of waste solids that uses earthworms (brandling or red worms) rather than bacteria to stabilize sludge [150, 149]. Sludge organics pass through the worm gut and emerge as dry, virtually odorless castings. Impetus exists for vermiculture, because there can be a substantially greater market for worm castings than for standard compost.

Wetlands planted on underdrained sand drying beds have been used at wastewater facilities for more than 20 years to dewater and stabilize sludges [152, 150, 149] and have been used to dewater and stabilize the biosolids from an aquaculture clarifier–backwash effluent [153]. Invervals between sludge addition allow for dewatering, drying, and stabilization of solids. Stabilized sludge is stored in these wetlands for as long as 10 years.

Hauling the collected sludge to the landfill is the most expensive alternative for sludge disposal. However, landfill disposal of waste aquaculture biosolids may be required in states in which they are classified as an industrial waste or when there are no other disposal options available.

13.6 Methods to Remove Dissolved and Colloidal Organic Matter

Dissolved organic matter (mostly nonbiodegradable) and color (i.e., refractory organic molecules) can become a problem in recirculating systems, if low water-exchange rates allow elevated levels to accumulate [142, 154].

There are three unit processes for reducing dissolved organic matter and color: ozonation, foam fractionation, and activated carbon adsorption. Activated carbon adsorption is not often used for treating color in large aquaculture production farms because of its relatively high cost. Use of activated carbon adsorption in industrial and municipal water and wastewater treatment industries has been described by many environmental engineering texts (Montgomery [130] and Metcalf and Eddy [147] to name a few).

13.7 Methods to Remove Carbon Dioxide

Accumulation of high levels of carbon dioxide can become a limiting toxicity factor with high fish densities and inadequate water exchange, that is, with high fish loadings [155, 156]. Carbon dioxide toxicity is more likely to occur in intensive aquaculture systems that inject pure oxygen because oxygen-injection processes use insufficient gas exchange to strip carbon dioxide [157]. Additionally, more carbon dioxide is usually produced in systems that inject pure oxygen because these systems have the oxygen to support higher fish loading rates. If aeration is used to supply oxygen to aquaculture systems, fish-loading levels are lower than can be obtained with pure oxygen, and enough air–water contact is generally provided to keep carbon dioxide from accumulating to toxic levels [59].

High levels of carbon dioxide can be reduced by stripping during aeration or by neutralization produced by alkaline addition.

13.7.1 Air Stripping

Air stripping and aeration are mass-transfer processes that occur together if water is contacted with air to bring its concentration of dissolved gases (such as nitrogen, carbon dioxide, and oxygen) into equilibrium with the partial pressures of these gases in the surrounding atmosphere.

If water is in contact with the atmosphere, the equilibrium concentration of carbon dioxide in water will depend on the amount of carbon dioxide in the air. At standard temperature and pressure, air contains about 0.032% carbon dioxide by volume (320 ppm), [158]—a mole fraction of about 0.00032 and a partial pressure of 0.00032 atm. Henry's law can be assumed to predict gas–liquid equilibrium because, for dilute solutions, the gas–liquid equilibrium relationship is linear. Henry's constant for carbon dioxide decreases with increasing temperature according to [159]:

$$K_H = 10^{\left(6.73 - \frac{1042}{T}\right)}, \tag{13.13}$$

where T is temperature in degrees K.

Water in contact with the atmosphere, therefore, would normally have an equilibrium concentration of 10^{-5} mol carbon dioxide per liter—about 0.5 mg/L at 20°C.

The rate of mass transfer (J), as defined by Fick's Law, is equal to the product of the overall mass-transfer coefficient (K_L), the total interfacial contact area per unit system volume (a), and the concentration gradient [160]:

$$J = K_L \cdot a \cdot (X - X^{eq}) \tag{13.14}$$

The concentration gradient is the driving force for mass transfer; that is, the rate of carbon dioxide stripping is greatest when the mole fraction of carbon dioxide in the water (X) is much greater than the carbon dioxide saturation mole fraction (X^{eq}).

Oxygen and carbon dioxide can be transferred to or from water with any nonclosed aeration system [155, 161, 162]. However, enormous quantities of air are required to achieve substantial carbon-dioxide transfer rates when compared with the airflow rates required for oxygen transfer alone. Therefore, aeration/stripping units that pass water through air are typically used to remove carbon dioxide, because in these units it is easier to provide the larger ratio of air to water volume needed for carbon-dioxide exchange [45].

The obvious way to let water fall through air is by a gravity drop, which can be over a weir, onto a splash board, through plastic media or stacked splash screens, down an inclined corrugated sheet (with or without holes), or down a stair-stepped surface. Grace and Piedrahita [163, 164] and Summerfelt [74] have reviewed and reported models and criteria that can be used to estimate carbon dioxide stripper performance and to design carbon-dioxide strippers for aquaculture systems. The general criteria for air stripping in forced ventilation columns are: a hydraulic fall of 1 to 1.5 m, a hydraulic loading of 1.0 to 1.4 $m^3/min/m^2$, and a volumetric air : water ratio of 6 : 1 to 10 : 1. High porosity packing or splash screens are needed to avoid flooding or gas hold-up. If high solids loadings are expected, a stripping tower with screens or media that resist biofouling should be used.

13.7.2 Chemical Addition

Alkalinity and pH play an important role in aquaculture. The pH of water controls acid/base chemistry, and alkalinity is a measure of the acid neutralizing capacity of the water. Alkalinity depends on the concentrations of the bicarbonate, carbonate, hydroxide, and hydrogen ions:

$$\frac{\text{Alk}}{50{,}000} = [\text{HCO}_3^-] + 2[\text{CO}_3^=] + [\text{OH}^-] - [\text{H}^+] \qquad (13.15)$$

Alkalinity is not a function of carbon dioxide, so alkalinity does not change when carbon dioxide is produced or removed. In recirculating systems, an alkalinity of at least 100 mg/L as calcium carbonate is recommended to maintain nitrification and reduce rapid swings in pH [80].

The influence of pH on the equilibrium of the carbonic acid and ammonia systems is important in aquaculture because ammonia (un-ionized) and carbon dioxide are much more toxic to aquatic organisms than ammonium and other species in the carbonic-acid system.

Dissolved carbon dioxide can combine with water in a hydrolysis reaction to form carbonic acid (Table 13.3). The equilibrium relationship that describes the hydrolysis of carbon dioxide indicates that there is about 633 times as much carbon dioxide in water as carbonic acid under equilibrium conditions.

The equilibrium alkalinity and carbon dioxide concentrations are related according to:

$$\frac{\text{Alk}}{50{,}000} = [\text{CO}_2] \cdot \left(\frac{K_0 K_1}{[\text{H}^+]} + \frac{K_0 K_1 K_2}{[\text{H}^+]^2} \right) + \frac{K_w}{[\text{H}^+]} + [\text{H}^+]; \qquad (13.16)$$

where the equilibrium constants K_0 and K_1 are defined in Table 13.3.

The equilibrium concentration of carbon dioxide in water can be estimated from alkalinity and pH using Eq. (13.16).

Table 13.3. Equilibrium relationships and constants in the inorganic carbon system

Equilibrium Reactions	Equilibrium Relationship	Equilibrium Constant
$CO_2\ (aq) + H_2O(l) \Leftrightarrow H_2CO_3\ (aq)$	$K_0 = \frac{[H_2CO_3]}{[CO_2]}$	1.58×10^{-3}@25°C
$H_2CO_3\ (aq) \Leftrightarrow H^+\ (aq) + HCO_3^-\ (aq)$	$K_1 = \frac{[H^+][HCO_3^-]}{[H_2CO_3]}$	2.56×10^{-4} mol/L@20°C[a]
$HCO_3^-\ (aq) \Leftrightarrow H^+\ (aq) + CO_3^=\ (aq)$	$K_2 = \frac{[H^+][CO_3^=]}{[HCO_3^-]}$	4.20×10^{-11} mol/L@20°C[a]
$H_2O\ (l) = H^+\ (aq) + OH^-\ (aq)$	$K_w = [H^+][OH^-]$	0.68×10^{-14} mol^2/L^2@20°C[a]

[a] Larson and Buswell [166].

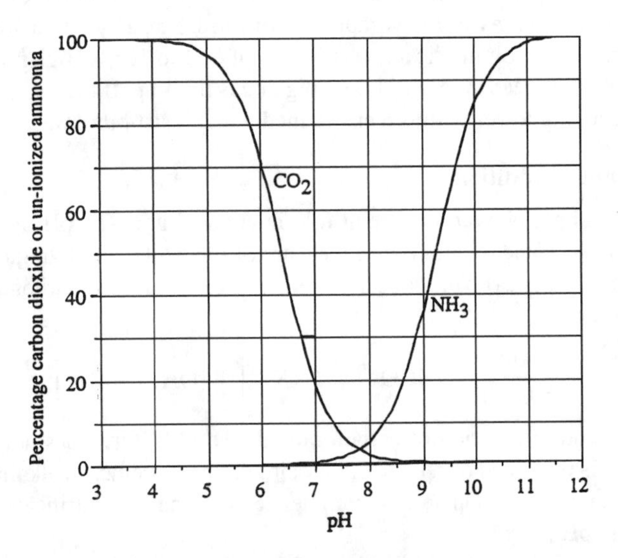

Figure 13.5. Percentage of carbon dioxide and ammonia as a function of pH at 25°C (after Summerfelt, [122]).

The range of pH values at which ammonia and carbon dioxide coexist indicates that the smallest fractions of both carbon dioxide and ammonia coexist at a pH of 7.5 to 8.2 (Fig. 13.5). Changing the system pH only one unit changes the corresponding equilibrium carbon dioxide or ammonia concentration ten-fold. Chemical treatment can be used to maintain a pH that will minimize the potentially toxic effects of ammonia and carbon dioxide in fish-culture systems. The treatment process consists of adding a supplemental source of alkalinity such as lime, caustic soda, soda ash, or sodium bicarbonate to the water [165]. Lime, caustic soda, and soda ash react with carbon dioxide to produce bicarbonate alkalinity and increase the pH of the water. Adding sodium bicarbonate is simply a source of bicarbonate alkalinity and a means to increase pH.

References

1. Phillips, M. J., M. C. M. Beveridge, and R. M. Clarke. 1991. Impact of aquaculture on water resources. In: D. E. Brune and J. R. Tomasso (eds.), Aquaculture and Water Quality. *Advances in World Aquaculture*, vol. 3. World Aquaculture Society, Baton Rouge, LA, pp. 568–591.
2. MacMillan, R. 1992. Economic implications of water quality management for a commercial trout farm. In: *National Livestock, Poultry, and Aquaculture Waste Management (Proceedings of the National Workshop)*. American Society of Agricultural Engineers, St. Joseph, MI.
3. Timmons, M. B. and T. M. Losordo. 1994. *Aquaculture Water Reuse Systems: Engineering Design and Management*. Elsevier Science B.V., New York.
4. Wade, E. M., S. T. Summerfelt, and J. A. Hankins. 1996. Economies of scale in recycle systems. In: G. Libey and M. Timmons (eds.), *Successes and Failures in Commercial Recirculating Aquaculture (Conference Proceedings)*, NRAES-98. Northeast Regional Agricultural Engineering Service, Ithaca, NY, pp. 575–588.
5. Summerfelt, S. T. and E. M. Wade. 1997. Recent advances in water treatment processes to intensify fish-production in large recirculating systems. In M. Timmons (ed.), *Advances in Aquacultural Engineering (Proceedings)*. Northeast Regional Agricultural Engineering Service, Ithaca, NY, pp. 350–367.
6. Bastian, R. K. 1992. Overview of federal regulations pertaining to aquaculture waste management and effluents. In: J. Blake, J. Donald, and W. Magette (eds.), *National Livestock, Poultry, and Aquaculture Waste Management*. American Society of Agricultural Engineers, St. Joseph, Michigan, pp. 220–226.
7. Ewart, J., J. A. Hankins, and D. Bullock. 1995. State Policies for Aquaculture Effluents and Solid Wastes in the Northeast Region. Northeast Regional Aquaculture Center, University of Massachusetts, Dartmouth, MA.
8. Beveridge, M. C. M., M. J. Phillips, R. M. Clarke. 1991. A quantitative assessment of wastes from aquatic animal production. In: D. E. Brune and J. R. Tomasso (eds.), Aquaculture and Water Quality. *Advances in World Aquaculture*, vol. 3. World Aquaculture Society, Baton Rouge, LA, pp. 506–533.
9. Wedemeyer, G. A. and G. W. Wood. 1974. Stress as a predisposing factor in fish disease. Fish Disease Leaflet No. 38. U.S. Department of Interior, Fish and Wildlife Service, Division of Fisheries Research, Washington, DC.
10. Wedemeyer, G. A. 1996. *Physiology of Fish in Intensive Culture*. International Thompson Publishing, New York.
11. Noble, A. C. and S. T. Summerfelt. 1996. Diseases encountered in rainbow trout reared in recirculating systems. *Annual Review of Fish Diseases* 6:65–92.
12. Piper, R. G., I. B. McElwain, L. E. Orme, J. P. McCraren, L. G. Fowler, and J. R. Leonard. 1982. *Fish Hatchery Management*. U.S. Fish and Wildlife Service, Washington, DC.
13. Liao, P. B. and R. D. Mayo. 1974. Intensified fish culture combining water reconditioning with pollution abatement. *Aquaculture* 3:61–85.

14. Kruner, G. and H. Rosenthal. 1987. Circadian periodicity of biological oxidation under three different operational conditions. *Aquacultural Engineering* 6:79–96.
15. Meade, J. W. 1985. Allowable ammonia for fish culture. *Progressive Fish-Culturist* 47:135.
16. Colt, J. E. and D. A. Armstrong. 1981. Nitrogen toxicity to fish, crustaceans and mollusks. In: L. J. Allen and E. C. Kinney (eds.), *Proceedings of the Bio-engineering Symposium for Fish Culture*. American Fisheries Society, Fish Culture Section, Bethesda, MD, pp. 39–42.
17. Russo, R. C. and R. V. Thurston. 1977. The acute toxicity of nitrite to fishes. In: R. A. Tubb (ed.), *Recent Advances in Fish Toxicology*. EPA-600/3-77-085. U.S. Environmental Protection Agency, Ecological Research Service, Corvallis, OR, pp. 118–131.
18. Westin, D. T. 1974. Nitrate and nitrite toxicity to salmonid fishes. *Progressive Fish-Culturist* 36(2):86.
19. Laird, L. M. and T. Needham. 1988. *Salmon and Trout Farming*. Ellis Horwood Limited, New York.
20. Losordo, T. M. 1991. Engineering considerations in closed recirculating systems. In: P. Giovannini (session chairman), *Aquaculture Systems Engineering*. American Society of Agricultural Engineers, St. Joseph, MI, pp. 58–69.
21. Wheaton, F., J. Hochheimer, and G. E. Kaiser. 1991. Fixed-film nitrification filters for aquaculture. In: D. E. Brune and J. R. Tomasso (eds.), *Aquaculture and Water Quality*. World Aquaculture Society, Baton Rouge, LA, pp. 272–303.
22. Muncy, R. J., G. J. Atchison, R. V. Bulkley, B. W. Menzel, L. G. Perry, and R. C. Summerfelt. 1979. Effects of suspended solids and sediment on reproduction and early life of warmwater fishes: A review. EPA-600/3-79-042. National Technical Information Service, Springfield, VA, p. 101.
23. Chen, S., D. Stechey, and R. F. Malone. 1994. Suspended solids control in recirculating aquaculture systems. In: M. B. Timmons, T. M. Losordo (eds.), *Aquaculture Water Reuse Systems: Engineering Design and Management*. Elsevier Science B.V., New York, pp. 61–100.
24. Chapman, P. E., J. D. Popham, J. Griffin, and J. Michaelson. 1987. Differentiation of physical from chemical toxicity in solid waste fish bioassay. *Water Air Soil Poll.* 33:295–308.
25. Klontz, G. W. 1993. Environmental requirements and environmental diseases of salmonids. In: M. K. Stoskopf (ed.), *Fish Medicine*. WB Saunders, Philadelphia, pp. 333–342.
26. Bullock, G. L., R. Herman, J. Heinen, A. Noble, A. Weber, and J. Hankins. 1994. Observations on the occurrence of bacterial gill disease and amoeba gill infestation in rainbow trout cultured in a water recirculation system. *J. Aquat. Anim. Health* 6:310–317.
27. Herbert, D. W. M., and J. C. Merkens. 1961. The effects of suspended mineral solids on the survival of trout. *Air Water Poll.* 5:46–50.
28. Juell, J. E., D. M. Furvik, and A. Bjordal. 1993. Demand feeding in salmon farming by hydroacoustic food detection. *Aquaculture Engineering* 12:155–167.

29. Hankins, J. A., S. T. Summerfelt, and M. D. Durant. 1995. Impacts of feeding and stock management strategies upon fish production within water recycle systems. In: M. B. Timmons (ed.), *Aquacultural Engineering and Waste Management*. Northeast Regional Agricultural Engineering Service, Ithaca, NY, pp. 70–86.
30. Mayer, I. and E. McLean. 1995. Bioengineering and Biotechnology Strategies for Reduced Waste Aquaculture. *Water Science and Technology* 31:85–102.
31. Thorpe, J. E. and C. Y. Cho. 1995. Minimizing waste through bioenergetically and behaviorally based feeding strategies. *Water Science and Technology* 31: 29–40.
32. Goddard, S. 1996. *Feed Management in Intensive Aquaculture*. Chapman and Hall, New York.
33. Clark, E. R., J. P. Harman, and J. R. M. Forster. 1985. Production of metabolic and waste products by intensively farmed rainbow trout (*Salmo gairdneri*) Richardson. *J. Fish Bio.* 27:381–393.
34. Chen, S., D. E. Coffin, and R. F. Malone. 1993. Production, characteristics, and modeling of aquaculture sludge from a recirculating aquaculture system using a granular media biofilter. In J.-K. Wang (ed.), *Techniques for Modern Aquaculture*. American Society of Agricultural Engineers, St. Joseph, MI, pp. 16–25.
35. Seymour, E. A. and A. Bergheim. 1991. Towards a reduction of pollution from intensive aquaculture with reference to the farming of salmonids in Norway. *Aquacultural Engineering* 10:73–88.
36. Mathieu, F. and M. B. Timmons. 1993. Time dependent BOD of aquacultural wastes in recirculating rainbow trout systems. In: J.-K. Wang (ed.), *Techniques for Modern Aquaculture*. American Society of Agricultural Engineers, St. Joseph, MI, pp. 44–47.
37. Garcia-Ruiz, R., and G. H. Hall. 1996. Phosphorus fractionation and mobility in the food and faeces of hatchery reared rainbow trout (*Onchorhynchus mykiss*). *Aquaculture* 145:183–193.
38. Muir, J. F. 1982. Recirculating systems in aquaculture. In: J. F. Muir and R. J. Roberts (eds.), *Recent Advances in Aquaculture*, vol. 1. Croom Helm and Westview Press, London, pp. 358–446.
39. Chen, S., M. B. Timmons, D. J. Aneshansley, and J. J. Bisogni, Jr. 1993. Suspended solids characteristics from recirculating aquaculture systems and design implications. *Aquaculture* 112:143–155.
40. Robertson, W. D. 1992. Assessment of filtration and sedimentation systems for total phosphorus removal at the Lake Utopia Fish Culture Station. Canadian Fish Culture Operations, Aquaculture Division, Connors Bros., Limited., Blacks Harbour, New Brunswick, Canada.
41. Timmons, M. B. and W. D. Young. 1991. Considerations on the design of raceways. In: *Aquaculture Systems Engineering*, Publication 02-91. American Society of Agricultural Engineers, S. Joseph, MI, pp. 34–46.
42. Randall, D. J. and C. Daxboeck. 1984. Oxygen and carbon dioxide transfer across fish gills. In: W. S. Hoar and D. J. Randall (eds.), *Fish Physiology*, vol. 10A. Academic Press, New York, pp. 263–314.

43. Walsh, P. J. and R. P. Henry. 1991. Carbon dioxide and ammonia metabolism and exchange. In: P. W. Hochachka and T. P. Mommsen (eds.), *Biochemistry and Molecular Biology of Fishes: Phylogenetic and Biochemical Perspectives*. Elsevier Science B.V., New York, pp. 181–207.
44. Kutty, M. N. 1968. Respiratory quotients in goldfish and rainbow trout. *J. Fish. Res. Board Can.* 25:1689–1728.
45. Colt, J. E. and K. Orwicz. 1991. Aeration in intensive culture. In: D. E. Brune and J. R. Tomasso (eds.), *Aquaculture and Water Quality*. World Aquaculture Society, Louisiana State University, Baton Rouge, LA, pp. 198–271.
46. Basu, S. P. 1959. Active respiration of fish in relation to ambient concentrations of oxygen and carbon dioxide. *J. Fish. Res. Board Can.* 16:175–212.
47. Smart, G. R., D. Knox, J. G. Harrison, J. A. Ralph, R. H. Richards, and C. B. Cowey. 1979. Nephrocalcinosis in rainbow trout *Salmo gairdneri* Richardson: The effect of exposure to elevated CO_2 concentrations. *J. Fish Dis.* 2:279–289.
48. Heinen, J. M., J. A. Hankins, A. L. Weber, and B. J. Watten. 1996. A semi-closed recirculating water system for high density culture of rainbow trout. *Progressive Fish-Culturist* 58:11–22.
49. SECL (Sigma Environmental Consultants, Limited). 1983. Summary of water quality for salmonid hatcheries, SECL 8067. Prepared for Department of Fisheries and Oceans, Vancouver, British Columbia, Canada.
50. Bergheim, A. and T. Âsgård. 1996. Waste production from aquaculture. In: D. Baird, M. C. M. Beveridge, L. A. Kelly, and J. F. Muir (eds.), *Aquaculture and Water Resource Management*. Blackwell Science, Osney Mead, Oxford, United Kingdoms, pp. 50–80.
51. Stewart, J. E. 1997. Environmental impacts of aquaculture. *World Aquaculture* 28(1):47–52.
52. Munro, A. L. S. and N. Fijan. 1981. Disease prevention and control. In: K. Tiews (ed.), *Aquaculture in Heated Effluents and Recirculation Systems*, vol. 2. Heenemann Verlagsgesellschaft mbH, Berlin, pp. 20–32.
53. NCC (Nature Conservancy Council). 1990. *Fish Farming and the Scottish Freshwater Environment*. Nature Conservancy Council, Edinburgh.
54. Bergheim, A., J. P. Aabel, and E. A. Seymour. 1991. Past and present approaches to aquaculture waste management in Norwegian net pen culture operations. In: C. B. Cowey and C. Y. Cho (eds.), *Nutritional Strategies for Aquaculture Waste*. University of Guelph, Guelph, Ontario, Canada, pp. 117–136.
55. Braaten, B. 1991. Impact of pollution from aquaculture in six Nordic countries: release of nutrients, effects, and waste water treatment. In: N. DePauw and J. Joyce (eds.), *Aquaculture and the Environment*. EAS Special Publication No. 16. European Aquaculture Society, Gent, Belgium, pp. 79–101.
56. Phillips, M. J., R. Clarke, and A. Mowat. 1993. Phosphorus leaching from Atlantic salmon diets. *Aquacultural Engineering* 12:47–54.
57. Cripps, S. J. and L. A. Kelly. 1996. Reductions in wastes from aquaculture. In: D. Baird, M. C. M. Beveridge, L. A. Kelly, and J. F. Muir (eds.), *Aquaculture and*

Water Resource Management. Blackwell Science, Osney Mead, Oxford, United Kingdoms, pp. 166–201.
58. Heinen, J. M., J. A. Hankins, and P. R. Adler. 1996. Water quality and waste production in a recirculating trout-culture system with feeding of a higher-energy or a lower-energy diet. *Aquaculture Research* 27:699–710.
59. Speece, R. E. 1973. Trout metabolism characteristics and the rational design of nitrification facilities for water reuse in hatcheries. *Trans. Amer. Fish. Soc.* 102(2):323–334.
60. Kaiser, G. E., and F. W. Wheaton. 1983. Nitrification filters for aquatic culture systems: State of the art. *J. World Maricult. Soc.* 14:302–324.
61. Wheaton, F. W., J. N. Hochheimer, G. E. Kaiser, M. J. Krones, G. S. Libey, and C. C. Easter. 1994. Nitrification filter principles. In: M. B. Timmons and T. M. Losordo (eds.), *Aquaculture Water Reuse Systems: Engineering Design and Management*. Elsevier, New York, pp. 101–126.
62. Schlegel, S. 1984. The use of submerged biological filters for nitrification. *Water Science Technology* 20:177.
63. Wanner, O. and W. Gujer. 1984. Competition in biofilms. *Water Science Technology* 17:27–44.
64. Harremoes, P. 1982. Criteria for nitrification in fixed film reactors. *Water Science Technology* 14:167–187.
65. Bovendeur, J., A. B. Zwaga, B. G. J. Lobee, and J. H. Blom. 1990. Fixed-biofilm reactors in aquacultural water recycle systems: Effect of organic matter elimination on nitrification kinetics. *Wat. Res.* 24(2):207–213.
66. Sharma, B. and R. C. Ahlert, 1977. Nitrification and nitrogen removal. *Water Research* 11:897–925.
67. Cutter, D. W. and L. M. Crump. 1983. Some aspects of the physiology of certain nitrite forming bacteria. *Ann. Appl. Biol.* 20:291–296.
68. EPA (Environmental Protection Agency). 1975. Process Design Manual for Nitrogen Control. Office of Technology Transfer, Washington, DC.
69. Alleman, J. E. and K. Preston. 1991. Behavior and physiology of nitrifying bacteria. In: L. Swann (ed.), *Proceedings of Regional Workshop on Commercial Fish Culture Using Water Recirculating Systems*. Illinois State University, Normal, IL, pp. 1–13.
70. Manem, J. A. and B. E. Rittman. 1992. The effects of fluctuations in biodegradable organic matter on nitrification filters. *Journal AWWA*. 84(4):147–151.
71. Eikebrokk, B. and R. H. Piedrahita. 1997. Nitrification performance and ammonia excretion from salmonids, calculated from nitrogen mass balances in water reuse systems under field and laboratory conditions. *Journal of Applied Aquaculture* 7(4): 15–32.
72. Haug, R. T. and P. L. McCarty. 1972. Nitrification with submerged filters. *J. Water Pollut. Control Fed.* 44:2086.
73. Lomax, K. M., and F. W. Wheaton. 1978. Pretreatment of fish culture wastewater for nitrification. *Trans. ASAE* 197–200.

74. Summerfelt, S. T. 1993. Low-Head Roughing Filters for Enhancing Recycle Water Treatment for Aquaculture. Doctoral diss., Iowa State University, Ames, IA.
75. Rusten, B. 1984. Wastewater treatment with aerated submerged biological filters. *J. Water Pollut. Control Fed.*, 56:424–431.
76. Anderson, D. 1974. Biological trickling filter system for water reuse in trout rearing. Investigational report no. 322. Minnesota Division of Fish and Wildlife, Section of Fisheries, St. Paul, MN.
77. Nijhof, M. 1995. Bacterial stratification and hydraulic loading effects in a plug-flow model for nitrifying trickling filters applied in recirculating fish culture systems. *Aquaculture* 134:49–64.
78. Tanaka, K., A. Oshima, and B. E. Rittmann. 1987. Performance evaluations of rotating biological contactor process. *Water Sci. Tech.* 19:483–494.
79. Tchobanoglous, G. and E. D. Schroeder. 1985. *Water Quality*. Addison-Wesley, Reading, MA, p. 168.
80. Malone, R. F., B. S. Chitta, and D. G. Drennan. 1993. Optimizing nitrification in bead filters for warmwater recirculating aquaculture systems. In: J.-K. Wang (ed.), *Techniques for Modern Aquaculture*. American Society of Agricultural Engineers, St. Joseph, MI, pp. 315–325.
81. Golz, W., K. A. Rusch, and R. F. Malone. 1996. Developing backwash protocols for floating bead filters: A model of solids-loading and biofilm retention effects on nitrification. In: G. Libey and M. Timmons (eds.), *Successes and Failures in Commercial Recirculating Aquaculture (Conference Proceedings)*, NRAES-98. Northeast Regional Agricultural Engineering Service, Ithaca, NY, pp. 196–205.
82. Summerfelt, S. T., J. A. Hankins, M. D. Durant, and J. N. Goldman. 1996. Removing obstructions: Modified pipe-lateral flow distribution mechanism reduces backflow in fluidized-sand biofilters. *Water Env. and Tech.* 8(11): 39–43.
83. Burden, D. G. 1988. Development and Design of a Fluidized Bed/Upflow Sand Filter Configuration for Use in Recirculating Aquaculture Systems. Master's thesis, Louisiana State University, Baton Rouge, LA.
84. Cooper, P. F. and B. Atkinson. 1981. *Biological Fluidized Bed Treatment of Water and Wastewater*. Water Research Centre, Ellis Horwood Ltd., Chichester, England.
85. Jewell, W. J. 1990. Fundamentals and advances in expanded bed reactors for waste-water treatment. In: R. D. Tyagi and K. Vembu (eds.), *Wastewater Treatment by Immobilized Cells*. CRC Press, Boca Raton, FL, pp. 223–252.
86. Sutton, P. M. and P. N. Mishra. 1991. Biological fluidized beds for wastewater treatment: A state-of-the-art review. *Water Environment Technology* 3(8): 52–56.
87. Summerfelt, S. T. and J. L. Cleasby. 1996. A review of hydraulics in fluidized-bed biological filters. *Trans. ASAE* 39:1161–1173.
88. Cleasby, J. L. 1990. Filtration. In: F. W. Pontius (ed). *Water Quality and Treatment*, American Water Works Association, 4th ed. McGraw-Hill, New York, pp. 455–560.

89. Cleasby, J. L. and K. S. Fan. 1981. Predicting fluidization and expansion of filter media. *Journal of the Environmental Engineering Division*, American Society of Civil Engineers 107:455–471.
90. Tsukuda., S. M., S. T. Summerfelt, T. K. Sawyer, G. L. Bullock, and C. P. Marshall. 1997. Effects of sand size on fluidized-bed biofilter performance in cold-water systems. In: *World Aquaculture '97 Book of Abstracts*. World Aquaculture Society, Baton Rouge, LA, pp. 467–468.
91. Monaghan, T. J., A. A. Delos Reyes, T. M. Jeansonne, and R. F. Malone. 1996. Effects of media size on nitrification in fluidized sand filters. In: *Aquaculture America '96 Book of Abstracts*. World Aquaculture Society, Baton Rouge, LA, p. 110.
92. Thomasson, M. P. 1991. Nitrification in Fluidized Bed Sand Filters for Use in Recirculating Aquaculture Systems. Master's thesis, Louisiana State University, Baton Rouge, LA.
93. Cooley, P. E. 1979. Nitrification of Fish-Hatchery Reuse Water Utilizing Low-Density Polyethylene Beads as a Fixed Film Media Type. Master's thesis, Idaho State University, Moscow, Idaho.
94. Wimberly, D. M. 1990. Development and Evaluation of a Low-Density Media Biofiltration Unit for Use in Recirculating Finfish Culture Systems. Master's thesis, Louisiana State University, Baton Rouge, LA.
95. Delos Reyes, A. A. and R. F. Malone. 1996. Design and evaluation of a commercial-scale, paddle-washed floating bead filter. In: G. Libey and M. Timmons (eds.), *Successes and Failures in Commercial Recirculating Aquaculture (Conference Proceedings)*, NRAES-98. Northeast Regional Agricultural Engineering Service, Ithaca, NY, pp. 183–195.
96. Chesness, J. L., W. H. Poole, and T. K. Hill. 1975. Settling basin design for raceway fish production systems. *Trans. ASAE* 18(5):159–162.
97. Warren-Hansen, I. 1982. Methods of treatment of waste water from trout farming. In: J. Alabaster (ed.), EIFAC Technical Paper No. 41. Report of the EIFAC Workshop on Fish-Farm Effluents, Silkeborg, Denmark, 26–28 May, 1981. FAO, Rome, pp. 113–121.
98. Camp, T. R. 1936. A study of the rational design of settling tanks. *Sewage Works Journal* 8:742–758.
99. Stechey, D. and Y. Trudell. 1990. Aquaculture waste water treatment: waste water characterization and development of appropriate treatment technologies for the Ontario trout production industry. PIBS 1319, Log # 90-2309-041. Ontario Ministry of the Environment, Toronto, Ontario, Canada.
100. Stechey, D. 1991. Build your own settling pond. *Northern Aquaculture* 7(5):22–29.
101. Henderson, J. P. and N. R. Bromage. 1988. Optimising the removal of suspended solids from aquaculture effluents in settlement lakes. *Aquacultural Engineering* 7:167–188.
102. Mudrak, V. A. 1981. Guidelines for economical commercial fish hatchery wastewater treatment systems. In: L. J. Allen and E. C. Kinney (eds.), *Proceedings*

of the Bio-engineering Symposium for Fish Culture. American Fisheries Society, Fish Culture Section, Bethesda, MD, pp. 174–182.

103. Michelsen, K. 1991. Past and present approaches to aquaculture waste management in danish pond culture operations. In: C. B. Cowey and C. Y. Cho (eds.), Nutritional Strategies and Aquaculture Waste. *Proceedings of the First International Symposium on Nutritional Strategies in Management of Aquaculture Waste*. University of Guelph, Guilph, Ontario, Canada, pp. 155–161.
104. Arceivala, S. J. 1983. Hydraulic modelling for waste stabilisation ponds. *Journal of Environmental Engineering* 109(5):265–268.
105. Paul, T. C., S. K. Sayal, V. S. Sakhuja, and G. S. Dhillon. 1991. Vortex-settling basin design considerations. *J. Hydraul. Eng.* 117:172–189.
106. Scott, K. R. and L. Allard. 1983. High-flowrate water recirculation system incorporating a hydrocyclone prefilter for rearing fish. *Progressive Fish-Culturist* 45:148–153.
107. Scott, K. R. and L. Allard. 1984. A four-tank water recirculating system with a hydrocyclone prefilter and a single water reconditioning unit. *Progressive Fish-Culturist* 46:254–261.
108. Mäkinen, T., S. Lindgren, and P. Eskelinen. 1988. Sieving as an effluent treatment method for aquaculture. *Aquacultural Engineering* 7:367–377.
109. Eikebrokk, B. and Y. Ulgenes. 1993. Characterization of treated effluents from landbased fish farms. In: H. Keinertsen, L. A. Dahle, L. Jorgensen, and K. Tvinnereim (eds.), *Fish Farming Technology*. Balkema, Rotterdam, pp. 361–366.
110. Jenkins, M. R., E. M. Wade, and C. W. Yohn. 1996. The opportunity for small-scale aquaculture as an alternative farm enterprise: An economic analysis of a small-scale partial reuse trout production system in West Virginia. *Journal of the ASFMRA* 81–89.
111. Yao, K. M. 1970. Theoretical study of high rate sedimentation. *Journal Water Pollution Control Federation* 42:218–228.
112. Yao, K. M. 1973. Design of high rate settlers. *J. Env. Engn. Div. ASCE* 99:621–636.
113. Smethurst, G. 1979. Settling basins: Practical considerations and choice. In: *Basic Water Treatment*. Published by Thomas Telford Ltd., available from the American Society of Civil Engineers, New York.
114. McLaughlin, T. W. 1981. Hatchery effluent treatment: U.S. fish and wildlife service. In: J. A. Lochie and E. C. Kinney (eds.), *Bioengineering Symposium for Fish Culture*. American Fisheries Society, Bethesda, MD, pp. 167–173.
115. Libey, G. S. 1993. Evaluation of a drum filter for removal of solids from a recirculating aquaculture system. In: J.-K. Wang (ed.), *Techniques for Modern Aquaculture*. American Society of Agricultural Engineers, S. Joseph, MI, pp. 519–532.
116. Liltvedt, H. and B. R. Hansen. 1990. Screening as a method for removal of parasites from inlet water to fish farms. *Aquacultural Engineering* 9:209–215.
117. Bergheim, A., R. Kristiansen, and L. Kelly. 1993. Treatment and utilization of sludge from landbased farms for salmon. In: J.-K. Wang (ed.), *Techniques for*

Modern Aquaculture. American Society of Agricultural Engineers, St. Joseph, MI, pp. 486–495.

118. Bergheim, A., S. Sanni, G. Indrevik, and P. Holland. 1993. Sludge removal from salmonid tank effluent using rotating microsieves. *Aquacultural Engineering* 12:97–109.
119. Cripps, S. J. 1995. Serial particle size fractionation and characterization of an aquaculture effluent. *Aquaculture* 133:323–339.
120. Kelly, L. A., A. Bergheim, and J. Stellwagen. 1997. Particle size distribution of wastes from freshwater fish farms. *Aquaculture International* 5:65–87.
121. Summerfelt, S. T., J. A. Hankins, J. M. Heinen, A. L. Weber, and J. D. Morton. 1994. Evaluation of the Triangel^{TM} microsieve filter in a water-reuse system. In: *World Aquaculture '94 Book of Abstracts*. World Aquaculture Society, Baton Rouge, LA, p. 45.
122. Summerfelt, S. T. 1996. Engineering design of a water reuse system. In: *The Walleye Culture Manual*. North Central Regional Aquaculture Center, Michigan State University, East Lansing, MI, pp. 277–309.
123. Summerfelt, S. T., J. A. Hankins, A. Weber, and M. D. Durant. 1997. Ozonation of a recirculating rainbow trout culture system: II. Effects on microscreen filtration and water quality. *Aquaculture*, 158:57–67.
124. Bergheim, A. and O. I. Forsberg. 1993. Attempts to reduce effluent loadings from salmon farms by varying feeding frequencies and mechanical effluent treatment. In: G. Barnabé and P. Kestemont (eds.), *Production, Environment and Quality: Proceedings of the International Conference Bordeaux Aquaculture '92*. Special Publication No. 18. European Aquaculture Society, Ghent, Belgium, pp. 115–124.
125. Singh, S. and Marsh, L. S. 1996. Modeling thermal environment of a recirculating aquaculture facility. *Aquaculture* 139:11–18.
126. Cryer, E. 1992. Recent applications of ozone in freshwater fish hatchery systems. In: W. J. Blogoslawski (ed.), *Proceedings of the 3rd International Symposium on the Use of Ozone in Aquatic Systems*. International Ozone Association, Pan American Committee, Stamford, CT, pp. 134–154.
127. Baeyens, J., I. Y. Mochtar, S. Liers, and H. De Wit. 1995. Plugflow dissolved air flotation. *Water Environment Research* 67:1027–1035.
128. Liers, S., J. Baeyens, and I. Mochtar. 1996. Modeling dissolved air flotation. *Water Environment Research* 68:1061–1075.
129. Kemmer, F. N. 1979. *The NALCO Water Handbook*. Nalco Chemical Company. McGraw-Hill Book Company, New York.
130. Montgomery, J. M. (Consulting Engineers, Inc.). 1985. *Water Treatment Principles and Design*. John Wiley and Sons, New York.
131. Chen, S., M. B. Timmons, J. J. Bisogni, Jr., and D. J. Aneshansley. 1993. Suspended-solids removal by foam fractionation. *Progressive Fish-Culturist* 55:69–75.
132. Weeks, N. C., M. B. Timmons, and S. Chen. 1992. Feasibility of using foam fractionation for the removal of dissolved and suspended solids from fish culture water. *Aquacultural Engineering* 11:251–265.

133. Chen, S., M. B. Timmons, J. J. Bisogni, Jr., and D. J. Aneshansley. 1993. Protein and its removal by foam fractionation. *Progressive Fish-Culturist* 55:69–75.
134. Chen, S., M. B. Timmons, J. J. Bisogni, Jr., and D. J. Aneshansley. 1994. Modeling surfactant removal in foam fractionation: I. Theoretical development. *Aquacultural Engineering* 13:163–181.
135. Chen, S., M. B. Timmons, J. J. Bisogni, Jr., and D. J. Aneshansley. 1994. Modeling surfactant removal in foam fractionation: II. Experimental investigations. *Aquacultural Engineering* 13:183–200.
136. Sander, E. and H. Rosenthal. 1975. Application of ozone in water treatment for home aquaria, public aquaria, and for aquaculture purposes. In: W. J. Blogoslawski and R. G. Rice (eds.), *Aquatic Applications of Ozone*. International Ozone Institute, Stamford, CT, pp. 103–114.
137. Otte, G. and H. Rosenthal. 1979. Management of a closed brackish water system for high density fish culture by biological and chemical treatment. *Aquaculture* 18: 169–181.
138. Williams, R. C., S. G. Hughes, and G. L. Rumsey. 1982. Use of ozone in a water reuse system for salmonids. *Progressive Fish-Culturist* 44:102–105.
139. Gargas, J. 1989. Fresh water ozonation and foam fractionation in a discus hatchery. *Tropical Fish Hobbyist* 38(3):114–122.
140. Rice, R. G., C. M. Robson, G. W. Miller, and A. G. Hill. 1981. Uses of ozone in drinking water treatment. *Journal of American Water Works Association* 73:1–44.
141. Bablon, G., W. D. Bellamy, M.-M. Bourbigot, F. B. Daniel, M. Doré, F. Erb, G. Gordon, B. Langlais, A. Laplanche, B. Legube, G. Martin, W. J. Masschelein, G. Pacey, D. A. Reckhow, and C. Ventresque. 1991. Fundamental aspects. In: B. Langlais, D. A. Reckhow, D. R. Brink (eds.), *Ozone in Water Treatment: Application and Engineering*. American Water Works Association Research Foundation. Denver, CO, pp. 11–132.
142. Rosenthal, H. and G. Otte. 1980. Ozonation in an intensive fish culture recycling system. *Ozone: Science and Engineering* 1:319–327.
143. Maier, D. 1984. Microflocculation by ozone. In: R. G. Rice and A. Netzer (eds.), *Handbook of Ozone Technology and Applications*, vol. 2. Butterworth Publishers, Boston, pp. 123–140.
144. Summerfelt, S. T. and J. N. Hochheimer. 1997. A review of ozone processes and applications as an oxidizing agent in aquaculture. *Progressive Fish-Culturist* 59:94–105.

145a. Bullock, G. L., S. T. Summerfelt, A. Noble, A. Weber, M. D. Durant, and J. A. Hankins. 1997. Ozonation of a recirculating rainbow trout culture system: I. Effects on bacterial gill disease and heterotrophic bacteria. *Aquaculture* 158:43–55.

145b. Westers, H. 1991. Operational waste management in aquacultural effluents. In: C. B. Cowey and C. Y. Cho (eds.), Nutritional Strategies and Aquaculture Waste. *Proceedings of the First International Symposium on Nutritional Strategies in Management of Aquaculture Waste*. University of Guelph, Guelph, Ontario, Canada, pp. 231–238.

146. EPA (Environmental Protection Agency). 1987. Dewatering Municipal Wastewater Sludges. Center for Environmental Resarch Information, Cincinnati, OH.
147. Metcalf and Eddy, Inc. 1991. *Wastewater Engineering: Treatment, Disposal, and Reuse*, 3rd ed. McGraw-Hill, New York.
148. Black and Veatch, Inc. 1995. Wastewater Biosolids and Water Residuals: Reference Manual on Conditioning, Thickening, Dewatering, and Drying. CEC Report CR-105603. The Electric Power Research Institute, Community Environment Center, Washington University, St. Louis, MO.
149. Reed, S. C., R. W. Crites, and E. J. Middlebrooks. 1995. *Natural Systems for Waste Management and Treatment*, 2nd ed. McGraw-Hill, New York.
150. Outwater, A. B. 1994. *Reuse of Sludge and Minor Wastewater Residuals*. Lewis Publishers, Boca Raton, FL.
151. Chen, S., D. E. Coffin, and R. F. Malone. 1991. Suspended solids control in recirculating aquaculture systems. In: *Engineering Aspects of Intensive Aquaculture*. Northeast Regional Agricultural Engineering Service, Cooperative Extension, Ithaca, NY, pp. 170–186.
152. Riggle, D. 1991. Reed bed system for sludge. *Biocycle* 32(12):64–66.
153. Summerfelt, S. T., P. R. Adler, D. M. Glenn, and R. N. Kretschman. 1996. Aquaculture sludge removal and stabilization within created wetlands. In: *5th IAWQ Conference on Constructed Wetland Systems for Water Pollution Control* (Vienna, Austria). Institute for Water Provision, Water Ecology and Waste Management, Unversitaet fuer Bodenkultur Wien, pp. 2–1 to 2–7.
154. Hirayama, K., H. Mizuma, and Y. Mizue. 1988. The accumulation of dissolved organic substances in closed recirculation culture systems. *Aquacultural Engineering* 7:73–87.
155. Colt, J. E. and V. Tchobanoglous. 1981. Design of aeration systems for Aquaculture. In: L. J. Allen and E. C. Kinney (eds.), *Proceedings of the Bio-engineering Symposium for Fish Culture*. American Fisheries Society, Bethesda, MD, pp. 138–148.
156. Colt, J. E., K. Orwicz, and G. Bouck. 1991. Water quality considerations and criteria for high-density fish culture with supplemental oxygen. In: J. Colt and R. J. White (eds.), *Fisheries Bioengineering Symposium 10*. American Fisheries Society, Bethesda, MD, pp. 372–385.
157. Watten, B. J., J. E. Colt, and C. E. Boyd. 1991. Modeling the effect of dissolved nitrogen and carbon dioxide on the performance of pure oxygen absorption systems. In: J. Colt and R. J. White (eds.), *American Fisheries Society Symposium 10: Fisheries Bioengineering Symposium*. Bethesda, MD, pp. 474–481.
158. Giddings, J. C. 1973. *Chemistry, Man and Environmental Change*. Canfield, San Francisco.
159. Cornwell, D. A. 1990. Air stripping and aeration. In: F. W. Pontius (ed.), *Water Quality and Treatment*. American Water Works Association, 4th ed. McGraw-Hill, New York, pp. 229–268.
160. Treybal, R. E. 1980. *Mass-Transfer Operations*, 3rd ed. McGraw-Hill Book Company, New York.

161. Speece, R. E. 1981. Management of dissolved oxygen and nitrogen in fish hatchery waters. In: L. J. Allen and E. C. Kinney (eds.), *Proceedings of the Bio-engineering Symposium for Fish Culture*. American Fisheries Society, Bethesda, Maryland, pp. 53–62.
162. Boyd, C. E. and B. J. Watten. 1989. Aeration systems in aquaculture. *Reviews in Aquatic Sciences* 1:425–473.
163. Grace, G. R. and R. H. Piedrahita. 1993. Carbon dioxide control with a packed column aerator. In: J. K. Wang (ed.), *Techniques for Modern Aquaculture*. American Society of Agricultural Engineers, St. Joseph, MI, pp. 496–505.
164. Grace, G. R. and R. H. Piedrahita. 1994. Carbon dioxide control. In M. B. Timmons and T. M. Losordo (eds.), *Aquaculture Water Systems: Engineering Design and Management*. Elsevier Science, New York, pp. 209–234.
165. Bisogni, J. J., Jr. and M. B. Timmons. 1991. Control of pH in closed cycle aquaculture systems. In: Engineering Aspects of Intensive Aquaculture: *Proceedings from the Aquaculture Symposium*. Northeast Regional Agricultural Engineering Service, Ithaca, NY, pp. 33–348.
166. Larson, T. E. and A. M. Buswell. 1942. Calcium carbonate saturation index and alkalinity interpretations. *Journal American of Water Works Association* 34:1667.

Index

acrylic, 241
activity, 50, 51, 52
admixtures, 5
adsorption, principle of, 216
advantages, 6
aeration, 213, 216, 309, 310, 320, 322, 326, 336–7
aerators
  diffusers, 216
  surface, 216
aerobic conditions, 165, 168
aerobic lagoon, 182, 183, 184
aerobic systems, 164, 182, 183, 184,
Africa, 200, 202, 206, 207
Africa donkeyes, 206
aggregate, 4
air distribution, 74
air quality, 167
air resource
  effect of manure on, 167–70
  odors, 168–9, 184
air stripping, 310, 336–7
air, solubility, 223
alfalfa, 147
algae, 213
alkalinity, 224
  carbon dioxide interactions, 315, 337–8
  control, 320, 337–8
  nitrification interactions, 320, 337
  supplements, 338
alley, 119
aluminum, 224, 239
ammonia, 94, 169, 170, 216, 224
  acid-base equilibrium, 312, 337–8
  ionized, 228
  production, 312, 313, 315
  removal methods (*see also* biofiltration, nitrification), 316–26, 332
  removal rates, 320, 322, 324
  suggested levels, 311, 312
  total, 224
  toxicity, 312, 337
  unionized, 228
ammonia loading, 253
ammonia, pond uptake, 249
anaerobic, 186
anaerobic conditions, 168, 169
anaerobic lagoon, 164, 168, 182, 183, 184, 186
anaerobic systems, 164, 169, 182, 183, 184, 186
animal resource-effect of manure on, 170
animal traction logger, 199, 203
aquacultural systems, 211–12
  classification, 212
aquaculture, 211
aquariums, 214
arches, 16
ard, 200, 201, 202
area units, 93
arsenic, 224
as-excreted, 176
as-excreted manure characteristics, 174
as-excreted swine manure, 177
as-excreted poultry manure, 178
Asia, 202, 207
attached, 167
attached phosphorus, 167
automatic feeders
  cattle, 118
autotrophs, 216

bacteria, 164, 167
bale wrap, 151
Bangladesh, 205
barium, 224
bass, hybrid striped, 211
bead filters, 313, 316, 318, 320, 326
beams, 16
beef, 176, 183
beef manures, 177
bentonite, 237
best available technology (BAT), 310–11
best management practices (BMPs), 310
biochemical oxygen demand (BOD), 310, 311, 316, 320
biofilm, 320, 322
biofilters, 215, 216, 316–26
  efficiency, 216
  media, 216, 320–1
  operation, 216
  oxygen/space limitations, 316, 320–1, 323, 326
  solids, and, 320, 321, 322, 326
  start-up, 320
  types, 216, 316–26
biofiltration (*see also* biofilters), 310, 316–22
biogas, 193
biological filters, 216
  operational parameters, 216
blacksmith, 200, 201
blocks, 9

body surface area, 35, 36
bolts, 20
bos indicus, 198
bovine, 197, 202
bowl, 133, 136
broilers, 128
building, 7
building systems, 96
bullock, 202, 203
butyl rubber, 241

cadmium, 224
cage culture, 211–12, 214
cages, 214
calcium, 224
camelid, 197, 202, 205
capacity, 188
carbon dioxide, 50, 52, 169, 170, 224
  acid-base equilibrium, 315, 337, 338
  production, 314–15
  removal (*see also* aeration, pH control, stripping columns), 322, 326, 336–8
  suggested levels, 311, 315
  toxicity, 315
carbon, activated, 216
carbon-to-nitrogen ratio, 174
carrying capacity (*see also* oxygen), 309
cart, 202, 205
cascade aeration equipment (*see* aeration)
catfish, 219
cement, 3, 4, 5
Centre for Tropical Veterinary Medicine, 199, 203
channel/racewace, cleaning velocity, 256
channel/raceway
  characteristics, 176, 177
  geometry raceway, 258
  roughness coefficient, 259
  side slope, 259
chemicals, 310, 311, 314, 315, 332, 333, 337–8
chemical composition, 147
chemical filters, 216
chemotherapeutants, 315
chlorine, 224
CIRAD-SAR, 203
circular opening, 68
clarifiers (*see* solids removal processes)
Clean Water Act, 310
closed, system, 212
coefficient for convection, 36
coefficient for radiation, 36
cold water, 223
coliforms, 167
columns, 16
concrete, 3, 4, 5, 6, 238
condition, 200, 203, 204, 206
conductivity, thermal, 220
constraints (in aquacultural systems), 219
construction material, 10
copper, 224, 238, 240
copper alloys, 235
corrosion, 8, 231
Costa Rica, 205
cow, 203, 206
critical level, oxygen concentration, 225
crushing, 202
cubes, 152
cultivator, 197, 200, 201

dairy, 181, 188
DAP, 200, 206, 207
density (of water), 220
dentrification processes, 165
detoxification, 216
diffuser, 216
discharge of jet, 74
disease, gas-bubble, 221, 225
disinfection, ozone, 216
dissolved air flotation (*see* flotation)
dissolved matter, 313, 314, 316
  methods to remove (*see also* ozonation, biofiltration), 321, 333–6
dissolved oxygen, 224
diurnal rhythm, 32, 50
donkey, 200, 202, 203, 205, 206
doors, 25
downflow, 216
draught animal power (DAP), 197, 200, 201, 206, 207
draught capability, 198, 203
drinking facility, 138
drinking system
  pig, 138
  poultry, 141
drinking water requirements, 138
  cattle, 133
  goats, 135
  poultry, 140
  turkeys, 141
dry weight basis, 174

Earth, 9–10
effective size (mesh), 233

effluent, 187, 190
  regulations, 309–11, 315, 334
  treatment technologies, 309–10, 316–36
energy, 193–4
energy consumption, 79, 84
Environmental Protection Agency (EPA), 310
equal-pressure system, 67
equid, 197, 198, 202
equilibrium moisture, 159
ergometer, 203
ergometer CTVM, 199, 203
ergonomics, 205
erosion, channel, 256
erosion velocity, 256
estuaries, 236
Europe, 202
evaporation losses, 212

fan capacity, 80, 82
farm buildings, 10
farming system, 200, 205, 206–7
fasteners, 19
fatigue, 202, 203
fatigue, fish swimming, 257
fecal coliform, 167
feed
  barriers, 116–17
  metabolite production relationships, 312, 313, 315
  waste, 313
feed requirements, 204
feeders
  automatic, 282, 284
  broilers, 128
  demand, 282, 287
  laying hens
    auger, 127
  pig
    dry feeding, 125
    liquid system, 126
  turkeys, 128
feeding alley, 115
feeding alley width
  cattle, 115
  goat, 122
  sheep, 119
feeding method, 3
feeding space, 122
feeding trailers, 130
feeds and feeding
  frequency, 282
  particle size, 282
  rate, 282, 297
  regime, 281
feeds barrier
  cattle, 116
fertilization, 189–90
fiberglass, 240
fibreboards, 21
fields, 188
filters, 216
filter media, 216
filtration (*see* granular media filters, microscreen filters)
finishing, 23
fish culture tank, 215
fixed feeding system, 132
floor, 6, 7
flotation, 314, 317, 332–4
flowrate, dimensioning, 83
fluidized-bed biofilter, 316, 319, 320, 322, 323–6
fluidized-bed biofilter applications, 113–16
fluidized bed filters, 216
foam fractionation, 327, 332–5
forage, 147
for maintenance, 204
fouling, 232
foundation, 6
free jet, 68–9
frequency, 72
freshwater production, 211
frost, 135
frost protection, 135
functional areas, 94
functional sector, 105

galvanization, 7, 8
galvanized steel, 235
Gambia, 205
gas transfer (*see* air stripping)
gas-bubble disease, 221, 225
glass, 243
glued joints, 20
glue-laminated timber, 13
goat, 122
granular media filters, 313, 326, 327, 332, 333
gravity sand filters (*see* granular media filters)
growth, fish, 255
Gujarat, 202
gunite, 243

hardness (of water), 220, 224
harness, 198

harnessing, 197, 198
harrow, 201, 205
harvest, 201, 206
harvesting (fish), 214
hay racks, 118
  cattle, 118, 128
  sheep, 121
heart rate, 203
heat balance, 32, 33, 34, 38, 40
heat balance, from a building, 56
heat balance house level, 46
heat of fusion (of water), 220
heat of vaporization (of water), 220
heterotrophic organisms, 316, 320, 321, 323
heterotrophs, 216
hoe, 200, 201
horse, 202, 203, 204, 205
human labor, 62–5, 197, 200, 201, 207
husbandry, 200
hydrogen cyanide, 224
hydrogen sulfide, 169, 171, 224
hyperthermia, 32
hypothermia, 32

ice-accumulation cooler, 143
incineration of solid manure, 193
India, 202
Indonesia, 205
inlet, 68–70
  porous, 71
inlet ducts, 73–4
intensity of, 2
ion exchange, 216
iron, 224

jet, circular openings, 69
jet, discharge, 74
jet, free, 68
jet, nonisothermal, 71
jet, obstructions, 74
jet, wall, 70
joints, 18–20

labile P, 167
labor, 201
labor, effiency, 246
lagoon, 184, 186
land preparation, 197, 200, 201, 205, 206
latent heat, 32, 34, 37, 38, 39, 44, 45, 46
Latin America, 200, 202, 207
layer and broiler, 178
LC_{50}, 228, 239
lead, 224
length, 116, 119
lethal temperature, 234
lift, 87
liquid, 182, 184
liquid manure 171, 183, 184, 185, 186
litter, 171, 174, 185, 186
livecar, 296, 298
livestock, 188
live weight, 197, 202
load, 6
loading capacity, 252
losse hay, 156
losses, 148
lower critical temperature, 33, 38, 39, 40, 41, 45

magnesium, 224
maintenance, 200, 204
maize, 202, 204
manganese, 224
manning equation, 259
manure, 9, 163, 164, 167, 168, 169, 171, 174, 175, 176, 178, 179, 180, 182, 183, 184, 187, 188, 189, 191, 192, 193
manure characteristics, 170–8
  beef-as excreted, 176–7
  beef-feedlot, 177
  dairy, 175–6
  litter, 171, 178, 185–6
  poultry-as excreted, 178
  swine, 177–8
  units of measure, 174–6
manure characterization, 175
manure consistency, 171, 179, 180
  liquid, 171, 182, 183, 184, 186
  semi-liquid (*see* slurry)
  semi-solid, 171, 182, 183, 184
  slurry, 171, 182, 183, 184
  solid, 171, 182, 183
manure management functions, 179–81
  collection, 180
  production, 179
  storage, 180–1
  transfer, 181
  treatment, 181
  utilization, 181
manure management systems, 171, 179, 180, 181–6
  beef, 183–4
  dairy, 181–3

poultry, 176, 185–6
swine, 184–5
manure-handling equipment, 171
manure-storage pond, 184
media, 316, 321, 322, 324, 326, 332
mercury, 224
metabolic activity, 204
metabolic by-products (*see* wastes, carbon dioxide, ammonia, nutrients, solids, dissolved matter)
metabolic rate, 202, 203, 204
metabolism, 32, 33
methane, 169, 185, 186
methemoglobinemia, 165
microscreen filters, 327, 330–2, 334
milk, 206
milk storage
direct-accumulation cooler, 143
direct-distension cooler, 143
refrigeration, 142
milking, 188
milling, 202
mixer trailers, 131
moisture production, 44, 46, 50
mold, 152
mortar, 9
mule, 202, 203

nails, 19
nail plates, 20
National Pollution Discharge Elimination System (NPDES) permit, 310
negative-pressure system, 67
Nepal, 205
net forces, 58
net pens, 214, 263
aesthetics, 265
construction, 265
current speed, 264
disadvantages, 214
flotation, 268
mooring design, 269
net forces, 268
net support, 266
operational aspects: dead fish removal, 270
operational aspects: feeding, 270
operational aspects: net cleaning, 270
permit, 214
site selection, 264–5
size, 214
water depth, 264
water temperature, 265
wave action, 264
wind, 265
net retaining structures, 214
nets, 237
cast, 297
gill, 297
lift, 297
seine, 297
net support, 56
nitrate, 14, 224, 311, 312, 315–16, 319–21
acid-base equilibrium, 313
production, 312, 319, 326
removal methods (*see also* biofiltration, nitrification, ozonation), 319–21, 334
suggested levels, 311, 312
toxicity, 312
nitrification (*see also* ammonia removal, biofilters), 165, 228, 312, 314, 316, 319–26, 337
nitrogen, 164, 165, 174, 187, 189, 224, 228,
nitrosomonas, 228
nitrobacter, 228, 319–21
nitrosomonas, 319–21
noise, 81
nutrition, 202
nutrients (*see also* phosphorus, nitrate), 163, 164, 174, 185, 188, 189, 216, 310, 315–16, 329, 335
nitrogen, 164–6, 171–5
phosphorus, 166–7, 171–5
nylon, 237

odors, 167, 168, 169
operator, 199, 200
optimum temperature (for common aquatic species), 225
outdoor systems, 220, 223
outlet, 68, 76–7
outlet diffusor, 82
ox, 200–206
oxygen, 202
carrying capacity, and, 309, 310, 315, 336
consumption (by fish), 225–7
nitrification, and, 319–22, 324, 326
pure, 223
solubility, 221–3
suggested levels, 311
oxygen consumption, 202
oxygen loading, 253, 254
oxygen transfer, 221
rate, 221

oxygenation (*see also* aeration, air stripping), 309, 310, 321
oxylog, 203
oyster production, 211
ozonation (*see also* ozone), 333–4, 335
ozone, 311, 314, 321, 333–4, 335
ozone treatment, 216, 235, 237

panel products, 20–3
particle boards, 21
particleboards and fibreboards, 21
particulates, 216
  removal devices, 216
pathogenic, 167
pathogens, 167–8
pathogens, fish, 311, 314, 315, 332, 335
paving, 7
PCB's, 224
pellets, 154
per unit area, 3
peristaltic, 82
pesticides, 228
pH, 224
  control, 310, 337–8
  interactions, 312, 315, 337–8
  suggested levels, 310, 311, 313, 315, 338
pH, range for production, 249
phosphorus, 164, 166, 167, 174, 187, 188, 189, 310, 311, 315–6, 332
photosynthesis, 37
pig breeding center, 105
pig housing, 101
piscilator, 300
plastic, 240
plough, 197, 198, 199, 200, 201, 202
plywood, 20–1
pollution, 187
polyculture, 248
polyethylene, 216, 237, 241
polypropylene, 241
polyvinyl chloride (PVC), 216, 241
ponds, 182, 212, 246, 281, 282, 283–4, 288, 292–3, 295–9
  depth, 212
  design types, 249
  dirurnal water quality, 247
  embankment, 213
  excavated, 213
  feeding method, 213
  photosynthesis, 247
  site selection, 213
  size, 212
  stocking, 213
  types, 213
  water losses, 212
  water supply, 212
porous, 68, 70
portal frames, 16
positive-pressure system, 68
post and beam construction, 16
potassium, 224
poultry, 185
poultry manure, 186
power, 199, 203, 205
power tiller, 205
preservatives, 27–8
pressure, 70
pressure loss, 68, 72
production of
  carp and tilapia, 211
  catfish, 211
  finfish, 211
  freshwater, 211
  intensity of, 212
  per unit area, 213
  salmon and trout, 211
  shrimp and prawn, 211
  world aquacultural, 211, 212
production intensification, 309, 310
productivity, 246
puddler, 201
pump
  airlift, 293
  centrifugal, 289
  peristaltic, 292
  reciprocating, 293
  rotary, 292

Q_{10} factor, 219

raceways, 212, 213, 252
  parallel, 213
  series, 213
  water supply, 213, 236
raceways, design, 254
racks, 121
recirculating aquacultural systems, 214, 215
  advantages, 215
  components, 215
  recycle ratio, 215
  replacement time, 215
  residence time, 215
  turnover time, 215
  wastewater discharge, 215

recirculating systems (*see* water-reuse systems)
recovery, 203
recycle ratio, 5
refrigeration, 142
regulators, 200
reinforcing, 6, 9
relative humidity, 158
replacement time, 5
residence time, 5
ridger, 201
roof coverings, 25–6
Rotating Biological Contactors (RBCs), 316, 318, 320, 322, 326
rotating biological contractors, 216
round bale, 150
roundwood, 13

salinity (of water), 212, 220, 221, 224
salmon, 213
sand, 216
  cost, 324, 325
  effective size (D_{10}), 323–5
  fluidized-bed biofilter applications, 323–6
  granular media filter applications, 332–3
  surface area, 324, 325
  uniformity coefficient (D_{60}/D_{10}), 323–4
sand filters, 213, 332, 333
sawn timber, 12–13
sealers (of pond), 213
sedimentation (*see* settling basins, swirl separators, tube or plate settlers)
sedimentation chamber, 216
seeder, 200, 201
seepage, 188, 212
seine reel, 298
selenium, 224
semi-solid, 182, 183, 184, 185
sensible heat, 32, 34, 35, 37, 44, 45, 46, 47, 48, 49
sensor
  ammonia, 304
  oxygen, 303
  pH, 303
  temperature, 303
settleable solids, 311, 313, 314, 316, 321, 329, 334
settling basins, 327, 328, 329, 330, 334
shear resistance, 103, 104
sheep housing, 89
Silsoe Research Institute, 203
silver, 224
single-pass (flow-through) systems, 309, 316, 332, 334
slopes, 213
slurry, 182, 183, 184, 185, 188
slurry manures, 182
smallholder farmer, 197, 200, 201, 202, 206
smallholders, 202
socioeconomic, 207
sodium, 224
soil, 213
soil quality, 200
solids, 182, 184, 185, 186
  total dissolved, 224
  total suspended, 224
solids characteristics
  chemical make-up, 313, 314
  dissolution, 313, 329, 334
  particulate size, 313, 314, 327, 332, 333
  settling rate, 327, 329
  shear resistance, 313, 314
  specific gravity and density, 314, 327, 330, 332
solids disposal, 310, 334–5
  composting, 335
  land application, 310, 335
  landfill, 335
  process options
  regulations (*see also* effluent regulations), 310, 335
  vermiculture, 335
  wetland, 335
solids effect on culture systems, 313, 320, 321, 322, 326, 337
solids effect on fish health, 313
solid manure, 171, 176, 182, 184
solid or slurry, 184
solids production, 313
solids removal
  options discussed, 313, 314, 329, 330, 334
  ozone, and, 321, 327, 333–4, 335
  processes, 310, 313, 314, 321, 327–34
solid, slurry, or liquid, 184
solids thickening (*see also* settling basins; microscreen filters), 334–5
soluble P, 167
soluble phosphorus, 166
spatial elements, 91
spatial module, 105
specific flow, 83
specific gravity and density, 104, 117, 120, 122
specific heat (of water), 219–20
spilled feed, 171, 175, 177, 182

spreading, 190, 191
spreading manure, 190
square bale, 148
stainless steel, 237–9
statistics, 149
steel, 216, 238, 239
Stoke's law, 327–8
storage, 188
storage capacity for livestock wastes, 188
storage facilities, 169
storage stability, 150
storing alfalfa, 156
stratification, 219
streams, 212
strength, 3, 4
structural systems, 13
submerged biofilters, 316, 317, 320, 321–2, 326
sulfate, 224
sulfur, 224
sunlight, 212
surface aerators, 216
swathing, 148
swine, 184
swine manure, 177, 184, 185
swirl separator, 327, 328, 329, 330
systems, 141
systems planning, 105

tank, 191, 192, 271, 287, 288, 292–3, 295–6, 299–300
  aquaculture, 214
  biomass loading, 271
  cleaning behavior, 272
  desirable characteristics, 216
  inlet and outlet design, 274, 275
  labor requirements, 271
  shapes, 216
  size and shape, 272, 273
tanks, 236
temperature, 93
temperature (of water), 219, 221, 225
temperature ranges (for common aquatic species), 225
thermal conductivity (of water), 220
thermal insulance, 35, 36
thermal insulation, 28–9
thermal resistance, 35, 37, 38
thermoneutral zone, 32, 33, 38, 39, 43
thermoregulatory, 33, 34, 37, 38
threshing, 202
tilapia, 211
timber, 11–12
timber connections, 18
tool bar, 201
topography, 213
total coliforons, 167
total dissolved, 14
total heat production, 33, 34, 39, 41, 43, 44, 45, 46, 47, 48, 49, 50
total kjeldehl nitrogen, 164
total phosphorus, 166
toxicity, 102, 105, 216, 226, 228
tractor, 197, 200, 201, 206, 207
trailer, 130
training, 200
transport, 202, 206
transportation, 157
trap, 298
treatment, 169
trickling biofilter, 316, 317, 320, 322, 325, 326
trickling filters, 216
trough, 116
trough length
  cattle, 116
  goat, 122
  sheep, 119
trout, 213
truss buildings, 16
tsetse, 200
tube or plate settler, 327, 328, 329, 330
turnover time, 215

u-tube, 216
ultraviolet (UV) light, 234
underfloor suction, 76–7
upflow, 216
upper critical temperature, 33, 45
uranium, 224

vacuum, 192
vanadium, 224
vasoconstriction, 37, 46
vasodilation, 37, 46
ventilation, automatically controlled natural, 55
ventilation, equal pressure, 68, 79
ventilation, forced, 68
ventilation, mechanical, 54
ventilation, natural, 54
  continuity of mass flow, 55
ventilation, natural, combined effects of thermal buoyancy and wind, 62–5
ventilation, natural, due to thermal buoyancy, 55–9
  neutral plane, 55

ventilation, natural, due to wind, 59–62
  pressure coefficients, 59–63
ventilation, negative pressure, 68, 78
ventilation, positive pressure, 68, 79
vinyl, 241
viscosity, 219
void fraction, 216
volume expansion, 158

wagon, 202
wall coverings, 23–8
walls, 7–8
warmwater, 223
waste (*see also* solids, ammonia, nitrate, phosphorus, chemotherapeutants, effluent regulations, solids removal processes), 103, 188, 189, 221
  defined, 311
  disposal, 334, 335
  minimization technologies, 309–10, 313
  treatment technologies, 310, 313, 314, 315–38
  types of, 310, 311
water, 5, 141
  properties of, 219
  temperature, 219
water bowl
  cattle, 133
  goats and sheep, 136, 137(F)
water depth, 54
water equipment, 138
water quality criteria, 224, 310–12, 315
water requirements, 133, 135
water resources, 309
water resources, effect of manure on, 163–7
water reuse (recirculating) systems, 309–16, 321–4, 327, 330, 332–5, 337
water supply, 212
water trough, 134
water use, 309, 310, 316
water-raising, 202
weather boardings, 23
weed control, 197, 200, 201, 206
weeder, 200
weeding, 201, 206
welfare code, 31
wet-weight basis, 174
wheeled tool-carrier, 201
windbreakers, 72
wood, 10–14
wood (as tank material), 216, 241
wood protection, 26
work, 202, 205
work output, 198, 202, 205, 206
work schedule, 205
working depth, 199
world aquacultural, 1, 2

yoke, 197, 198

Zimbabwe, 202
zinc, 224
zirconium, 224